SPRINGER SERIES
IN PERCEPTION ENGINEERING

Series Editor: Ramesh C. Jain

Ichiro Masaki
Editor

Vision-based Vehicle Guidance

With 236 Illustrations

Springer-Verlag
New York Berlin Heidelberg London Paris
Tokyo Hong Kong Barcelona Budapest

Ichiro Masaki
General Motors Research Laboratories
Computer Science Department
Warren, MI 48090-9055
USA

Series Editor
Ramesh C. Jain
Electrical Engineering and
    Computer Science Department
University of Michigan
Ann Arbor, MI 48109
USA

Library of Congress Cataloging-in-Publication Data
Vision-based vehicle guidance / [edited by] Ichiro Masaki.
        p.   cm.—(Springer series in perception engineering)
    Includes bibliographical references and index.
    ISBN 0-387-97553-5 (alk. paper)
        1. Automobile driving—Automation.   2. Mobile robots.   3. Computer
    vision.   I. Masaki, Ichiro.   II. Series.
    TL152.5.V55   1991
    629.28′3—dc20   91-487

Printed on acid-free paper.

Production managed by Bill Imbornoni. Manufacturing supervised by Genieve Shaw.
Typeset by Asco Trade Typesetting Ltd., Hong Kong.
Printed and bound by Edwards Brothers, Inc., Ann Arbor, MI.
Printed in the United States of America.

9 8 7 6 5 4 3 2 1

ISBN 0-387-97553-5 Springer-Verlag New York Berlin Heidelberg
ISBN 3-540-97553-5 Springer-Verlag Berlin Heidelberg New York

Preface

There is a growing social interest in developing vision-based vehicle guidance systems for improving traffic safety and efficiency and the environment. Examples of vision-based vehicle guidance systems include collision warning systems, steering control systems for tracking painted lane marks, and speed control systems for preventing rear-end collisions.

Like other guidance systems for aircraft and trains, these systems are expected to increase traffic safety significantly. For example, safety improvements of aircraft landing processes after the introduction of automatic guidance systems have been reported to be 100 times better than prior to installment. Although the safety of human lives is beyond price, the cost for automatic guidance could be compensated by decreased insurance costs. It is becoming more important to increase traffic safety by decreasing the human driver's load in our society, especially with an increasing population of senior people who continue to drive.

The second potential social benefit is the improvement of traffic efficiency by decreasing the spacing between vehicles without sacrificing safety. It is reported, for example, that four times the efficiency is expected if the spacing between cars is controlled automatically at 90 cm with a speed of 100 km/h compared to today's typical manual driving. Although there are a lot of technical, psychological, and social issues to be solved before realizing the high-density/high-speed traffic systems described here, highly efficient highways are becoming more important because of increasing traffic congestion.

Vision-based vehicle guidance systems are expected, as a third potential benefit, to improve our environment. The time required to run the vehicle engine to move a given distance would be decreased if the vehicle could get to the destination in a shorter time because of higher traffic efficiency. The shorter engine running time would contribute to cleaner air.

In addition to the social interests described, vision-based vehicle guidance is an exciting field for technical research. Vision systems for vehicle guidance must be able to deal with less structured environments, shorter processing time, lower processing cost, and higher reliability, compared to conventional machine vision systems for factory automation. Operating under various weather changes, for example, is a great challenge for conventional vision systems developed for controlled lighting conditions.

This book is based on the IEEE round-table discussion held on July 2, 1990. The round-table discussion on vision-based vehicle guidance was held as an activity of the Intelligent Vehicle Subcommittee in the IEEE Industrial Electronics Society. We decided to publish this book because there are no books dedicated to the young, rapidly growing vision-based vehicle guidance field. Please feel free to contact me for your comments, suggestions, or questions.

Ichiro Masaki
Warren, MI

Acknowledgments

I would like to express my appreciation to all of the participants who contributed to the discussions on July 2, 1990, at the IEEE round-table discussion on "Vision-based Vehicle Guidance." Special thanks are owed to the organizing committee members listed here in alphabetic order.

J.M. Brady	University of Oxford
R.A. Brooks	Massachusetts Institute of Technology
P.J. Burt	Stanford Research Institute
J.L. Crowly	LIFIA
L.S. Davis	University of Maryland
O.D. Faugeras	INRIA
M. Fujie	Hitachi
T. Fukuda	University of Nagoya
V. Graefe	Universität der Bw München
F. Harashima	University of Tokyo
H. Hosokai	Science University of Tokyo
R. Jain	University of Michigan
T. Kanade	Carnegie–Mellon University
J.C. Latombe	Stanford University
G.F. McClure	Martin Marietta
Y. Shirai	University of Osaka
T. Tomita	MITI
S. Tsuji	University of Osaka
T. Tsumura	University of Osaka Prefecture
M. Yoshida	Fujitsu
S. Yuta	University of Tsukuba

I am indebted to G.G. Dodd and others at the General Motors Research Laboratories for their encouragement and assistance. Financial support from the Foundation for Promotion of Advanced Automation Technology contributed to a successful round-table discussion.

Contents

Contributors

Philippe Bobet   LIFIA

Michel Buffa   INRIA

Stefan Carlsson   Royal Institute of Technology

James L. Crowley   LIFIA

Larry S. Davis   University of Maryland

Daniel DeMenthon   University of Maryland

Sven Dickinson   University of Maryland

Jan-Olof Eklundh   Royal Institute of Technology

Olivier D. Faugeras   INRIA

Kenji Fujita   Mazda Motor Corporation

Volker Graefe   Universität der Bundeswehr München

Tomoyuki Hamada   Hitachi Ltd.

Martial Hebert   Carnegie–Mellon University

Hiroshi Ishiguro   University of Osaka

Toshio Ito   Daihatsu Motor Co. Ltd.

Ramesh Jain   University of Michigan

Hiroshi Kamada   Fujitsu Laboratories Ltd.

Kohji Kamejima   Hitachi, Ltd.

Takeo Kanade   Carnegie–Mellon University

Shiro Kawakatsu   Daihatsu Motor Co. Ltd.

Klaus-Dieter Kuhnert   Universität der Bundeswehr München

Atsushi Kutami   Mazda Motor Corporation

Hideo Mori   University of Yamanashi

Akihiro Okuno   Mazda Motor Corporation

Roland Pesty   University of Osaka

Amy Polk   University of Michigan

Marie de Saint Blancard   Peugeot S.A.

Karen Sarachik   MIT

L.T. Schaaser   University of Bristol

Steven Shafer   Carnegie–Mellon University

Patrick Stelmaszyk   ITMI

B.T. Thomas   University of Bristol

Charles Thorpe   Carnegie–Mellon University

Masahiro Tsuchiya   Hitachi Ltd.

Saburo Tsuji   University of Osaka

Philip Veatch   University of Maryland

Yuriko C. Watanabe   Hitachi Ltd.

Masumi Yoshida   Fujitsu Laboratories Ltd.

Zhengyou Zhang   INRIA

1
Vision-based Autonomous Road Vehicles

VOLKER GRAEFE AND KLAUS-DIETER KUHNERT

1.1 Abstract

Autonomous road vehicles, guided by computer vision systems, are a topic of research in numerous places in the world. Experimental vehicles have already been driven automatically on various types of roads. Some of these vehicles are briefly introduced, and one is described in more detail. Its dynamic vision system has enabled it to reach speeds of about 100 km/h on highways and 50 km/h on secondary roads.

The field has advanced rapidly in recent years. Nevertheless, many problems remain to be solved before such vehicles may be introduced into ordinary road traffic. Some of the problems and approaches to their solutions are discussed. There are good prospects that in the future driving robots or highway autopilots will help to make driving easier and safer.

1.2 Introduction

One goal of the European research program PROMETHEUS (<u>Pro</u>gram for a <u>E</u>uropean <u>T</u>raffic with <u>H</u>ighest <u>E</u>fficiency and <u>U</u>nprecedented <u>S</u>afety) is the development of a robot with the ability to drive cars and trucks on any normal freeway without human intervention. Such a robot (also called a highway autopilot) could either relieve the human driver completely from the task of driving or, if the driver prefers to remain in charge, act as a proficient copilot, assisting the driver and warning him in the case of an emerging dangerous situation.

This article is an abbreviated version of the paper by V. Graefe and K.-D. Kuhnert (1988), "Toward a Vision-Based Robot with a Driver's License," *Proceedings, IEEE International Workshop on Intelligent Robots and Systems, IROS '88, Tokyo*, pp. 627–632. © 1988 IEEE.

Since many traffic accidents are caused by distracted, careless, or tired drivers, the introduction of driving robots will, hopefully, help to *improve traffic safety*, which is the main goal of PROMETHEUS.

The robot must be fully autonomous in the sense that no special installations on the road will be provided (e.g., buried cables or radio beacons) and that no external source of computing power will be available. The robot may only utilize those markings and traffic signs that exist for the convenience of human drivers, and it must coexist with conventional vehicles driven by humans.

Considering the recent progress in microelectronics, artificial intelligence, and, particularly, computer vision, it appears no longer impossible to construct such a "robot with a driver's license." A fully functional prototype of the driving robot is, in fact, planned to be demonstrated in 1994. Its main sensor will be a dynamic vision system.

Highways, rather than secondary roads or city streets, will be the initial domain for automatic driving. The reason is that, in spite of high speeds (in Germany typically between 100 km/h and 200 km/h), driving on highways is, in fact, easier than driving on other roads.

1.3 Some Experimental Autonomous Road Vehicles[1]

1.3.1 Japan

The first report on an autonomous road vehicle came from Japan [39]. A passenger car, equipped with two TV cameras, some signal processing electronics, and a small computer, was able to drive autonomously on well-marked roads at speeds up to 30 km/h. The particular methods used depended largely on specialized hardware and were not easily extendable to a greater variety of environments. After adding a component for obstacle detection [40], the project was apparently discontinued.

Another Japanese autonomous vehicle tested on roads is Harunobu-4 (Yamanashi University). It is a small cart, equipped with a TV camera, some ultrasonic sensors, and a Motorola microcomputer [31]. It has a very flexible

---

[1] This section reflects the state of the various projects as it was known by the authors at the time of writing the original article (early 1988). By spring 1991 we heard that the ALV project has apparently been discontinued, and that the speed of NavLab has been increased to 20 or 30 km/h, with the WARP computers removed from the vehicles. The Japanese personal vehicle was demonstrated at 30 km/h on a test course and at 60 km/h on a straight highway using two largely independent vision systems simultaneously, one for road following and one for obstacle avoidance. Many automobile companies in Europe, Japan, and the United States are now operating autonomous or semi-autonomous road vehicles, none of them able to run autonomously in ordinary traffic. The speed of all fully autonomous road vehicles, except VaMoRs, is still limited by their vision systems.

architecture, but because of its limited computer power, its speed is only about 1 km/h.

In 1987, the Personal Vehicle System project was started by the Ministry of International Trade and Industry, Nissan, and Fujitsu [32]. Experiments involving an autonomous automobile on a test track were scheduled for 1988. The authors are not aware of any detailed information on this project.

1.3.2 United States of America

In the United States, several laboratories are involved in the development of autonomous land vehicles. The projects are financed by the Department of Defense, and the announced goal is cross-country driving. Two of these vehicles, the ALV (Autonomous Land Vehicle; Martin Marietta Denver) and the NavLab (Carnegie–Mellon University, Pittsburgh), may, however, be considered autonomous road vehicles since they have been extensively tested on roads [2, 22, 43]. Both vehicles are equipped with a multitude of sensors (color TV cameras, imaging laser range finders, sonars, and others) and with several very powerful computers. The ALV has reached speeds of 20 km/h on an unobstructed blacktop road [41], while the NavLab has been limited to speeds of less than 1 km/h on driveways on the CMU campus [38]. Higher speeds were not reached because of the long cycle times of the vision systems (ALV, ~ 2 s; NavLab, ~ 20 s). Recently, WARP computers have been installed in both vehicles; they have 10-processor pipelines with theoretical performances of 10^8 floating-point operations per second [1]. It is not obvious, however, that the architecture of the WARP is particularly well adapted to typical vision tasks; therefore, probably a fraction of the theoretical performance may actually be used in this type of application.

1.3.3 United Kingdom

The Royal Armament Research and Development Establishment (RARDE) is operating an experimental vehicle, ROVA (robot vehicle, autonomous). It is a camper of about 2 t, equipped with actuators and sensors for autonomous driving, including 3 TV cameras with lenses of different focal lengths on a common pan and tilt head. A multiprocessor vision system is under development [42].

1.3.4 Germany

In Germany, there are 3 institutions that operate experimental road vehicles for research in autonomous mobility. The Optopilot of Volkswagen AG [45] employs a TV camera as a sensor. Selected parts of the image are analyzed to find the locations of lane markers relative to the vehicle. This information is used to control the steering of the vehicle. The speed is controlled by a human driver, therefore, the Optopilot does not provide the ability of completely autonomous driving. The automatic lateral control has been tested at speeds up to 120 km/h on well-marked roads with large radii of curvature. Like Tsugawa's system, the Optopilot is probably not easily applicable to less constrained environments.

For a number of years, the Daimler-Benz AG has operated semiautonomous buses that may automatically follow a predetermined track, guided laterally either mechanically by rails or electronically by buried cables. Such buses are used as test beds for experiments in autonomous mobility. One of them has been equipped with cameras and computers and is now used in experiments involving automatic road following and obstacle avoidance.

The world's fastest fully autonomous road vehicle was designed and is operated by the Universität der Bundeswehr München. It is described in more detail in the next section.

The Experimental Vehicle of the Universität der Bw München

A commercial 5-t van has been converted into a test vehicle for autonomous mobility and computer vision, VaMoRs (Figure 1.1) by the Institute of System Dynamics and Flight Mechanics of the Universität der Bw [9, 44]. It is, in effect, a moving laboratory equipped with an electric generator (220 V, 6 kVA) and computer-controllable actuators for accelerating, braking, and steering. It has a number of sensors, for example, for velocity, angular velocities, and accelerations. Two monochrome TV cameras are mounted on a computer-controllable platform, one with a short focal length yielding a wide field of view and one with a telescopic lens for higher resolution at a greater distance.

There are two computers in the vehicle, an IBM IC (industrial version of the PC/AT) for vehicle control and high-level vision, and a real-time image

FIGURE 1.1. VaMoRs, the world's fastest autonomous road vehicle.

processing system, "BVV 2," a multiprocessor system designed as an efficient hardware basis for dynamic vision.

The Vision System of VaMoRs

The vision system fully implements the concepts for dynamic machine vision developed at the Universität der Bw [6, 7], including (1) a hardware architecture that is well matched to the task of dynamic vision, (2) an efficient feature extraction method based on correlation, and (3) an internal 4-D model world for feature aggregation and control of system behavior.

Hardware. The vision system of the vehicle is based on the multi-microcomputer system BVV 2, which is described in the appendix in detail [12, 16].

The system architecture of the BVV 2 is optimized for real-time image processing and, in particular, for feature extraction. The freely programmable parallel processors (PPs) may concentrate their entire computing power on those parts of the scene that, at any given moment, yield the most relevant information. They make it possible to partition the task of image interpretation in such a way that almost no loss of efficiency occurs when these subtasks are processed in parallel. The image memories in the videobus interfaces enable the PPs to access pixel data independently and simultaneously without any need for coordination.

Feature Extraction. The first major step in scene interpretation is the recognition and localization of suitable features in the image sequence. Interesting points, e.g., corners and borderlines between regions of different gray levels, so-called edges, may be used as elementary features. Edge elements have been found particularly useful for recognizing and locating the road, even when no actual markers are present.

The method of controlled correlation, an efficient generalization of ordinary correlation [23, 25, 26], is used to find and track edges in the image sequence. It employs knowledge of the expected position of a feature to reduce the size of the search space and thus the time required to locate the feature. Also, the maxima of the correlation function are computed only along suitably chosen search paths. Using a very efficient implementation, an algorithm based on this method requires only 15 ms of time on one of the parallel processors of the BVV 2 to localize and track 3 elements of a road border, allowing every video frame to be analyzed. Several processors may be employed to track numerous elements of both edges of the road simultaneously.

Feature Aggregation. When a human driver perceives his environment and assesses a situation he depends heavily on his past experience, and if he is a novice, his reactions tend to be slow and clumsy. Similarly, a driving robot should—at least—be able to utilize knowledge of spatial structures, objects, and processes when analyzing the video signals produced by its camera for building an internal representation of the environment. Knowing the present state of the environment (and of itself) is crucial for computing appropriate control signals or for issuing sensible warnings.

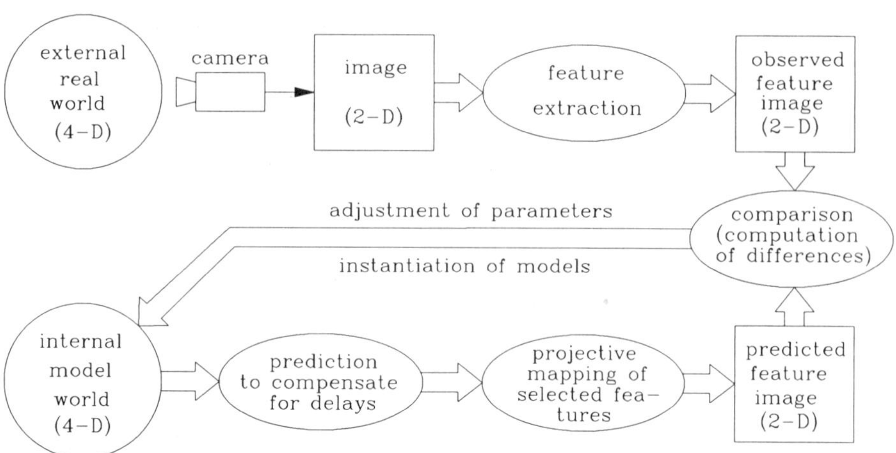

FIGURE 1.2. Scene interpretation using a 4-D model world.

Because of its limited abilities (compared to the human brain), a computer must form its internal model world using only a reduced or simplified image of the external world based on a limited number of visible features. In another respect, the inner world should, however, not be reduced: just as the external world is four-dimensional (space and time), the internal model world should, according to a concept developed by Dickmanns, be four-dimensional, too [4, 6, 7]. It incorporates the physical laws of motion (in the form of differential equations) and combines them with the data measured in the image sequence. At certain points in time, 2-D projections of both worlds are generated (Figure 1.2). For the external world, this is done by capturing an image with a TV camera and extracting the relevant features from it. For the internal model world, an *equivalent* perspective projection (not the ill-conditioned, non-unique inverse!) is performed by computation. To compensate for the delays of the process (about 80 ms), an equivalent prediction of the state of the model world is included in the generation of the internal image. The most recent image produced by the camera (reduced to the relevant features) is compared to a predicted image derived from the model world; the two images compared relate to the same instant in time. Differences between them are used to adjust parameters of the model world. No inversion of perspective transformation is required.

There are then two 2-D feature images, both reflecting the state of the world at the same instant in time. Usually, the two images are similar, but not identical. The differences between them are evaluated and used to adjust the parameters of the model world in such a way as to reduce the differences. Comparison and adjustment are performed many times per second, thus causing the internal world to follow closely all changes of the external world.

Results Obtained with VaMoRs

Presently, VaMoRs is able to follow an unobstructed road and to adjust its speed according to the curvature of the road ahead of the vehicle. In 1987, numerous tests were successfully completed where it ran autonomously along the entire length of 24 km of a newly constructed highway that was not yet open to public traffic. During the experiments, the vehicle automatically entered the freeway, accelerated, reached its top speed of 96 km/h, and autonomously drove the entire available distance of 24 km. The experiments were successfully repeated numerous times under various weather conditions, including light rain and direct sunshine causing extreme shadows, mainly under bridges [44].

In 1988, experiments on campus roads were performed. These roads are without painted markers, and many shadows of trees and buildings make image interpretation difficult. Nevertheless, speeds of 50 km/h were reached even under adverse conditions of sunshine or rain. Powerful methods for *dynamic* computer vision have been developed to cope with such complex scenes in real time [4, 6, 7]. In both cases, highway and campus road, the smoothness and accuracy of steering were comparable to that of a human driver.

Avoiding collisions with obstacles is a topic of current research. Detection and tracking of large, well-recognizable obstacles were accomplished from within the vehicle while it was running at a speed of about 40 km/h [17, 18] (for more recent results see [35] and [37]). Approaching an obstacle and stopping in a predetermined distance in front of it has also been demonstrated [5].

1.4 Future Developments

The next long-term goal is to demonstrate the feasibility of a highway autopilot for motor vehicles to be used in public traffic. This means that a vision-guided vehicle, able to run fully automatically on ordinary highways without any special installations or markings, should be designed and tested. Of course, it must be designed in such a way that it conforms with all traffic rules and never annoys or endangers drivers of other vehicles.

At present, VaMoRs is limited to relatively simple scenes (e.g., empty roads). At the end of the project, it should be able to drive autonomously under all traffic situations on freeways. This includes driving in dense traffic, understanding signals and traffic signs, avoiding obstacles, and tolerating occasional violations of rules by drivers of other vehicles.

This is a very ambitious goal, indeed, and only achievable in a joint research effort by several institutions. Some of the numerous problem areas that must be addressed are

camera systems with a high dynamic range and the ability to adapt to widely
  varying illumination levels,

more powerful hardware for low- and intermediate-level real-time vision,
feature extraction under complex situations and difficult viewing conditions,
real-time knowledge bases for recognizing shapes and situations,
definition and implementation of all the behavioral competence required for
    safe driving in ordinary traffic, and
hardware and software concepts for implementing systems that are by orders
    of magnitude more complex than the present ones.

These topics are discussed in more detail in the following.

1.4.1 Camera Systems

Typical solid-state cameras (CCD cameras) have a dynamic range (referring
to the light intensities falling on different pixels of one image) of about 20 dB.
In natural scenes, much larger differences in brightness exist, sometimes up to
50 dB, which exceeds the dynamic range of the camera by a factor of 1000.
This may cause parts of a scene to appear totally black while other parts of
the same scene are extremely overexposed causing severe blooming.

The range over which the sensitivity of a camera can be adjusted is insufficient, too. If an iris is varied between $f/1$ and $f/32$, this corresponds to a
change of sensitivity over 3 orders of magnitude. This may be compared to the
adaptability of the human eye, which extends over 11 orders of magnitude
[33]. Much better sensor elements and also better signal processing are,
therefore, required for computer vision systems that are to function even under adverse lighting conditions.

The number of pixels in a typical camera is not sufficient to provide a wide
field of view and a high local resolution at great distances at the same time.
Therefore, it is necessary to use several cameras fitted with lenses with different focal lengths simultaneously. Experimental studies are necessary to decide
how many cameras will be required, where they should be placed, and how
well they should be stabilized.

1.4.2 Hardware for Low- and Intermediate-Level Real-Time Vision

To accomplish the stated goals, scenes of great complexity must be understood by the system. One prerequisite for handling such complex scenes is
very powerful hardware. Based on the well-proven real-time vision system
BVV 2, a successor has been designed. This system, the BVV 3, implements
the same architecture, although its parallel processors are of a more advanced
and more powerful type (Intel 80286 and 80386 instead of 8086). Also, the
concept of distributed image memories and the method of interprocesor communication were adopted from the predecessor. Thus, the global structure of
the system remains unchanged. (The BVV 3 is described in more detail in the
appendix.)

1.4.3 Feature Extraction

LANE BOUNDARIES

Presently, it is possible to recognize the markers of the lane being used with reasonable reliability if weather conditions are not too bad. Mandatory is the need to recognize adjacent lanes [46], also, and to classify different types of lane markers (e.g., broken, solid, branching or joining lines, different line widths). Furthermore, the ability of recognizing lanes under more difficult conditions, which sometimes occur, must be developed (e.g., construction areas with ambiguous markings or roads covered by water or snow).

OTHER VEHICLES AND OBSTACLES

How well external objects can be recognized depends, among other things, on their size, the distance, weather, and illumination conditions. At the moment, it is possible to recognize large objects, such as cars, from a distance of a few dozen meters. Desired is the recognition of objects with a diameter of a few centimeters from a distance of about 100 m. To accomplish this, lenses with long focal lengths must be used and should be stabilized. Considering the cost of an effective stabilization, algorithms should be developed that are able to analyze images even in the presence of motion blur [20].

SIGNS

Most of the signs posted along highways require for their interpretation the ability to read. Mainly, these are signs to assist the driver in finding the right direction. Probably reading systems that have been developed for other applications will be used for interpreting such signs. For other signs, requiring or forbidding certain actions, suitable pattern recognition methods must be developed.

1.4.4 Model Banks and Situation Recognition

To recognize 2-D and 3-D shapes, objects, and situations, it is necessary to provide dynamical models of moving objects as well as environment schemes that may be instantiated in a situation-dependent manner. Methods must be found to organize such types of knowledge in such a way that it is easily extended and accessible in real time.

To recognize situations, relationships between observed objects and their behavior on one hand, and stored models and schemes on the other hand, must be established. An internal spatio–temporal world must be constructed, and it must be updated according to the measured data from the real external world. Compared to the existing approaches, the size and the complexity of the model world must be drastically increased.

1.4.5 Behavioral Competence

The simplest behavior is driving along an empty lane with a suitably chosen speed, for example, according to the curvature of the road [8, 9]. Upon this basis more complex behaviors must be developed, including

stopping in front of an obstacle;
driving in convoys;
changing over to another empty lane, passing obstacles, and entering and leaving a highway; and
passing slower vehicles.

1.4.6 Hardware and Software Concepts

The complexity of the system to be developed will exceed that of the present one by orders of magnitude. This is true for all areas, including feature extraction, modeling, situation recognition, knowledge representation, and behavioral competence. To cope with this complexity, far more powerful computer systems and software concepts are required than presently available.

An improved hardware architecture for feature extraction has already been designed and built (see appendix) as a partial prototype. Similar, suitable hardware architectures for the higher levels of the system (world model, situation recognition, knowledge base) should be developed that are as efficient as the BVV 2 and BVV 3 for the lower levels.

For the representation and efficient processing of many different types of knowledge, software structures will have to be found. They should be as uniform as possible, should lend themselves to an efficient implementation, and should combine the advantages of typical artificial intelligence (AI) techniques with the numerical efficiency of the methods known from modern control theory and system dynamics.

1.5 Applications

There are many industrial applications for autonomous mobile systems, for example, mobile robots and transportation vehicles. Here, however, only those applications will be discussed in which such systems assist the drivers of motor vehicles on highways.

It should be understood that the availablity of an autopilot will never stop a driver from actively driving the car, either completely or partially, that is, in cooperation with the autopilot if he so desires [3]. It is always the human driver who selects the mode of operation. On the other hand, even if a system is intended to merely warn, monitor, or advise a driver, in order to be accepted, it must still have the demonstrated ability to drive by itself, because most human drivers will resent taking advice from a person or a machine not competent of driving.

To meet the requirements of different drivers and different situations, several different modes of operation of the autopilot should be available to the driver. A few of them are briefly discussed here.

1.5.1 Warning and Monitoring

In this mode, the driver operates the vehicle. The autopilot may, however, warn the driver in certain situations, for example, when he is violating a rule, when he follows too closely, when the speed is not adequate, or when an obstacle exists that he has apparently failed to notice. In addition, the autopilot could recognize if the driver drives unsafely, for example, being tired, intoxicated, or distracted. In such cases the driver could be warned in an unemotional way.

It is difficult to determine how the autopilot is supposed to react in this mode once it realizes that the driver has made a mistake and is about to cause an accident. Should it take control of the vehicle against the will of the driver? If it did, this would raise difficult questions of legal responsibility in the actual case of an accident. Besides, it would be an extremely difficult technical problem to provide the autopilot with the ability of assuming control of the vehicle in a situation that was already out of hand. This is certainly much more difficult than driving automatically from the beginning and avoiding such situations altogether.

1.5.2 Intelligent Cruise Control

This mode is just one example for a number of conceivable modes in which the driving is neither fully automatic nor fully manual. All of these modes have in common the requirement of a very careful design of the man–machine interface to avoid confusion of the driver or a situation in which driver and autopilot counteract each other. Also, the problem of legal responsibility must be carefully resolved.

The driver might choose this mode if he wants to be driven most of the time but feels that he masters some of the more complex (and more interesting) situations better than the autopilot. The vehicle would normally follow its lane automatically, selecting an adequate speed. The driver may at any time set upper limits for speed and acceleration. The autopilot is not allowed to change lanes; therefore, it will always stay behind slower traffic in its lane, and if the lane were blocked (perhaps by a broken down car), it would stop. If the driver wants to pass other vehicles or drive around an obstacle, he may do so by temporarily taking control. After the maneuver, he may, of course, reactivate the autopilot. This mode may well have the greatest potential for contributing to improve traffic safety.

1.5.3 Fully Automatic Driving

Finally, there is the fully automatic mode, similar to what is now common practice in commercial aviation. The driver would enter the highway manually, then switch on the autopilot, and relax. For those who do not consider highway driving their hobby and who cannot afford a chauffeur, this should be an attractive perspective.

Fully automatic driving might at first glance appear to be the most difficult mode to realize, but, in fact, it is the basis of the other modes. A warning system, for instance, must have the same full capacity to understand traffic situations as a complete autopilot, and in addition, it must understand the intentions of the driver. Semiautomatic systems, such as the intelligent cruise control, would probably not be approved by the authorities as long as they cannot act as a full autopilot in a sudden emergency caused, for example, by illegal actions of other drivers.

1.6 Conclusions

It will certainly take a long time before autonomous vehicles will be an everyday appearance on public roads, and certainly many difficult problems, in addition to the mentioned ones, will have to be solved. But, the first step has been taken, and there are approaches to solutions, some of them already validated experimentally. There is no indication why it should be impossible in the long run to construct autonomous highway vehicles that are less accident prone than vehicles controlled by human drivers.

Acknowledgment. Part of this work has been funded by the German Ministry of Research and Technology (BMFT) and by the Daimler-Benz AG.

Appendix. Dynamic Vision Systems: The BVV Family[2]

Adequate computer hardware is a key to practical robot vision. General-purpose computers, even very powerful ones, are definitely not adequate, neither are certain highly parallel computers designed for demanding numerical computations or for the simulation of neural networks. Surprisingly, multiprocessor systems containing a small number of processing elements, each of them based on a standard microprocessor of moderate performance, have been demonstrated to outperform much more expensive computer systems in robot vision applications. In the following sections, three generations of such vision systems are described, and the underlying principles guiding their design are explained. The high performance of these vision systems is

---

[2] This appendix is an abbreviated version of the paper by V. Graefe (1990), "The BVV Family of Robot Vision Systems." In *Proceedings of the IEEE Workshop on Intelligent Motion Control, Istanbul,* O. Kaynak, ed. pp. IP55–IP65. © IEEE.

due to their system architecture, rather than to the exceptional speed of their hardware elements.

Flexibility is a key issue. It is important in several respects, including the flexiblity of random access to pixel data by the processing elements, flexibility in dynamically concentrating the computing power of the system on those parts of an image containing the most relevant information at any particular moment, and flexibility in restructuring the system under software control to match the inherent structure of the vision task.

The decision to design hardware for real-time vision systems resulted from the desire to study vision and motion control in combination and to base this study on real-world experiments in which the motions of some mechanical device were to be controlled by vision. It was obvious that such an approach would involve tremendous experimental difficulties, but there was hope that it might lead to insights and an understanding of the essence of vision, which may not be gained otherwise. This hope was mainly based on the observation that in the organic world vision always seems to be linked to the ability of motion control. Organisms that cannot move, also cannot see. Another motivation for studying vision and motion control together originated from the expectation that sooner or later some vision-controlled robots or other sophisticated machines would find practical application and that eventually there may be a time when only the simplest machines and vehicles would not have a sense of vision.

A.1. Concept of Dynamic Vision

Vision for motion control, or robot vision, must always be real-time vision. Real-time can have many meanings. In the context of robot vision, it means that the vision system must enable the robot to perceive changes in its environment while they are occurring and soon enough to influence these changes by interacting with the environment or to react to them in some purposeful manner. The interpretation thereof depends on the environment and the type of motion to be controlled. A snail, for instance, would have little advantage from a very fast perception system because this animal is limited to sluggish motions, anyway. A fly, however, being capable of very brisk motions, sometimes depends on a correspondingly fast vision system for survival. For robots or other technical systems intended to interact with humans or to operate in an environment shaped for humans, a speed of reaction similar to that of a human would be desirable [16]. In order to meet this goal, a robot vision system should not introduce a delay of more than 100 ms in reporting an event in the environment or in providing data on some visible motion. The words real-time and robot vision will be used in this sense in the following.

Initially, the required speed was the most obvious problem in real-time vision. It was commonly known that the interpretation of a single static image took from a few minutes to a few hours on a powerful computer [30, 34, 36]. Reddy [34] had estimated that the segmentation of an image alone required

1000 operations per pixel. It was difficult to imagine how real-time vision should be feasible if it meant doing similar things as in static image processing, but somehow doing them faster by 3 to 5 orders of magnitude. Delaying the start of experimental work in real-time vision until more powerful computers would become available did not seem to be a solution for three reasons: (1) in the past, the complexity of algorithms had always grown at least as fast as the speed of available hardware; (2) it was felt that "hard" knowledge and fresh ideas gained from experiments were needed more than additional results from computer simulations; and (3) considering the fact that the tiny brains of some primitive insects are sufficient for real-time vision, it was not likely that the processing power of, for example, existing minicomputers was, in principle, insufficient.

Finally, the hypothesis evolved that vision for motion control, or as it is now called, dynamic vision, is something totally different from static image processing. This was the key to a solution. Attempts to verify this hypothesis led to the discovery of significant differences between the two fields, a new understanding of the nature of dynamic vision, and the design of practical robot vision systems.

A.2 Architecture for Dynamic Vision

The design of vision hardware was never considered an end in itself but rather a necessary prerequisite for studying dynamic vision in real-world experiments. The initial approach was, therefore, to minimize the design effort by using available minicomputers to the greatest possible extent and build only a peripheral device enabling such a computer to process image data in real time. Accordingly, the first member of the BVV family of vision systems, the BVV 1, was initially intended to be such a peripheral device, and it was often referred to as a preprocessor in relation to an associated minicomputer.

Figure A.1 shows the overall structure of a vision system based on these considerations. An image sequence processor is inserted between the camera and a conventional computer. From a general point of view, the task of the image sequence processor is data reduction. This is done by extracting relevant information from the pixel data and converting it into a compact symbolical form, well suited for transmitting it to the master computer. According to this concept, the bulk of the computation was to be performed by a conven-

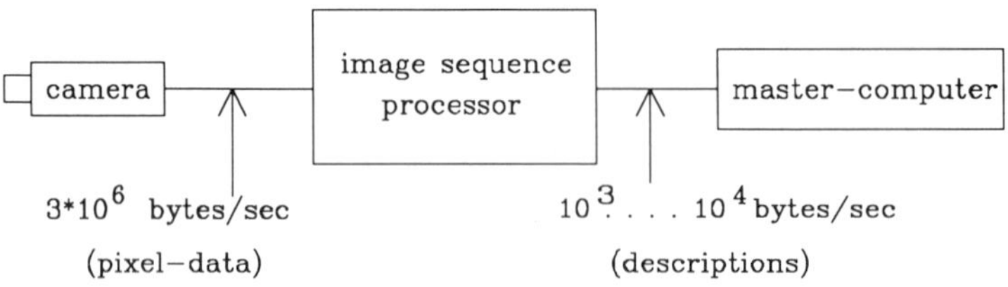

FIGURE A.1. Hierarchical robot vision system.

tional minicomputer (master computer), while a specialized image sequence processor was used only for relieving the master computer from the task of inputting pixel data at high speed. The master computer is, thus, free to do what it can do best: computation on relatively small sets of data.

The underlying idea behind this structure was that the minicomputer, which was to be used as the master computer, was a much more powerful computer than the image sequence processor, which was based on one of the 8-bit microprocessors available at that time.

With the advent of ever stronger microprocessors, and eventually the PC, this view has changed. Later BVV systems, while still utilizing commercial hardware as much as possible (e.g., single board computers), are completely self-contained vision systems, except that an additional PC is used for mass storage and as a human interface. The reason is that, at least in not-too-complex situations as they have been studied up to date, the bulk of the processing actually occurs in the lower levels of a vision system (feature extraction, 2-D object modeling [14]) for which the BVV 1 was originally intended. The task originally assigned to the master computer (dynamical object model and assessment of a relatively simple situation) is comparatively small and may, therefore, be performed by the image sequence processor as an additional task, using only a fraction of that system's resources. In fact, it was soon discovered that even the BVV 1, in spite of its weak processing elements, could be used quite effectively without a master computer. General-purpose computers, including minicomputers, by themselves are not suited for dynamic vision. One obvious reason is the fact that they do not have an input channel capable of feeding them continuously with data from a TV camera. If a camera signal is digitized, a data stream of 4×10^6 to 10×10^6 bytes per second results. Even though some modern computers may have direct memory access, DMA, channels capable of handling such data streams, the continuous inputting of video data at such rates would absorb a large part of the memory access bandwidth, thus reducing the remaining computing power significantly. This was even more evident 13 years ago when the image sequence processor BVV 1 was conceived and when memory access times were about 10 times longer than they are today.

A system, as shown in Figure A.1, was actually built and used for experiments in robot vision. The master computer was, at different times, a DEC PDP 11/60, a VAX 750, and a Perkin Elmer 3220. The image sequence processor was a specially designed multi-microprocessor system, later named BVV 1. It was the first member of the BVV family.

A.3. Features of Dynamic Vision

The design of the BVV 1 was based on four concepts.

1. To control motion, it is sufficient to interpret only those parts of an image sequence that contain relevant features; it is unnecessary to analyze all other parts of the images, for instance, the background. Since, typically,

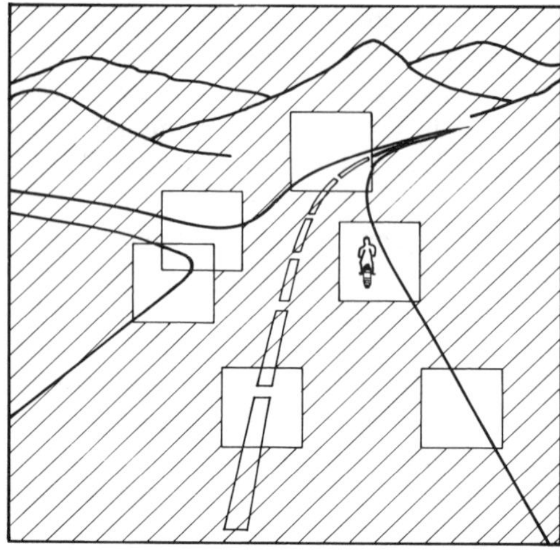

FIGURE A.2. A few small regions of an image contain almost all information relevant for motion control.

only small areas within each image contain relevant features, concentrating all the available processing power on the most relevant parts of each image leads to a significant gain in efficiency. Figure A.2 illustrates this concept, using the task of guiding an autonomous vehicle on a road as an example. Usually, in vision for motion control, the combined area of all regions of interest is less than 10% of the image area. Limiting all processing to these regions thus leads to a very efficient use of resources. Stating it differently, vision systems not limiting their processing to the relevant parts of a scene waste at least 90% of their resources.

2. Controlling motion in real time requires the highest possible processing speed. The task of image interpretation is, therefore, partitioned into several subtasks that are assigned to an equal number of subprocessors (parallel processors) within the system.

3. To prevent communication between subprocessors from becoming a bottleneck for the performance, communication channels with the necessary bandwidth must be provided. Also, the system should be structured in such a way as to minimize the need for interprocessor communication altogether. This may be done by defining the subtasks properly, making them, and thus the subprocessors associated with them, largely independent of each other.

4. Flexibility is a key to the high performance of a vision system. The subprocessors in the BVV 1 have, therefore, free random access to each pixel within their region of interest and they may freely position this region anywhere in the image. Moreover, the internal structure of the image sequence processor is, instead of being fixed, definable by the application programs. It is thus possible for appropriately written application pro-

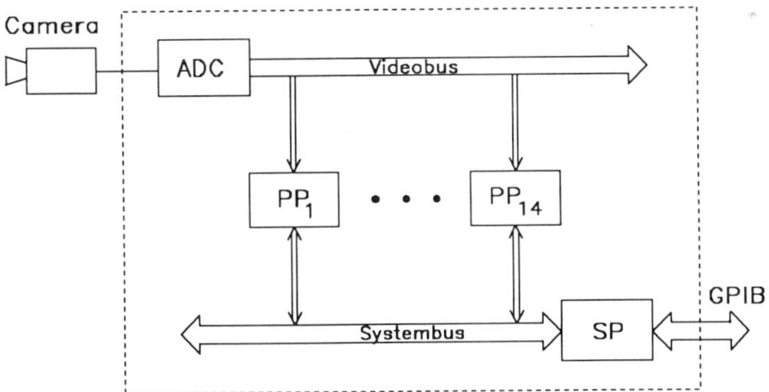

FIGURE A.3. Block diagram of the image sequence processor BVV 1.

grams to configure the BVV 1 as a parallel, hierarchical, or mixed, system of subprocessors.

Figure A.3 shows the block diagram of the BVV 1. It may contain up to 15 microcomputers: one system processor, SP, and up to 14 parallel processors, PP. The processors are custom-made single-board computers using the Intel 8-bit microprocessor 8085 A as a central processing unit, CPU. They have 16 kBytes of memory each (when the design was started in 1977, this seemed to be more than enough). Each PP is a self-contained, freely programmable computer, intended and equipped to execute its own task.

An analog-to-digital converter, ADC, digitizes the video signal and distributes it in digital form via the videobus to all parallel processors. Each PP copies the part of the image that constitutes its momentary region of interest into an internal private memory. Having a private image memory in each PP avoids delays that would exist if a central image memory would be shared by all PPs.

The private image memories of the BVV 1 are rather small (1 kByte each) because of the formerly high cost of fast memory chips. The region of interest of a PP is thus determined by the hardware to be contained in a rectangle of up to 1024 pixels. Location, size, shape, and sparseness of these rectangular "windows" are dynamically controlled by the microcomputer at run time.

Messages can be exchanged between any two PPs via the systembus. An external host computer is connected to the BVV 1 via a general purpose interface bus, GPIB, and may also communicate with any PP. The SP transports all messages, relieving the PPs completely from this task; they merely have to deposit outgoing messages into their private outbasket and periodically check their private inbasket. The "baskets" are FIFO (first in, first out) buffers within each parallel processor. This intermediate storage of all messages eliminates the necessity of synchronizing the sender of a message with its receiver, thus making the internal communication within the vision system both easy to program and efficient.

Typically, each PP will produce one result message per video cycle, and it will occasionally receive a message from higher levels, instructing it, for instance, to enter a different mode of operation. The total message traffic is, therefore, so weak that the systembus is never a bottleneck that might slow down the operation of the system. The videobus does not create a bottleneck either, since all PPs can receive video data from it simultaneously without disturbing each other.

The structure of the BVV 1 is clear and simple. It is, however, more flexible than might be apparent at first glance. Although all PPs are equal and are connected to the systembus in parallel, it is, nevertheless, possible to give the system a hierarchical structure by merely letting the application programs in the PPs communicate with each other in the appropriate way. For instance, PP_1 through PP_4 could send their results to PP_5, and PP_6 through PP_8 could send results to PP_9. PP_5 and PP_9 could then be considered to be on a higher hierarchical level. If they, in turn, produced results and sent them to, for example, PP_{12}, this would constitute still another level in the hierarchy. The interesting point is that the grouping of the parallel processors into a parallel, hierarchical, or mixed structure is perfectly flexible. The structure can be freely rearranged at any time by the software, even during run time.

In a typical robot vision application, each parallel processor tracks an object (or a part of an object) that has been assigned to it by a higher level process, running, for example, on the master computer.

It then produces a description of those features within its window for which it has been instructed to look. The intermediate results generated by the feature extracting PPs are usually in the form of coordinates of a feature and some parameters related to it. They are passed on to the system processor, from where they are transmitted to the master computer or, for example, in a system without a master computer, to another PP. The transformation from the video data to the sequence of result messages constitutes a data reduction by more than 3 orders of magnitude. These results are a very compact description of the relevant aspects of the scene. They are not sufficient to reconstruct the original image (which is not the goal), but they are sufficient to build and update an abstract description of the scene, allowing a robot to perform its task.

The image sequence processor BVV 1 was successfully tested in experiments involving motion control by computer vision [11, 19]. In connection with an algorithm developed by Meissner [29], it was demonstrated that the BVV 1 could provide, by visual observation of the motions of an unstable vertical rod ("inverted pendulum"), the data necessary to stabilize the rod. Even a rod that was only 30 cm long, and therefore, too agile to be stabilized by a person, could be stabilized.

During such experiments, it was realized that the advantage of having a minicomputer as a master computer had been overestimated. In fact, the inherent flexibility of the BVV 1 made it easy to let one of its parallel processors assume the role of the master computer. Figure A.4 shows such an

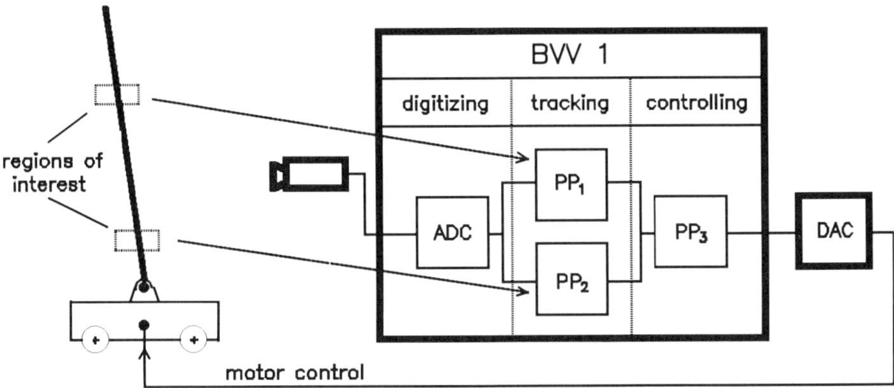

FIGURE A.4. Stabilizing an inverted pendulum using only the BVV 1 without any master computer.

application, an inverted pendulum stabilized by the BVV 1 without a master computer. What had begun as a data reduction unit thus turned into a complete robot vision system.

An important and somewhat surprising lesson learned by working with the BVV 1 was that, *if the system architecture is correct,* four standard 8-bit microprocessors (1 for communication and 3 for image processing) are sufficient for a robot vision task as demanding as the stabilization of an agile mechanical system.

Nevertheless, it became obvious that the computing power of the 8085 A microprocessor was sufficient for the real-time interpretation of visually simple scenes only. In order to be able to interpret more complex scenes, it was decided to develop a more powerful, but otherwise similar, image sequence processor, the BVV 2 [10].

Another vision system, also based on the concepts that had guided the design of the BVV 1, and remarkably similar to the BVV 2, was later built and described by an unrelated Japanese group [21].

A.4. The BVV 2

The block diagram (Figure A.5) clearly shows the architectural similarity between the vision system BVV 2 and its predecessor BVV 1. The main difference in a conceptual sense is a videobus system with originally 2, and later 4, channels allowing the simultaneous transmission of up to 4 independent image sequences instead of the simple videobus of the BVV 1. Other differences, not visible in the block diagram, are more powerful microprocessors and much larger memories in all subprocessors.

The multichannel videobus system may be used, for instance, in the following ways:

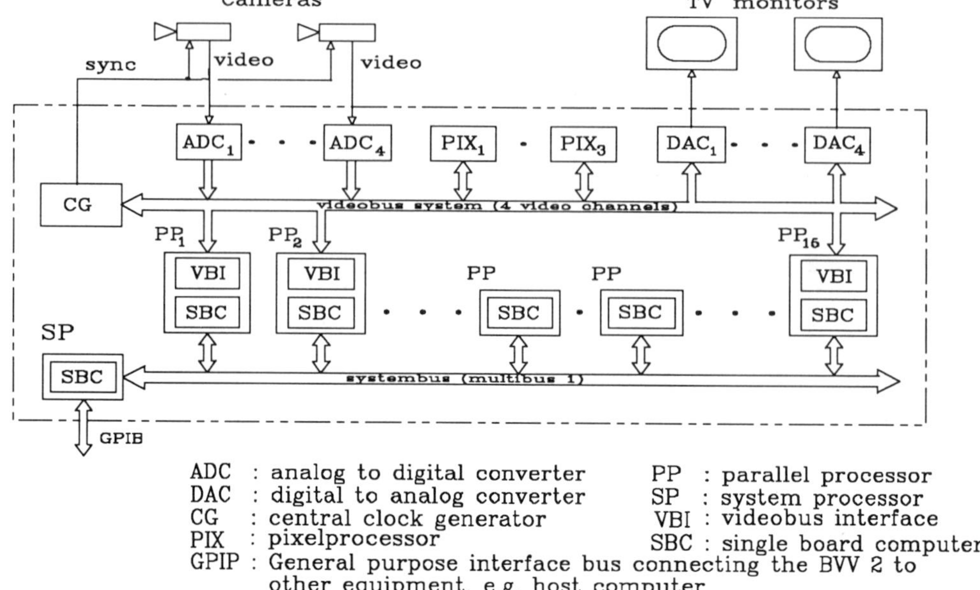

FIGURE A.5. The real-time vision system BVV 2.

employing two identical cameras simultaneously for stereo vision;

simulating the foveal and the peripheral vision of the eye by using two, or more, identical cameras equipped with lenses of different focal lengths and mounted on a common movable platform;

using an RGB camera for full color vision;

using several cameras pointing into different directions, for instance, the front and rear of a vehicle; and

using fewer than four cameras and, in addition, one or more preprocessing units ("pixelprocessors") to derive "on the fly" from an original image a new image containing, for instance, only edges or only interesting points.

If more than one camera is connected to the system, they must run synchronously. A central clock generator in the BVV 2 provides the control signals necessary to synchronize the cameras and all other parts of the system. In the BVV 1, where only one camera could be used, the camera provided the synchronization signals.

Each parallel processor consists typically of a single-board computer, SBC, and a videobus interface, VBI. The SBCs are standardized commercial products. They were originally based on Intel's 16-bit microprocessor 8086. This processor is generally more powerful and faster than the 8085 A, but, what is most important in the context of image processing, it is much more efficient in addressing two-dimensional arrays of data. In this respect, the 8085 A is quite weak because of its insufficient number of index registers. For typical tasks in feature extraction the 8086 is about 10 times faster than the 8085 A.

Each one of the original parallel processors has between 40 and 148 kBytes of memory. The greater speed of the 8086, in combination with the larger memory of each parallel processor, makes it possible to run more complex programs in real time and to use more refined methods for the extraction of features from the image than with the older BVV 1.

Communication between the subprocessors within the BVV 2 is organized in a manner similar to the one used in the BVV 1, using FIFO buffers to decouple the subprocessors from each other. However, since the SBCs of the BVV 2 are provided with dual-port memories, they have direct access to each other's memory. Messages within the BVV 2 are, therefore, transferred by the system processor reading them from, or writing them into, a parallel processor's memory. The FIFO buffers, which physically exist in the BVV 1, are thus simulated by software in the BVV 2. This made the interprocessor communication slightly slower than with hardware FIFOs, but it allowed commercially available single board computers to be used for all PPs and the SP without modification. This had the advantage that later some of the SBCs could be replaced by even more powerful ones based on the microprocessors 80286 and 80386.

The VBI is a custom-made printed circuit board containing all those parts of the PP that are unavailable on a standard SBC. Its main task is to select from the videobus those pixels that belong to the selected window and make them available to the parallel processor. Since the data rate on the videobus is much higher than on the local processorbus, the video data must be temporarily stored in a buffer memory before being placed in the parallel processor's main memory.

In the BVV 1, a 32-byte FIFO buffer had been used for this purpose (Figure A.6). Therefore, the number of pixels within one row of the window has been limited to 32. This restriction turned out to be inconvenient at times, and a

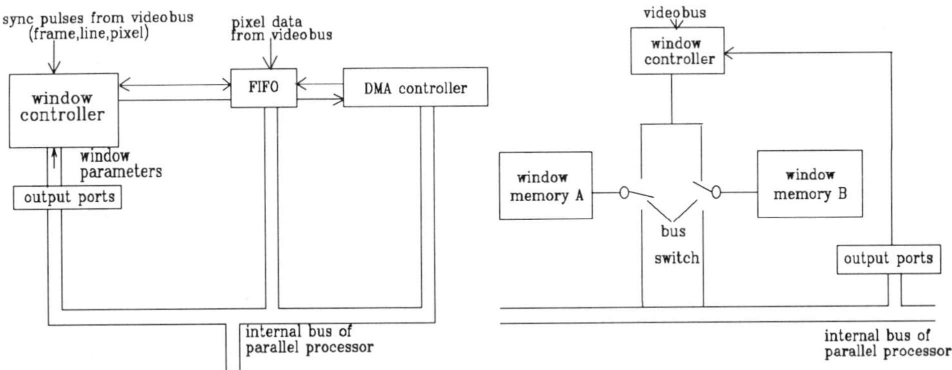

FIGURE A.6. Videobus interfaces of the BVV 1 (left) and the BVV 2 (right).

better solution was desired for the BVV 2. Another restriction was that the window parameters controlling shape, size, and position of the window could only be changed during the interframe gap of the video signal. This could cause an unnecessary increase of the vision system's response time by as much as 20 ms in the worst case.

The videobus interface of the BVV 2 avoids these restrictions. It contains two window memories with a capacity of 4 kBytes each. The window memories are connected through bus switches to the videobus and to the local processor bus in such a way that always one of them is connected to the videobus and the other one to the processorbus. The switches are controlled by the parallel processor's CPU. In operation, the window controller selects, according to the window parameters set by the processor, those pixels that belong to the window and stores their gray levels in the window memory connected to the videobus. As soon as all data belonging to the window have been stored, the processor may, by sending a command to the bus switches, exchange the two window memories. This means that the window memory, which has just been loaded with pixel data, is now connected to the local bus, thus becoming part of the processor's main memory, while the other image memory, which had been connected to the processor up to this moment, is now connected to the videobus, ready to receive the next set of pixel data. In this way, no time is lost in transferring data from the window memory to the parallel processor's main memory.

The BVV 2 has proven to be an efficient hardware basis for dynamic computer vision. Among the applications studied with it are autonomous vehicles, including a highway vehicle [6, 7, 44], and a vision-guided autonomous indoor vehicle to be used in factories and warehouses [27].

The ability to process image sequences at a high rate (50 to 60 images per second) was essential for allowing the experimental vehicle VaMoRs to run autonomously at its maximum speed of 96 km/h on highways, limited only by the power of its engine, not by the vision system.

Again, the main reason for the relatively good system performance, as evidenced in the applications, is not the sheer computing power of the processors of the BVV 2 but rather its flexibility, which allows the entire power of the system to be concentrated on the relevant regions of the image.

A.5. The BVV 3

After having worked with the BVV 2 for some time, it became clear that certain types of algorithms were particularly useful for feature extraction in dynamic scenes (e.g., controlled correlation [23]) and that the parallel processors spent a significant amount of time executing a fairly limited variety of operations (e.g., addition, subtraction, maximum detection, and, surprisingly, data conversion from the 8-bit format used on the videobus to the 16-bit format used in computing). Also, it was found that in a typical program algorithms from two distinct classes changed over and over again: (1) schematic application of operators on groups of pixels and (2) more or less complex

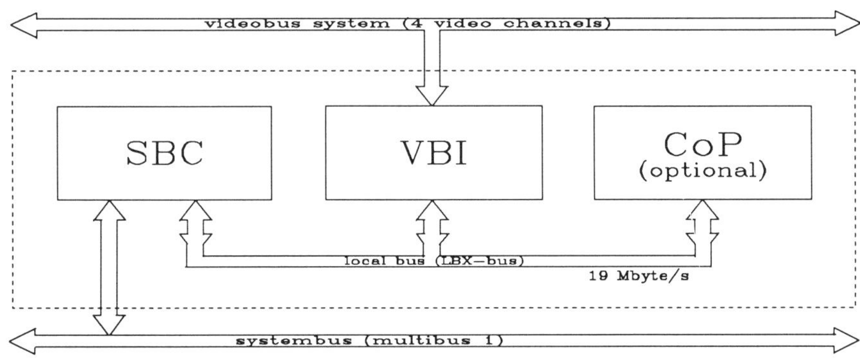

FIGURE A.7. The parallel processor of the BVV 3.

decisions in analyzing the results of the more schematic processing and determining which operator to apply next [28].

A standard microprocessor appeared to be well suited for the analysis and decision part, but it seemed likely that special hardware, similar to a signal processor, would be much more effective for the schematic part. It was, therefore, decided to develop the special hardware and implement it as a coprocessor, to be tightly coupled to an associated standard microprocessor.

If such a coprocessor is used in combination with a microcomputer, a very powerful device for feature extraction results. In fact, its performance would probably be limited by the quantity of pixel data ($4K$) available in a window of the videobus interface of the BVV 2. Fortunately, in the meantime, prices for memory chips were reduced to a point where it seemed justified to design a new videobus interface that would store an entire image rather than only a selected window of limited size.

Figure A.7 shows the structure of the new parallel processor. It connects to the same videobus and to the same systembus as its predecessor in the BVV 2. In fact, the overall architecture of both vision systems is exactly the same; they differ only by the improved performance in feature extraction of the parallel processors of the BVV 3.

The adherence to commercial standards had been very beneficial in the past, having permitted some of the single-board computers of the BVV 2 to be replaced by more modern ones with the newer and more powerful microprocessors Intel 286 or 386 as they became available. These same SBCs are initially being used in the BVV 3, possibly to be replaced again by even more powerful ones at a later time. They are connected to the VBI and coprocessor (COP) via a semistandardized LBX bus having a theoretical bandwidth of 19 MB/s.

The videobus interface is similar in its function to the one of the BVV 2, but instead of the two window memories, it has two full image memories. Among the advantages of such large memories are the facts that large objects, which would not fit into a single window, are easier to handle and that tracking objects is simpler because the need of repositioning a window in each tracking step is eliminated. This may save some time (up to one video cycle) in certain situations.

The smaller windows did, however, have one advantage: it was easy to visibly mark the position of a window on a TV screen, allowing the user or programmer of the BVV 2 to verify at all times that each parallel processor was operating on the right parts of the image and was correctly tracking its assigned object. This has proven to be an indispensable aid in debugging programs and for demonstrating results. Naturally, this simple and effective method cannot be used any more if the size of the "window" is equal to the size of the entire image. As a replacement, the new VBI contains a "marking memory" that records accesses of a processor to each pixel in the video memory. The contents of the marking memory may be graphically overlaid on the video data on a TV screen, allowing the operator to observe which pixels have actually been accessed in the last video cycle.

Figure A.8 shows the internal structure of the coprocessor. It consists of

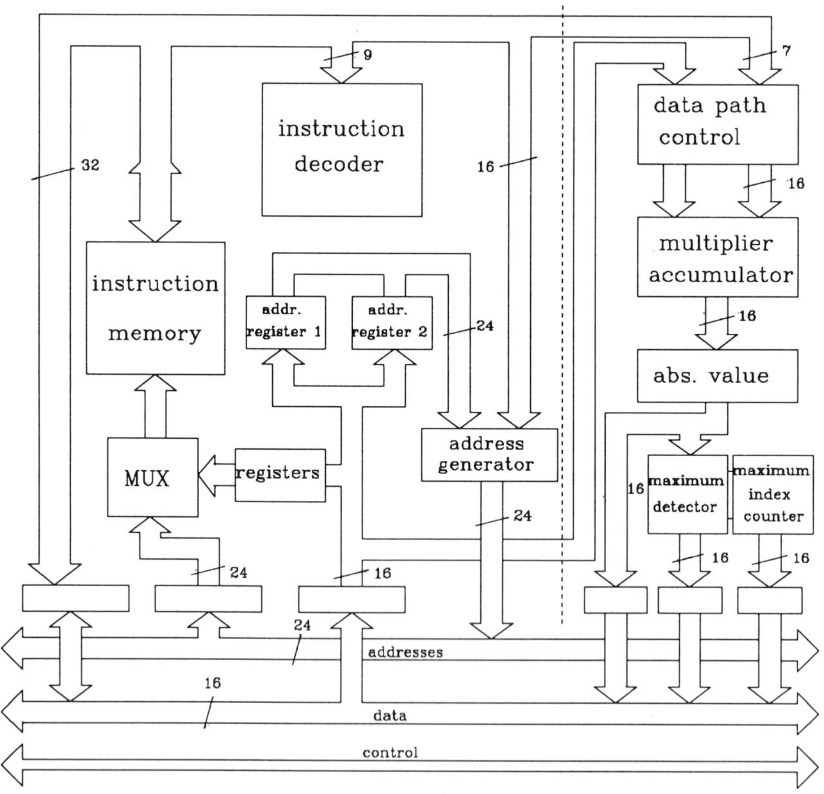

FIGURE A.8. Internal structure of the coprocessor.

two main components: the control unit (left part) and the arithmetic unit (right part). The control unit comprises the instruction and constant-data memory (presently $128K \times 32$ bit), addressed via a separate internal bus; three main registers; an LBX address generator; the internal address generator, the program sequencer; and the instruction decoder. The arithmetic unit consists mainly of a multiplier/accumulator, a complex data-type-conversion hardware, and a parallel maximum detector with index counter. Normally, 16-bit signed fixed-point numbers in the interval $(-1, +1)$ are used.

The coprocessor has a pipeline architecture with seven stages. One basic operation of the coprocessor, being executed in *one* cycle, may consist of the following steps: fetch the instruction and a constant; compute addresses; fetch the data; convert types; compute multiplication, addition, and absolute value; detect a maximum in the stream of intermediate results; record its position in the stream. As many as 10^7 such complex operations may be executed in 1 s.

The coprocessor was conceived by Kuhnert [24]. Additional details on the coprocessor are given in Kuhnert [26] and Kuhnert and Graefe [28].

The coprocessor's architecture resembles, in many ways, a typical signal processor. There are, however, some important differences that made it worthwhile to develop the coprocessor, rather than using one of the commercially available signal processors. The most important one is the limited addressing ability of all available signal processors (a few dozen K bytes at most). A large program memory is, however, essential for feature extraction in dynamic vision for at least two reasons. (1) In writing programs, a trade-off exists between speed and program size; for instance, if all loops are eliminated, a long program results that may, however, execute very fast, because no time is spent in computing addresses and checking loop termination conditions. (2) Many different feature extraction operators (programs) should reside in the memory at the same time to be selected at random by the microcomputer at run time. The coprocessor is able to address 16 MB of program memory. Some special functions that are unusual, but quite useful for the intended application, constitute another difference. Examples are an address generator allowing an algorithm to be shifted to any part of the image or a maximum detector that keeps track of the maximum in a series of intermediate results without requiring additional computing time.

A processor as complex as the coprocessor would normally be very difficult to program. To relieve the user from this task, a program generator was developed that allows the user to define the intended image processing algorithm interactively on a graphic screen of a computer. The machine program for the coprocessor is then generated automatically [13]. In addition, an emulator exists that lets the user test and debug the algorithms on a PC.

Some measurements were performed on a preliminary version of the coprocessor running at a reduced speed of 8.8 MHz instead of the design speed of 10 MHz [28]. Convolution of a 256×256 image with a 3×3 mask took 82 ms. The evaluation of a low-pass pyramid by computing the mean of 2×2 pixels and subsampling the resulting image by a factor of 2 took 16 ms. It should be kept in mind that this speed is reached by a programmable device and not by hardwired special hardware.

These figures are, however, given for comparison purposes only. Schematic application of an algorithm on all pixels of an image is of little interest in dynamic vision. The full power of the coprocessor becomes obvious only if its built-in flexibility is utilized. For instance, a pyramid over any 128×128 pixel region may be computed in 4 ms, and a pyramid over any 64×64 pixel region in 1 ms. Some typical algorithms used for feature extraction on the BVV 2 ran more than 100 times faster after being ported to the coprocessor.

On the other hand, it must be understood that there are other algorithms that contain many decisions and branches and, therefore, cannot be sped up nearly as much by using the coprocessor. A relatively complex road tracking program was tested on a single parallel processor of the BVV 3 [26, 28]. It used 12 regions of interest (6 for tracking sections of both road edges and 6 for following the road edges in the image, starting at the tracked sections) and one geometrical model for each of the two road edges to make the tracking robust in the presence of shadows from trees and of wet spots on the road. This entire program ran with a cycle time of 4 ms.

A.6. Conclusions

Looking for the inherent structure and relative simplicity of the problem of dynamic vision has led to a computer architecture that is well matched to the task of visual motion control [14, 15].

Over a period of 12 years, a family of three such systems has been built and tested in experiments relating to motion control by real-time computer vision.

All three are coarsely grained multiprocessor systems, comprising about a dozen loosely coupled microcomputers. Special attention has been paid to a careful design of the internal communication structure according to the requirements of the task. Two main communication channels exist in each of the BVV systems: a videobus for distributing pixel data and a systembus for interprocessor communication.

The architecture of these robot vision systems reflects the inherent structure of dynamic vision. It is very different from that of a typical image processing system. In spite of their relative simplicity, the BVV systems have proven to be a very powerful hardware basis for various real-world experiments in which fast-moving mechanical systems or vehicles were controlled by dynamic vision.

References

[1] Annaratone, M., Arnould, E., Gross, T., Kung, H. T., Lam, M. S., Menzilcioglu, O., Sarocky, K., and Webb, J. A. (1986). "Warp Architecture and Implementation." *13th Annual International Symposium on Computer Architecture, Tokyo, 1986.*

[2] Davis, L. S., Le Moigne, J., and Waxman, A. M. (1986). "Visual Navigation of Roadways." Preprints. *Conference on Intelligent Autonomous Systems, Amsterdam, December 1986*, pp. 21–30.

[3] Dickmanns, E. D. (1986). "Computer Vision in Road Vehicles Chances and Problems." Preprint. *ICTS-Symposium on Human Factors Technology for Next-Generation Transportation Vehicles, Amalfi, Italy.*

[4] Dickmanns, E. D. (1987). "4D-Dynamic Scene Analysis with Integral Spatio-Temporal Models." In R. Bolles, ed., *4-th International Symposium on Robotics Research, Santa Cruz.* MIT Press, Cambridge, Massachusetts.

[5] Dickmanns, E. D., and Christians, Th. (1989). "Relative 3-D State Estimation for Autonomous Visual Guidance of Road Vehicles." In T. Kanade et al., eds., *Intelligent Autonomous Systems 2*, Amsterdam, pp. 683–693.

[6] Dickmanns, E. D., and Graefe, V. (1988). "Dynamic Monocular Machine Vision." *Machine Vision and Applications* **1**, 223–240.

[7] Dickmanns, E. D., and Graefe, V. (1988). "Applications of Dynamic Monocular Machine Vision." *Machine Vision and Applications* **1**, 241–261.

[8] Dickmanns, E. D., and Zapp, A. (1986)."A Curvature-Based Scheme for Improving Road Vehicle Guidance by Computer Vision." In N. Marquino and J. H. Wolfe, eds., *Mobile Robots. Proceedings of the SPIE,* **727**, 161–168.

[9] Dickmanns, E. D., and Zapp, A. (1987): "Autonomous High-Speed Road Vehicle Guidance by Computer Vision." Preprint. *10th IFAC-Congress, Munich* **4**, 232–237.

[10] Graefe, V. (1983). Ein Bildvorverarbeitungsrechner für die Bewegungssteuerung durch Rechnersehen." In H. Kazmierczak, ed., *Mustererkennung 1983*, NTG Fachberichte, VDE-Verlag, pp. 203–208.

[11] Graefe, V. (1983). "A Preprocessor for the Real-Time Interpretation of Dynamic Scenes." In T. S. Huang, ed., *Image Sequence Processing and Dynamic Scene Analysis*, Springer-Verlag, Berlin and New York, pp. 519–531.

[12] Graefe, V. (1984): "Two Multiprocessor Systems for Low-Level Real-Time Vision." In J. M. Brady, L. A. Gerhardt, and H. F. Davidson, eds., *Robotics and Artifical Intelligence*, Springer-Verlag, Berlin and New York, pp. 301–308.

[13] Graefe, V. (1989a). "A Flexible Semiautomatic Program Generator for Dynamic Vision Systems." *Proceedings, International Workshop on Industrial Applications of Machine Intelligence and Vision—MIV 89, Tokyo,* pp. 100–105.

[14] Graefe, V. (1989b). "Dynamic Vision Systems for Autonomous Mobile Robots." *Proceedings, IEEE/RSJ International Workshop on Intelligent Robots and Systems (IROS '89), Tsukuba,* pp. 12–23.

[15] Graefe, V. (1990). "On the Design of Robot Vision Systems." *NATO ASI Active Perception and Robot Vision. Maratea, July 89.* To appear in NATO ASI Series, Springer-Verlag, Berlin and New York.

[16] Graefe, V., and Kuhnert, K.-D. (1988). "A High-Speed Image Processing System Utilized in Autonomous Vehicle Guidance." *Proceedings of the IAPR Workshop on Computer Vision, Tokyo,* pp. 10–13.

[17] Graefe, V., and Regensburger, U. 1988. "Analysis and Measurement of Objects in the Path of a Vision-Guided Mobile Robot." *International Advanced Robotics Program—Proceedings of the Second Workshop on Manipulators, Sensors, and Steps Toward Mobility, Manchester, October 1988.*

[18] Graefe, V., and Solder, U. (1988). "Detection of Objects in the Path of a Vision-Guided Mobile Robot." *International Advanced Robotics Program—Proceedings*

of the Second Workshop on Manipulators, Sensors, and Steps Toward Mobility, Manchester, October 1988.

[19] Haas, G. (1982). "Meßwertgewinnung durch Echtzeitauswertung von Bildfolgen." Dissertation, Fakultät für Luft- und Raumfahrttechnik der Universität der Bundeswehr München.

[20] Haas, G., and Graefe, V. (1983): "Locating Fast-Moving Objects in TV-Images in the Presence of Motion Blur." In A. Oosterlinck and A. G. Tescher, eds., *Applications of Digital Image Processing V., Proceedings of the SPIE* **397**, 440–446.

[21] Inoue, H., and Mizoguchi, H. (1984). "A Flexible Multiwindow Vision System for Robots." *Proc. 2nd. Int. Symposium of Robotics Research, Kyoto*, pp. 42–49.

[22] Kanade, T., Thorpe, C. E., and Whittacker, W. (1986). "Autonomous Land Vehicle Project at CMU." *Proceedings of the ACM Computer Conference, Cincinnati.*

[23] Kuhnert, K.-D. (1986). "A Model-Driven Image Analysis System for Vehicle Guidance in Real Time." *Proceedings of the Second International Electronic Image Week, CESTA, Nice*, pp. 216–221.

[24] Kuhnert, K.-D. (1986). "A Vision System for Real-Time Road and Object Recognition for Vehicle Guidance." In N. Marquino and W. J. Wolfe, eds., *Mobile Robots. Proceedings of the SPIE* **727**, 267–272.

[25] Kuhnert, K.-D. (1986). "Comparison of Intelligent Real-Time Algorithms for Guiding an Autonomous Vehicle." In L. O. Hertzberger, ed.. *Proceedings: Intelligent Autonomous Systems, Amsterdam.*

[26] Kuhnert, K.-D. (1988). "Zur Echtzeit-Bildfolgenanalyse mit Vorwissen." Dissertation. Fakultät für Luft- und Raumfahrttechnik der Universität der Bundeswehr München.

[27] Kuhnert, K.-D. (1990). "Dynamic Vision Guides the Autonomous Vehicle ATHENE." *Japan–USA Symposium on Flexible Automation, Kyoto, July 90*, pp. 507–510.

[28] Kuhnert, K.-D., and Graefe, V. (1988). "Vision Systems for Autonomous Mobility" *Proceedings, IEEE International Workshop on Intelligent Robots and Systems, IROS '88, Tokyo*, pp. 477–482.

[29] Meissner, H.-G. (1982). "Steuerung dynamischer Systeme aufgrund bildhafter Informationen." Dissertation, Fakultät für Luft- und Raumfahrttechnik der Universität der Bundeswehr München.

[30] Moravec, H. P. (1980). *Obstacle Avoidance and Navigation in the Real World by a Seeing Robot Rover.* Robotics Institute, Carnegie–Mellon University, Pittsburgh.

[31] Mori, H., Ishiguro, H., Kotani, S., Yasutoml, S., and Chino, Y. (1988). "A Mobile Robot Strategy Applied to Harunobu-4." *9th International Conference on Pattern Recognition, Rome, November 1988.*

[32] Niepold, R. (1988). Personal communication.

[33] Pohl, R. W. (1958): *Optik und Atomphysik. 10. Auflage*, Göttingen, Heidelberg, p. 335.

[34] Reddy, R. (1978). "Pragmatic Aspects of Machine Vision." In A. Hanson, and E. Riseman, eds., *Computer Vision Systems.* Academic Press, San Diego, pp. 89–98.

[35] Regensburger, U., and Graefe, V. (1990). "Object Classification for Obstacle Avoidance." *SPIE Symposium on Advances in Intelligent Systems. Boston, November 1990*, pp. 112–119.

[36] Shirai, Y. (1979). "On Application of 3-Dimensional Computer Vision." *Bul. Electrotech. Lab.* **43** (6), 358–377.

[37] Solder, U., and Graefe, V. (1990). "Object Detection in Real Time." *Proceedings of the SPIE Symposium on Advances in Intelligent Systems.* Boston, November 1990, pp. 104–111.

[38] Thorpe, C. E., and Kanade, E. (86). "Vision and Navigation for the CMU Navlab." In N. Marquino and W. J. Wolfe, eds., *Mobile Robots. Proceedings of the SPIE* **727**, 261–266.

[39] Tsugawa, S., Yatabe, T., Hirose, T., and Matsumoto, S. (1979). "An Automobile with Artificial Intelligence." *6th International Joint Conference on Artificial Intelligence, Tokyo,* pp. 893–895.

[40] Tsugawa, S., Hirose, T., and Yatabe, T. (1984). "An Intelligent Vehicle with Obstacle Detection and Navigation Functions." *Proceedings IECON '84, Tokyo* **1**, 303–308.

[41] Turk, A., Morgenthaler, G., Gremban, D., and Marra, M. (1987). "Video Road-Following for the Autonomous Land Vehicle." *Proceedings IEEE Intern. Conf. on Robotics and Automation, Raleigh, North Carolina,* pp. 273–280.

[42] Waddington, J. F. (1988). Personal communication.

[43] Waxman, A. M., Le Moigne, J., and Davis, L. S. (1986). "Maryland Autonomous Land Vehicles Project." *Proceedings of the Second International Electronic Image Week, CESTA, Nice,* pp. 208–215.

[44] Zapp, A. (1988). "Automatische Straßenfahrzeugführung durch Rechnersehen." Dissertation, Fakultät für Luft- und Raumfahrttechnik der Universität der Bundeswehr München.

[45] Zimdahl, W., Rackow, I., and Wilm, T. (1986). "OPTOPILOT—ein Forschungsansatz zur Spurerkennung und Spurführung bei Straßenfahrzeugen." *VDI Berichte* **162**, 49–60.

[46] Wershofen, K. P., Graefe, V. (1991). "A Real-Time Multiple Lane Tracker for an Autonomous Road Vehicle." Proceedings, EURISCON, Korfu, July 1991.

2
The New Generation System for the CMU Navlab

CHARLES THORPE, MARTIAL HEBERT, TAKEO KANADE, AND STEVEN SHAFER

2.1 Abstract

The Carnegie–Mellon University Navigational Laboratory (the CMU Navlab) project integrates sensing, image understanding, planning, control, and software systems architectures onto a self-contained mobile robot. The Navlab drives autonomously along a variety of roads (dirt, gravel unmarked bicycle paths, city streets, rural roads) and cross-country. This chapter describes the Navlab and its contributions in color vision, neural nets, 3-D perception, planning, and robot architectures.

2.2 Introduction

The goal of the CMU Navlab project is to build autonomous systems capable of outdoor navigation, both on roads and cross-country. Since the outset of the project in 1984, we have held two main tenets: the importance of complete systems and the importance of focusing on bottlenecks. Our emphasis on complete systems has meant that, since the beginning, we have closed the loop from sensing to action in realistic outdoor scenarios. We have been forced to deal with the vagaries of natural illumination, bright sunlight and clouds and rain and snow; and we have had to confront the problems of camera calibration, path planning, real-time computing power, and software systems architectures. While the logistical costs of performing such real experiments have sometimes been significant, our resulting algorithms and systems are calibrated to reality. Our second principle, of focusing on the bottlenecks, has pushed us to work on the most difficult problems first. For outdoor navigation, the biggest challenge, and our main area of research, has been in image understanding in difficult conditions. Instead of running at high speeds on clearly marked expressways, we have worked on unstructured roads (includ-

---

This chapter is based on "Towards Autonomous Driving: The CMU Navlab" by Charles Thorpe et al. which first appeared in *IEEE Expert* **6** (4), August 1991. © 1991 by IEEE.

ing dirt roads and a winding asphalt bicycle path), with the changing appearance of structured roads in dappled shadows and at intersections, and in off-road navigation over rough terrain. Once we had the first versions of reliable perception software, we also developed novel planning methods for rough terrain, and we have designed and built systems software to forge the separate perception and planning modules into integrated systems. Other technologies, such as vehicle design, high-speed computing, and control theory, are not the main bottlenecks. While important components, they have been or are being developed by other groups, often outside the mobile robotics community. By directly confronting the central areas of perception, planning, and system building for mobile robots, we are completing the missing links that will enable us to build the reliable high-speed mobile robots of the future.

We now have significant results in many of those areas. Our Navlab robot van (shown in Figure 2.1) drives itself at slow speeds along unmarked, unmapped trails, locating and traversing intersections. On more typical structured roads, the Navlab drives up to its mechanical limit of 28 km/h. It can run without a map, or use maps it has built, along with information from previous runs, to select different behaviors at different locations. Off road, the Navlab can move slowly over moderately rough terrain, and it can map large areas as it drives. The resulting software has been transferred to other projects, including the Defense Advanced Research Projects Agency (DARPA) Martin Marietta ALV (autonomous land vehicle) and our own NASA-sponsored AMBLER, a walking machine for planetary exploration.

FIGURE 2.1. The Navlab.

2.2.1 Systems

Our research on perception, planning, control, and architectures is being integrated into two main systems. The Autonomous Mail Vehicle work is the test bed for road following and landmark recognition, while the Generic Cross-Country Vehicle emphasizes 3-D perception and trajectory planning.

THE AUTONOMOUS MAIL VEHICLE

The road following system under construction for the Navlab is the Autonomous Mail Vehicle, or AMV. This system draws its inspiration from postal deliveries in suburban or rural areas, which follow the same route day after day, undeterred by "rain nor snow nor dark of stormy night". The mail carriers drive at relatively slow speeds, often on many different kinds of roads. They do gross navigation through a network of roads and intersections and fine position servoing to mail boxes.

This type of system is an example of a broader class of applications that focuses on map building and reuse, positioning, road following, and object recognition. Our AMV project is investigating those issues, including strategies for using different sensors and different image understanding operators for the perception components.

On an initial mapping run, the AMV system finds 3-D objects using range data, finds roads using color images and the reflectance data from our laser scanner, and enters them into the annotated map . After the annotated map has been built, the Navlab retraverses the same route more efficiently. Instead of processing range images continuously, Navlab object detection only runs when a trigger fires to tell it of a landmark, and it only looks where the annotation says the object should appear. Each time the system sees the road, it generates a two-parameter update to its dead-reckoned position estimate, updating its cross-track position and heading relative to the road but not its along-track position. Each time the system sees a landmark, it gets a full three-parameter position update and corrects its cross-track position, along-track position, and heading. Rather than plan paths anew with each image, the system plans its desired path for the entire run before it begins and only does minor servo modifications when the vehicle position is updated.

GENERIC CROSS-COUNTRY VEHICLE

The Generic Cross-Country Vehicle (GX-CV) packages 3-D perception and local trajectory planning into a solid foundation for off-road navigation. The basic GX-CV system enables the Navlab to travel in a general direction, planning and executing vehicle trajectories around obstacles and across rough terrain. Hooks in the GX-CV planner will enable missions such as cross-country traversing with global navigation or cross-country mapping.

The heart of the GX-CV consists of three parts: the EDDIE (Efficient Decentralized Database and Interface Experiment) architecture and controller, medium-resolution 3-D mapping, and an explicit vehicle model planner. We have prototypes of each of these modules. We are currently refining the modules, particularly to increase efficiency.

The GX-CV illustrates the importance of modules agreeing on task requirements and the importance of model explicitness. The planner includes detailed, explicit models of the vehicle and its capabilities. This, in turn, drives the design of perception and the selection of parameters governing resolution and coverage.

2.2.2 Context

Our work is part of the broader framework of DARPA's Strategic Computing Initiative, including the ALV project that began in 1984. Several of the contractors from the Strategic Computing Initiative worked on perception and planning for autonomous navigation. Among others, the University of Massachusetts and Honeywell developed motion tracking software [2, 5], SRI developed tracking using 3-D data [6], and ADS built qualitative navigation [24]. Martin Marietta built and operated the ALV vehicle itself and developed its own road-following software [32]. Hughes and the University of Maryland contributed off-road and on-road navigation, respectively, directly to the ALV test bed [18, 33].

The role of CMU was to build a new generation vision system. We were asked to look beyond the immediate problems of getting the ALV through its first demonstrations and to address the issues of more difficult perception and integration. Thus, our charter to look at more difficult vision problems matched our belief that solving those problems was the right first step in building capable systems. Also, the request that we address integration issues matched our concern for complete systems and led to much of our architectural research.

Beyond the DARPA community, the past five years have been several other outdoor mobile robot projects. In the United States, Texas A&M has begun work on visual tracking for convoy following and obstacle avoidance [17]. General Motors is working on lane following at high speeds under relatively constant illumination [20, 21]. In Germany, Professors Dickmanns and Graefe [25] have built an elegant control formulation for driving on autobahns. Fujitsu and Nissan in Japan have built prototype road-following software [26]. The research at CMU, and within the DARPA community, is distinguished from all of these by its concentration on the more difficult vision problems of bad weather, bad lighting, and bad or changing roads.

2.2.3 Evolution of Experimental Robots at CMU

Our early outdoor vision experiments used the Terregator, a desk-sized mobile robot designed for rough terrain operation. The Terregator was a rugged and versatile platform, serving as a "smart peripheral" for vision research. It carried an on-board 68000 computer, which interpreted commands such as "follow an arc of radius X at velocity Y for distance Z and acknowledge when complete". The Terragator was capable of reliable operation in rough terrain. We have driven it up and down steps, on steep slopes, even in an underground coal mine. But, it was hampered by a lack of on-board space. We could not carry the computers we needed for completely autonomous operation, especially for vision processing, so we always had to have a link to the laboratory. The link took several forms: a coaxial cable, 1200-baud radio modems, a microwave video link, or a UHF video (we own channels 24 and 46). None of these was completely satisfactory. The video signal was often degraded, particularly when operating among trees, exactly when it was most crucial to get good processing results. More important, the researchers sitting in the computer room had a hard time visualizing what the Terregator was doing.

The limitations of the Terregator led to the construction of our current vehicle, the Navlab. From the outside, it looks like a standard commecial van, with a rooftop air conditioner, plus one or more video cameras and a laser range finder mounted over the cab. On the inside, it looks more like a machine room, with five electronics racks, 20 kW of on-board power, and miscellaneous consoles and monitors. Over time, the Navlab has carried Sun 3's, Sun 4's, several generations of the CMU/GE Warp supercomputer, various specialized real-time controllers, gyrocompasses, an inertial navigation system, and a satellite positioning system. At the beginning of the Navlab's history, in 1986, the racks were mostly empty. Over the next three years, the insatiable desire of all vision researchers for more computing power had all the racks filled and made us add special rack-mounted air conditioning to keep some of our more power-hungry machines cool and content. Now, as computers are shrinking and as we understand better which algorithms we want to run, our rack space demands have decreased. Our current real-time controller occupies four slots in a VME cage, and our general-purpose computing consists of three Sun 3's in a single cage, and two Sun 4's in another cage. Even counting miscellaneous disk and tape drives, power conditioning, and inertial navigation, our racks are less than half full.

The most important payload of the Navlab is the researchers. There is always a safety driver in the driver's seat, watching over the Navlab's autonomous runs. In the back, there is room for five researchers plus observers. The quality of our mobile robot software increased greatly when the graduate students and engineers were able to ride along on autonomous runs, partly out of self-preservation, but mainly because they could see and feel how their code worked. We run standard SunOs Unix[1] on the Navlab, so we have a standard programming environment and tools to find and fix bugs in our programs during an experimental run (debuggers, editors, compilers, etc.). The Navlab is truly a laboratory on wheels.

---

[1] SunOs is a trademark of Sun Microsystems, and Unix is a trademark of Bell Labs.

2.3 Color Vision for Road Following

Roads that are nearly straight, evenly illuminated, and well marked, can be tracked easily in color or monochrome video images. Finding the edges of a clean sidewalk and tracking freshly painted white stripes on an empty expressway are both straightforward. Road following becomes much more difficult when the road runs through dappled shadows, illumination suddenly changes as the sun goes behind a cloud, or the "road" is a meandering bicycle path with no lines or stripes and with broken and uneven borders. Our first road-following software ran a simple edge detector (Roberts' operator, followed by thresholding) over the image and looked for edge fragments that had strong contrasts, were parallel, and pointed in roughly the correct direction. This worked very well for campus sidewalks. When we took our robot onto the bicycle path behind campus on Flagstaff Hill, the highest contrast edges in the scene were shadow edges. At the right time of day, the shadows of tree trunks fell along the road, producing strong, straight, parallel edges at nearly the predicted direction. Our road-following software turned into tree-shadow-following software.

We have taken several approaches to coping with difficult visual conditions. Our SCARF (Supervised Classification Applied to Road Following) system uses adaptive color classification. It deals with changing illumination and changing road appearance by updating its color models for each new image. It handles poorly defined roads by classifying all the pixels in the image and by using a simple road model in a voting scheme to find the most probable road in the image. YARF (Yet Another Road Follower), our second vision system, takes a different approach. It takes advantage of the lines and stripes of structured roads and uses an explicit model of those features both to guide individual trackers and filter and validate its detected road model. ALVINN (Autonomous Land Vehicle in a Neural Net), the third main color-vision system currently running on the Navlab, uses a connectionist architecture. It achieves its power by being trained directly on the current road and processing quickly so that small imperfections tend to be smoothed out.

2.3.1 SCARF

SCARF tracks roads by adaptive color classification [9]. SCARF runs in a loop: classify image pixels, find the road model that best matches the classified data, and update the color models for classification. The simple models of road color and geometry make very few assumptions about the road and make SCARF run robustly even when following unstructured roads.

The first strength of SCARF comes from representing multiple color classes as Gaussian distributions in full RGB (Red, Green, and Blue) color and calculating probabilities instead of using binary thresholds. SCARF typically uses four color classes to describe road appearance and four to describe off-road objects. In the classification step, each pixel is compared to all eight classes.

The output of classification is both the label of the most probable class and its probability. Having multiple classes allows SCARF to represent the different colors of the road (for instance, asphalt, wet patches, shadowed pavement, and leaves) and off-road objects (trees, sunlit grass, shaded grass, and leaves). Using full color, instead of monochrome images or some combination of colors, keeps all the image information that may be useful in discrimination. The Gaussian representation of each color class determines, intuitively, whether a particular variation in color is significant. Sunlit asphalt tends to be homogeneously colored and is represented by a class with small variance; grass has more variety and is represented with correspondingly larger variances. Having Gaussian representations of the colors for each class makes it possible to calculate the likelihood that a given pixel belongs to a particular class. While most other navigation systems simply use binary thresholds, SCARF gives the probability for each classified point. This is especially important for cases such as dry leaves that occur both on and off road. A particular pixel may somewhat more closely resemble off-road leaves than on-road leaves, but the confidence that it should be classified as off-road will be very small, so that pixel will (correctly) contribute very little to the overall road location determination.

Classified pixels vote for all road locations that would contain them, with votes weighted by classification confidence. The road with the most votes is used both for steering and recalculating the color classes using nearest mean clustering to collect new road and off-road color statistics. SCARF uses a very simple model of road geometry. Roads are represented as triangles in the image. The apex is constrained to lie on a particular image row, corresponding to the horizon, and the base of the triangle has a fixed width, dependent on road width and camera calibration. There are two free parameters: the column in which the apex appears and the skew of the triangle in the image. While this simple two -parameter model does not represent curves or hills or road-width variations, it does approximate the road shape well enough to allow reliable driving. It is especially effective because the voting procedure uses all pixels, not just those on the edges, and is therefore relatively insensitive to misclassifications. A model with more free parameters could represent more potential road shapes, but it would often be led astray and would find curves or branches where in fact all that exists is noise. Furthermore, the simple model allows for fast voting and functions with small amounts of data, so SCARF can process highly reduced images (typically 60 by 64 or 30 by 32) at high rates (approximately 1 frame per second). Processing images closely spaced along the road means that small errors in road representations are corrected before the vehicle arrives at the mistaken locations. Processing images closely spaced in time means that even the drastic illumination changes caused by clouds covering the sun appear as gradual shifts in road appearance and so do not derail SCARF.

We have built several versions of SCARF. Faster versions divide the image into strips and process each strip on a separate cell of the Warp parallel supercomputer. More capable versions of SCARF add to the road model by checking for intersections, as well as for the main road. The various versions of SCARF have driven the Navlab along bicycle paths, dirt roads, gravel roads, and suburban streets. SCARF has been integrated into several of our Navlab systems. Figure 2.2 shows SCARF correctly finding the road through deep shadows, where the road is not obvious even to a human observer. Figure 2.3 shows SCARF finding an intersection in a series of images as the vehicle approaches the branch point.

2.3.2 YARF

The YARF system explicitly models as many aspects of road following as possible for driving on structured roads [23, 3]. Highways, freeways, rural roads, and even suburban streets have strong constraints. Modeling these explicitly makes reasoning easier and more reliable. When a line tracker fails, for instance, an explicit model of road and shoulder colors adjacent to the line helps in deciding whether the line disappeared, became occluded, turned at an intersection, or entered a shadow. This kind of geometric and photometric reasoning is vital for building reliable and general road trackers.

YARF has individual knowledge sources that know how to model and track specific features, such as road edge markings (white stripes), road center lines (yellow stripes), and shoulders. YARF also uses as explicit geometry model of the road, consisting of the location of the vehicle on the road, the location of stripes, the type of stripes (e.g., broken or solid); and the maximum and current road curvature. Other features, which are not yet modeled but which may be helpful, include locations of shadows, 3-D effects as the road goes through valleys and over hills, and global illumination changes.

The yellow line tracker, for instance, uses the hue of the lines for segmentation. The hue is calculated for all pixels within a window around the predicted line location. Pixels with a hue between 40 (reddish-yellow) and 100 (greenish-yellow) are set to 1, others to 0. The results of this thresholding are quite noisy. Pixels that are very close to gray have an unstable hue value, while yellow lines in dark shadows are often so dark that digitization noise nearly swamps their yellow hue. As a result, the images after hue thresholding often have isolated noise points, both false positives and false negatives. We clean up the output with a "shrink and grow" operator. The resulting image is normally dominated by one or two blobs, corresponding to the yellow line or lines. The blob descriptors are returned as the line locations. Figure 2.4 shows the yellow line tracker and a separate white line tracker finding the road lines even in complex shadows.

YARF is designed for higher speeds than SCARF and runs in a more predictable environment. This requires and allows a more complex road model that encodes curvature as well as position. The equation of a circle

$$radius^2 = (x - xcenter)^2 + (y - ycenter)^2$$

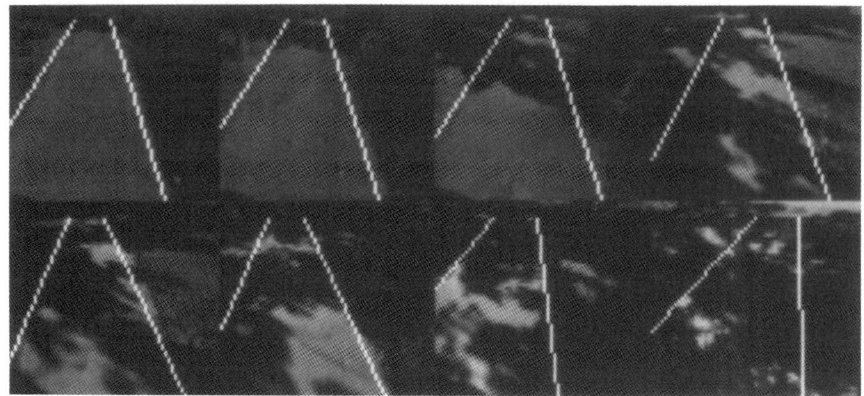

FIGURE 2.2. SCARF correctly finding the road in difficult shadows.

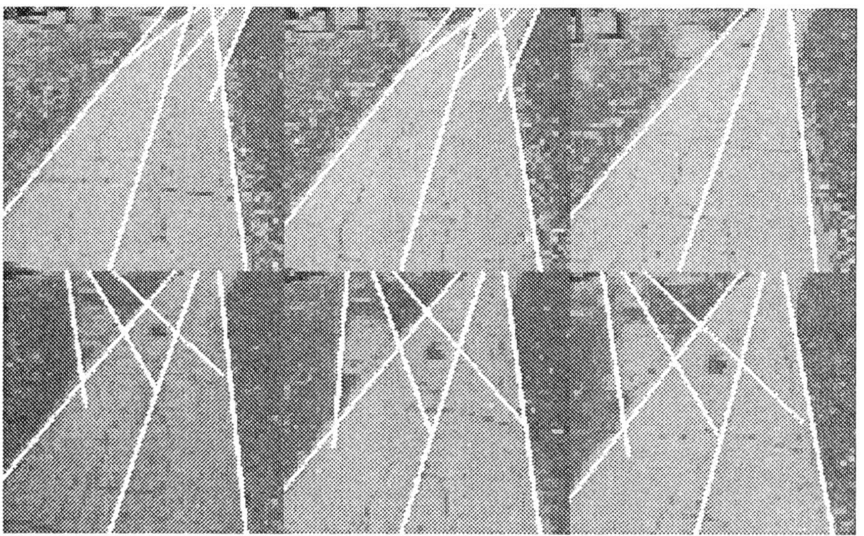

FIGURE 2.3. SCARF finding an intersection.

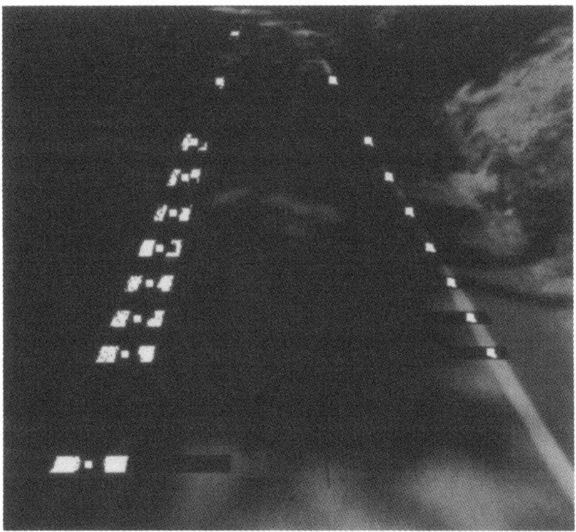

FIGURE 2.4. YARF tracking yellow and white lines in complex shadows.

is nonlinear and somewhat hard to solve in the presence of noise. We can approximate a circular arc with a parabola, similar to the approach of Dickmanns [25]:

$$x = \text{half\_curvature} \times y^2 + \text{slope} \times y + \text{lateral\_offset}.$$

The least-squares values of curvature, slope, and lateral offset are easily computed using the matrix pseudo-inverse. Since this is done in vehicle coordinates, with the vehicle pointing approximately along the road, the parameters calculated are good approximations to those of the best-fit circle. In practice, this approximation is adequate for the sorts of curvatures and slopes of roads within the Navlab's field of view.

Occasionally, features are found incorrectly. YARF detects these mistakes both locally, based on the results of a single operator, and globally, checking for consistency Local error detection depends on the specific operator. Some operators, such as the oriented window tracker, can only report a correlation measure as a confidence. Others, such as the two-color blob detector, can provide a little more information. The blob detector usually finds a light blob (the white line) against a dark background (asphalt). It carries statistics of the mean, variance, and covariances of red, green, and blue for both feature and background colors. If all pixels in its prediction window are the background color, the color blob detector reports a missing feature. If a light-colored blob is found, but only at the edge of the window, it reports a clipped feature. If all pixels are much lighter or darker than modeled by either color, it reports an illumination shift. It is up to the higher level calling program to decide whether the road is wider, the white stripe is temporarily missing, or the lighting really is changing.

Global error detection uses the output from all feature detectors in a single frame. There are many ways to check for data consistency. The simplest, performing a least-squares fit and examining the residual, gives some cues as to whether there is an outlier, but does not reliably indicate which point is in error. Better approaches come from the "robust statistics" literature.

Noise in measured positions or missing features is a separate but related issue. Since our model has three parameters (curvature, slope, and lateral_offset), we need, in general, three measured points to calculate the model. If one of our features is missing, not detected, or out of the field of view, we may not have enough measured points. More generally, if we have only a few points, our solution can be unstable and sensitive to noise, even using least-squares solutions. The right solution is to filter the results from successive frames. The intuition behind a Kalman filter is that the current estimate of state is a combination of current measurements and the previous state estimate, transformed to account for system dynamics. The weight given to current measurements depends on their believed accuracy. The weight given to the previous estimate depends both on how accurate the previous estimate was thought to be and on how accurately it can be transformed into current coordinates. In the case of road following, the weight for current detected features comes from the accuracy of feature detection and camera calibration, checked by measuring internal consistency (the residual in the least-squares fit of current measurements to the model). The weight given to the prior estimate is reduced by the uncertainty in dead-reckoned vehicle motion between the previous and current positions, and it is further reduced by extrapolation uncertainty in assuming a constant road model. Vehicle motion errors can be reduced by inertial sensing. Road model errors will depend on the situation. For the gentle curves and smoothly varying curvature of an interstate, prior estimates can be extrapolated for long distances, and they can carry the vehicle through shadows and other visually confusing areas where current tracking fails. For the winding turns of back roads in Pennsylvania, however, the road model can change radically over very short distances, and the weight given to prior models in the filter equations must decrease rapidly.

In practice, the complete Kalman filter may not be needed. We have achieved good results by fitting curves to the points detected over the three most current frames, with no error weighting or filtering at all. In Figure 2.5, the squares show tracked left and right lane lines, the asterisks show requested vehicle paths, and the diamonds show estimated road position and heading. For the relatively slow speeds (up to approximately 15 km/h) of YARF and with the Navlab's accurate dead reckoning, the errors in detected positions are probably dominated by image processing noise rather than vehicle motion noise. Future runs, at higher speeds, will probably require more elaborate filtering schemes than we currently need.

YARF works because of the integration of all the constraints. Explicitly modeling tracker performance and feature appearance and using specialized trackers allows high speeds, high accuracy, and local failure detection. Explic-

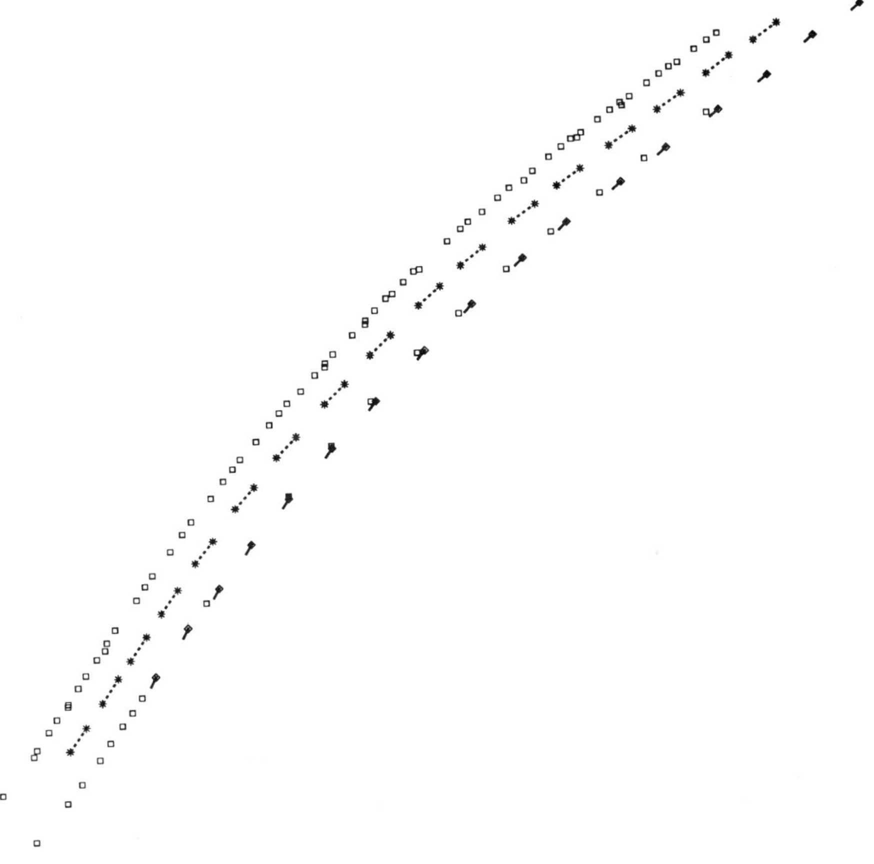

FIGURE 2.5. YARF tracking result.

itly modeling road geometry allows accurate predictions and enables global error detection. Explicitly modeling errors and uncertainty allows YARF to correctly size its prediction window. Explicitly modeling changes in road geometry gives YARF the capability to handle urban streets, with intersections and other discontinuities in lane structure and appearance, as well as the cleaner environment of highway lane following.

2.3.3 ALVINN

ALVINN is the newest of our road following programs, built by the CMU Connectionist group [27]. The weights in ALVINN's neural network hidden units are trained by driving the Navlab by hand. During the training phase, ALVINN inputs the camera image and the steering angle from the human driver at each time step. The image is preprocessed to enhance road contrast. The enhanced image and the steering angle are fed to a back-propagation algorithm that adjust weights in the hidden units until the weights settle to values that give the correct steering response for each input image. Typically, training takes less than 100 input images and uses less than 5 min.

With this training scheme, ALVINN directly learns how to follow roads. It is more difficult to train the network to recover from errors when it is not quite aligned on the road. To provide examples of images from slightly different vantage points and the proper steering commands, each input image is reused in several positions. The images are shifted to simulate a variety of errors, and the steering command is shifted to generate the command that would bring the Navlab back on the road.

When ALVINN runs, it preprocesses the input images and gives them to the net. ALVINN then directly outputs the steering wheel angles as dictated by the network, with no reasoning about road location. ALVINN uses reduced-resolution images (typically 30 by 32 or 45 by 48 pixels) and runs in about a quarter of a second per image.

One characterization of ALVINN is that it uses a compiled representation, going straight from images to steering with no intermediate geometric or symbolic representation. During its learning phase, the backpropagation algorithm automatically compiles this knowledge by selecting the features that discriminate between different steering angles, which correspond to different road locations. Since ALVINN starts with no preconceived idea of what the road looks like, it learns different sets of weights to follow many different types of roads with no change in the underlying algorithms.

The disadvantage of a compiled representation such as that used by ALVINN is that it cannot take advantage of geometric or symbolic input. If ALVINN is trained to run on a particular road, it is impossible to tell it that a second road is just like the first, only twice as wide. Since there is no explicit representation of "road width," or even "road," there are no symbolic parameters to be changed or manipulated. The advantages of such a representation are that it is fast and it is easy to train for a particular road. The weights learned by ALVINN tend to be large, low-frequency edge masks or matched filters that look for the road in general locations. Thus, local imperfections in the road or in lighting do not greatly distort the output steering direction. Figure 2.6 shows the weights for one of ALVINN's hidden units. The square shows the weights coming in to one hidden unit from the input image, and the line at the top shows the weights going out to different steering angles. White means positive weights, and dark means negative. This unit mainly looks for a road on the left edge of the image and mainly votes for turning left. There is also a secondary pattern that would match a road further to the right and slightly positive weights supporting straight ahead steering.

ALVINN is the fastest of our current road following systems because of its compiled representation. It has also been the most difficult to integrate into systems because it does not output detected and locations. While SCARF and

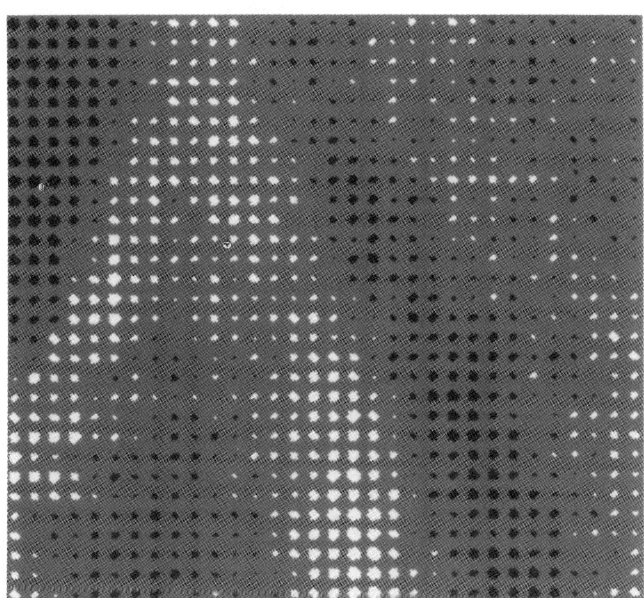

FIGURE 2.6. ALVINN's weights for one hidden unit.

YARF report the location and orientation of the road, ALVINN only produces steering commands. It is, however, possible to reason geometrically about ALVINN's steering output to infer some information about road location. The output arc will bring the Navlab onto the center of the road at some distance along the arc. Thid distance would in general be unknown, depending on the particular driving style used during training, except that the artificially shifted images used in training the Navlab to return to the road use a particular specified geometry. This geometry, used for shifting the roads to provide training examples, can also be used to measure how far along the steering arc the center of the road lies. The intersection of the arc and the road center gives a single point known to lie on the road center line. Observing several of these points over time allows us to approximate the apparent road position and shape, which allows the map-based reasoning needed to integrate ALVINN with other systems.

2.4 3-D Perception

An outdoor mobile robot needs information derived from appearance (*e.g.*, road location in a color image or terrain type), but it also needs to know the geometry of the observed environment. In some tasks, such as cross-country navigation, the most important information is the geometry of the *terrain*, the set of 3-D surfaces observed or traversed by the vehicle. The first step toward building geometrical representations of a terrain is to choose a suitable sensor. Clearly, a single-color camera is not suitable for collecting 3-D data. An alternative is to use passive techniques for recovering 3-D data, such as stereo vision. There are significant drawbacks to these techniques. Instead, we use an active sensor, a laser range scanner, which can generate a high-resolution depth image of the terrain in front of the vehicle. Using such a sensor alleviates the need for inferring 3-D information from 2-D information, and being active, it is also less sensitive to outside illumination. The technology used in the sensor is very recent and is not used very widely yet. We discuss the drawbacks, advantages, and prospects for future improvements in the technology in Section 3.1.

The terrain can be described at different levels of resolution depending on the task, the environment, and the amount of computation time allocated for 3-D perception in the system. For example, a system that follows roads that are known to be locally flat and mostly obstacle free requires a completely different representation than a system that navigates through rugged, open terrain. In the latter case, vehicle safety becomes the overwhelming issue, while vehicle speed becomes much less important. This is consistent with the general approach that the components of a mobile robot must be tailored to environment and task.

In addition to several types of terrain representations, we also distinguish between techniques that involve building a representation from a single image and techniques that put together representations from several images to build a consistent *map* of the terrain traversed by the vehicle. An example of the former is fast obstacle detection in which the goal is to detect unexpected objects in the current image as fast as possible. An example of the latter is building a 3-D map of a long stretch of terrain so that the system can later use the map to retraverse that area. Depending on the environment and the task, we should be able to build maps at all levels of resolution.

We have identified three types of representations that we discuss in detail in the following sections. For each type of representation, we describe two types of processing: range data processing for building a terrain representation from a single range image and matching techniques for building consistent maps from several observations.

1. *Discrete objects and obstacle detection:* A coarse description of the environment is one in which only discrete objects are represented without an explicit representation of the surrounding terrain. This representation is appropriate when navigating in mild terrain with clearly defined objects. There are two applications of this level of representation: fast detection of obstacles during road following and building maps of the objects observed as the vehicle travels on a network of roads. Such a map is used in the AMV system (Section 1.1) to correct the vehicle position by matching observed objects with predicted objects from the map.

2. *Feature-based terrain modeling:* A finer description of the environment involves describing the terrain by a set of features (regions and edges) in addition to discrete objects. This representation is used for cross-country navigation, in which the vehicle, in order to navigate safely, must take into account the local shape of the terrain as well as discrete objects. The terrain features can be used in conjunction with discrete objects to build more reliable terrain maps over many images. This is particularly true in mild open terrain in which not enough discrete objects are observed to reliably match observations.

3. *High-resolution terrain models:* The highest resolution representation is achieved by building elevation maps, the spatial resolution of which may be as fine as 10 cm. Large terrain maps can be built from individual high-resolution maps by correlation-type matching. The advantage over feature- or object-based representations is that information about the local shape of the terrain is preserved everywhere. The price to pay is much higher computation time. Therefore, high-resolution terrain maps are most useful in applications in which the computations can be done off-line or in applications in which stop-and-go motion of the vehicle is acceptable. The latter situation occurs when navigating through very difficult terrain, in which case the best terrain description is necessary to ensure the safety of the vehicle.

2.4.1 Range Sensing

The sensor that we use on the Navlab is the Environmental Research Institute of Michigan (ERIM) scanner (Table 2.1). This scanner is typical of the class of

TABLE 2.1. Relative performance of example range scanners.

Characteristics	ERIM	Perceptron
Eye safety	yes (?)	yes
Field of view	80 h by 30 w	60 h by 60 w (programmable tilt)
Pixels	256 by 64	256 by 256
Ammbiguity interval	20 m	40 m
Depth	8 bits (8 cm)	12 bits (1 cm)
Intensity	8 bits	8 bits
Maximum range	40 m (?)	50 m
Scan rate	2 frames/s	2 frames/s
Scan direction	top to bottom	programmable
Interface	VME to Sun	VME to Sun
Temperature	narrow range	"Pittsburgh"
Construction	wire wrap	printed circuit
Components	all custom	most off the shelf
Size	90 w by 35 h by 45 d, cm	45 w by 35 h by 35 d, cm
Weight	50 kg	<25 kg
Power	26 VDC	110 VAC

laser range finders based on the measurement of the phase shift of an amplitude modulated laser. Table 2.1 lists characteristics of both the ERIM, one of the earliest scanners, and the Perceptron, a later model.

Using an active laser range finder has considerable advantages over more traditional techniques, such as stereo vision. It is insensitive to outside illumination, it is fast compared to the computation time required by standard passive techniques, and it provides a high-resolution range map as opposed to the sparse map produced by most passive techniques. However, we have found a number of problems with this laser ranging technology that should be addressed in order for it to be widely used.

Mixed points: At the boundary between two objects, one part of the laser spot, or footprint, is on one surface while the other is on the other surface. Since the sensor integrates over the entire footprint, the resulting measured point does not lie on either surface but is between the two surfaces. Such a point is a mixed point, because it is measured from a mixture of reflections from both surfaces. Mixed points are inevitable with the current technology. Most mixed points can be removed through median filtering. However, the mixed point problem implies that edges in range images, especially edges between distant objects, are highly unreliable. This should be taken into account in the choice of range image processing algorithms.

Acquisition rate: The typical acquisition rate of 2 images/s is too slow. The motion of the vehicle can be significant while the image is being scanned, thus leading to a distorted range image. This is not a problem at very low speeds, such as in cross country navigation, but may preclude the use of this type of sensing at higher speeds, such as in highway driving and for tracking moving obstacles.

Sensitivity to surface material: In early scanners, the measured range varied with surface type. More recent scanners can adjust for different uniform surfaces, yet they still have problems with edges between surfaces that are at the same depth but have different reflectances. The change in reflectance causes changes in internal sensor gains, which upset the phase detection, which produces a spurious depth edge.

2.4.2 Discrete Objects and Obstacle Detection

The lowest resolution terrain representation is an object map that contains a small number of objects represented by their trace on the ground plane. Several techniques have been proposed for obstacle detection. The Martin Marietta ALV [12, 13, 32] detects obstacles by computing the difference between the observed range image and precomputed images of ideal ground planes at several different slope angles. Points that are far from the ideal ground planes are grouped into regions that are reported as obstacles to a path planner. A very fast implementation of this technique is possible since it requires only image differences and region grouping. It makes, however, very

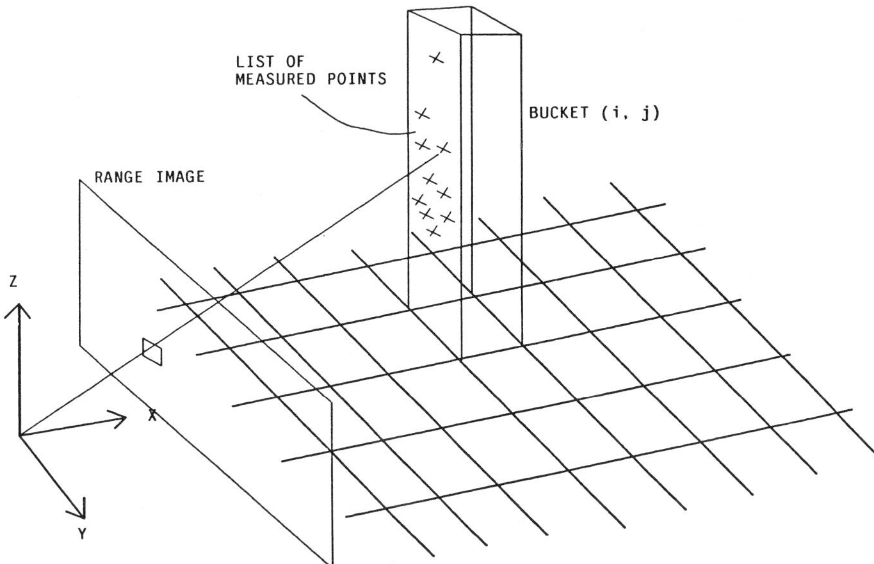

FIGURE 2.7. Building the obstacle map.

strong assumptions on the shape of the terrain. It also takes into account only the absolute positions of the potential obstacle points, not relative positions and slopes. As a result a short, sharp ridge or step would be overlooked, even though it may be an obstacle. Another approach proposed by the Hughes AI group [10] is to detect the obstacles by thresholding the normalized range gradient, $\Delta D/D$, and by thresholding the radial slope, $D\,\Delta\phi/\Delta D$. The first test detects the discontinuities in range, while the second test detects the portion of the terrain with high slope. This approach has the advantage of taking a vehicle model into account when deciding whether a point is part of an obstacle.

We use a terrain map approach to detect obstacles for the Navlab. Each cell of the terrain contains the set of data points that falls within its field. We can then estimate the surface normal at each elevation map cell by fitting a reference surface to the corresponding set of data points. Cells that have a surface normal far from the vehicle's idea of the vertical direction are reported as part of the projection of an obstacle (Figure 2.7). Obstacle cells are then grouped into regions corresponding to individual obstacles. The final product of the obstacle detection algorithm is a set of 2-D polygonal approximations of the boundaries of the detected obstacles that is sent to an A*-type path planner. In addition, we can roughly classify the obstacles into holes or bumps according to the shape of the surfaces inside the polygons.

Figure 2.8 shows the result of applying the obstacle detection algorithm to a sequence of ERIM images. The figure shows the original range images (top), the range pixels projected in the elevation map (left), and the resulting polygo-

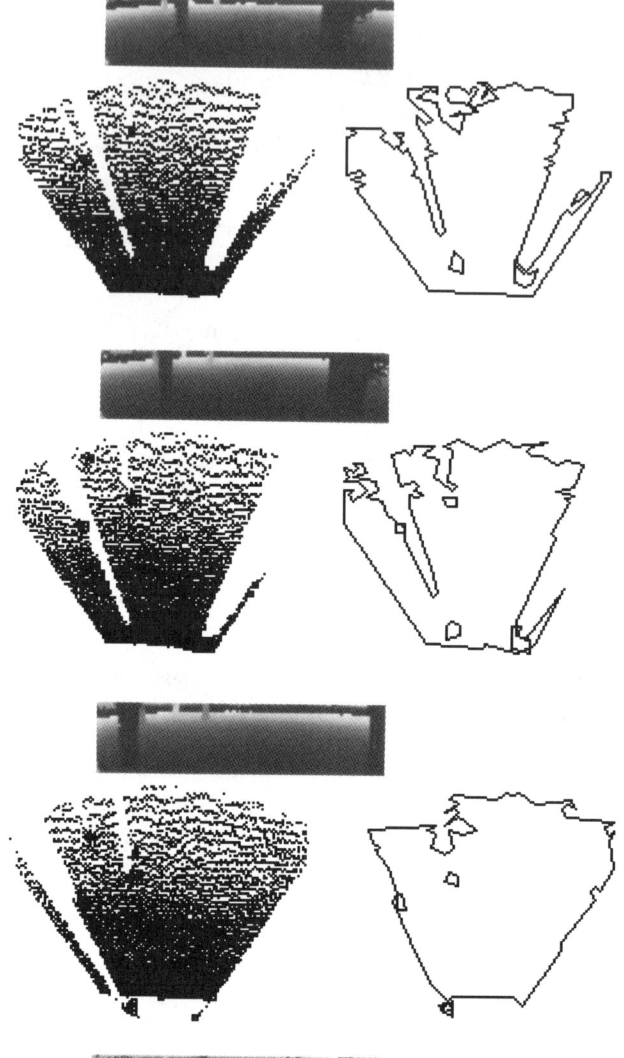

FIGURE 2.8. Obstacle detection on a sequence of images.

nal obstacle map (right). The large enclosing polygon in the obstacle map is
the limit of the visible portion of the world.

The obstacle detection algorithm does not make assumptions on the position of the ground plane, in that it only assumes that the plane is roughly
horizontal with respect to the vehicle. Computing the slopes within each cell
has a smoothing effect that may cause real obstacles to be undetected. Therefore, the resolution of the elevation map must be chosen so that each cell is

significantly smaller than the typical expected obstacles. In the case of Figure 2.8, the resolution is 20 cm. The size of the detectable obstacle also varies with the distance from the vehicle due to sparser range pixels at longer distances.

In the fast obstacle detection mode, several improvements can be used to decrease the computation time. We need to look for obstacles only in a narrow stripe in front of the vehicle. We do not need to detect all the objects, it is sufficient to raise an alarm as soon as one object is found. We do not need the high spatial resolution at close range, therefore the data can be subsampled close to the vehicle. Taking into account those improvements, we can achieve fast obstacle detection that runs in 600 ms on a Sun Microsystems SPARC workstation, which is fast enough at the current speeds, considering the fact that the acquisition time is still 500 ms on average.

Another application of object detection is to build object maps by combining many observations. The resulting map is used for navigating through the same region, as in the AMV system. Matching objects is not very expensive in our case because we have only a few objects to match in each frame and because we can assume that we have a reasonable estimate of the displacement between frames from INS or dead reckoning so that the locations of the objects detected in one image can be easily predicted in the next image. The main issues are to remove spurious objects and to compute the location of the objects as accurately as possible. Spurious objects can be detected in two cases: noise in the range image may cause the object detection program to hallucinate and moving objects (*e.g.*, people) crossing the field of view are detected as objects even though they should not be included in a map. Spurious objects must be eliminated because they may lead to disastrous results when they are later used to correct the position of the vehicle. The position of the objects must be computed as accurately as possible so that the position corrections that are computed using the object map are also accurate.

The problem of spurious objects is solved by calculating a confidence measure for each object. Once an object has been seen in one image, it should appear in subsequent images, as predicted by vehicle motion, object position, and sensor field of view. If it appears as predicted, its confidence is increased, otherwise its confidence decreases. Objects with low confidence are discarded. Accurate object locations are obtained by updating the uncertainty on object location (mean and covariance matrix) each time an object is observed in a new image. The initial uncertainty is based on a sensor model and depends mostly on the distance between the object and the vehicle. The uncertainty also takes into account the fact that only a small part of the object surface is observed. The uncertainties are combined using standard maximum likelihood techniques. Figure 2.9 shows a sequence of 14 images. The images are separated by about 50 cm. The white lines connect the objects that are matched images. The white dots indicate the locations of the detected objects in the images. Spurious objects are detected in images 13, 18, 20, and 22. Since they are not matched, their confidence is low, and they are eventually discarded from the map.

FIGURE 2.9. Matching objects in a sequence of range images.

2.4.3 Feature-based Terrain Modelling

Obstacle detection is sufficient for navigation in flat terrain with discrete obstacles, such as following a road bordered by trees. We need a more detailed description when the terrain is uneven as in the case of cross-country navigation. For that purpose, an elevation map could be used directly [11] by a path planner. This approach is costly because of the amount of data to be handled by the planner that does not need such a highresolution description to do the job in most cases. A better alternative is to group smooth portions of the terrain into regions and edges that are the basic units manipulated by the planner. This set of features provides a compact representation of the terrain, thus allowing for more efficient planning, as described in Section 4 [29].

The features used are of two types: smooth regions and sharp terrain discontinuities. The terrain discontinuities are either discontinuities of the elevation of the terrain, as in the case of a hole, or discontinuities of the surface normals, as in the case of the shoulder of a road [4]. We detect both types of discontinuities by using an edge detector over the elevation map and the surface normal map. The edges correspond to small regions on the terrain surface. Once we have detected the discontinuties, we segment the terrain into smooth regions. The segmentation uses a region-growing algorithm that first identifies the smoothest locations in the terrain based on the behavior of the surface normals and then grows regions around those locations. The result of the processing is a covering of the terrain by regions corresponding either to smooth portions or edges.

Figure 2.10 shows the interpolated elevation map. Figure 2.11 shows the polygonal boundaries of the regions extracted from the image of Figure 2.10. In this implementation, the resolution of the elevation map is 20 cm. Since we need a dense map in order to extract edges, we interpolated linearly between the sparse points of the elevation map.

More recent implementations of the terrain modeling for cross-country planning do not use a region representation, but instead directly use a set of maps that describe the underlying data, including elevation, slope, terrain edges, etc. The maps are organized in quad-trees for fast access. Using this representation saves a considerable amount of computation time both in building the terrain representation and using it for path planning. A complete terrain representation can now be built in 2 s on a SPARC work station.

As in the case of object descriptions, composite maps can be built from terrain descriptions. The basic problem is to match terrain features between successive images and compute the transformation between features. In this case, the features are the regions that describe the terrain parameterized by their areas, the equation of the underlying surface, the center of the region, and the main directions of the region. If objects are detected, they are also used in the matching in the same way as before. Finally, if the vehicle is traveling on a road, the edges of the road can also be used for the matching. As in the case of object matching, an initial estimate of the displacement

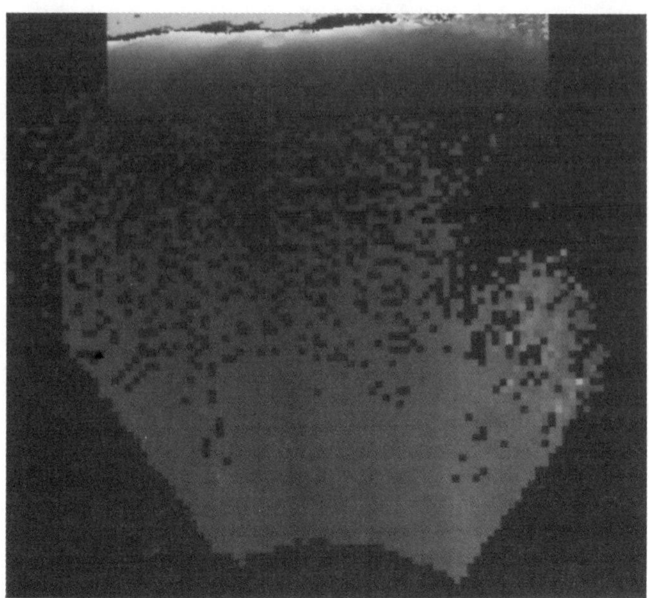

FIGURE 2.10. Range image and elevation map.

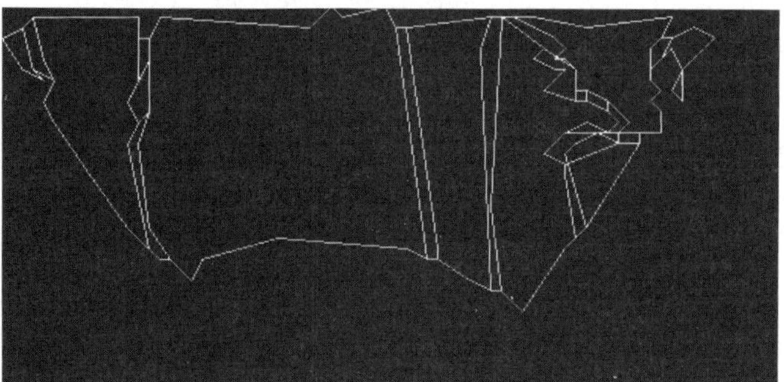

FIGURE 2.11. Polygonal boundaries of terrain regions.

between successive frames is used to predict the matching features. A search procedure is used to find the most consistent set of matches. As before, the search is actually very fast because of the small number of features and the fact that the initial guess of the transformation between images is usually quite close to the actual value. The features are weighted in the search according to how reliably they can be detected. The reliability of a feature depends on its type: discrete objects are more reliable than terrain regions and road edges. Once a set of consistent matches is found, the transformation between frames is recomputed, and the common features are merged.

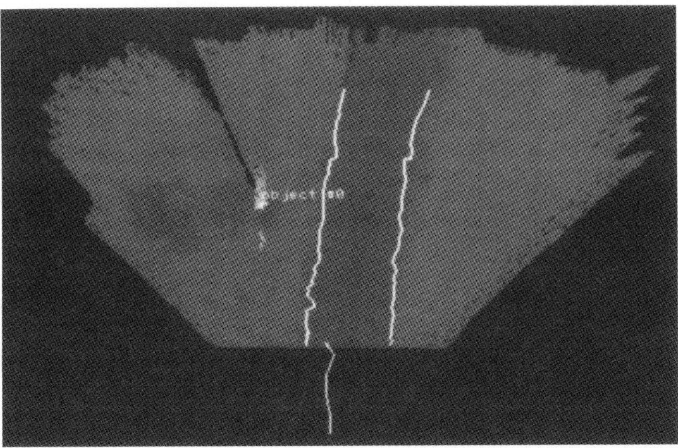

FIGURE 2.12. Map building using feature matching.

This map building approach has been tested on sequences of images with errors in position estimation of up to 1 m in translation and 20° in rotation. For example, Figure 2.12 shows a sequence of 5 maps that are merged into a composite map using feature matching.

2.4.4 High-Resolution Terrain Models

The high-resolution terrain representation is an elevation map, which is a function $z = f(x, y)$ represented by a regular grid of values (x_i, y_i). The most straightforward way to convert a range image to an elevation map representation would be to map each pixel (*row, column, range*) of the range image to an (x, y, z) location in map coordinates. There are a number of problems with this approach.

Sampling: Since this approach is similar to image warping, the distribution of data points in the elevation map is not uniform. The map gets sparser farther from the sensor.

Shadows: Objects create range shadows, that is, regions of space that are not visible even though they lie within the sensor's field of view. Shadowed regions must be explicitly identified and represented separately since no information is available in those regions. The difficulty here is to distinguish between genuine range shadows and regions of the map with no information because of sparse sampling.

Uncertainty: Range measurements are corrupted by noise due to electronic noise, surface material, laser footprints, etc. The noise can be modeled by a variance, σ, for each measurement. As a first approximation, σ depends only on the measured range. This uncertainty is represented in sensor space and must be converted into a representation of the uncertainty in the elevation map.

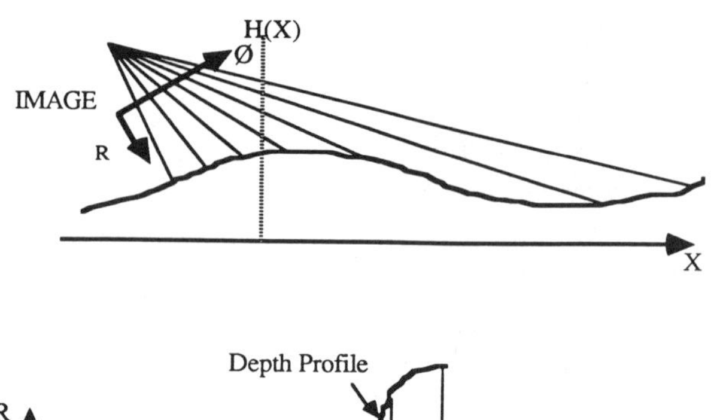

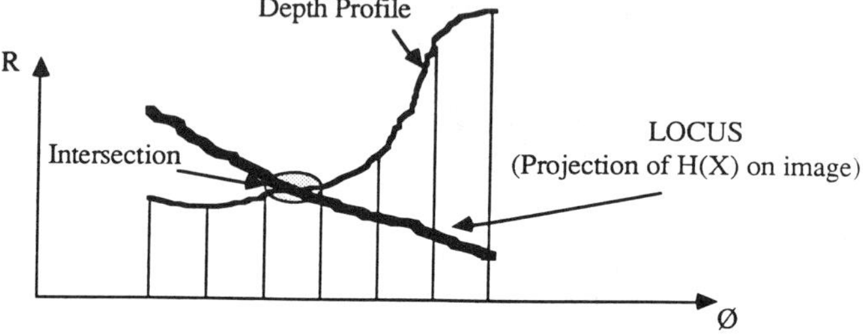

FIGURE 2.13. The locus method: intersection of scanned surface with vertical line in world space (top) and same intersection in image space (bottom).

Those problems could be solved by applying a standard interpolation technique to the sparse elevation map. This would provide a dense elevation that is a reasonable interpolation of the sparse input. However, such an interpolation technique would not take into account the geometry of the sensor, thus making it difficult to identify shadows or to convert sensor uncertainty to map uncertainty.

The *locus* algorithm overcomes many of these problems by explicitly taking into account the sensor geometry in building a dense elevation map. The idea is illustrated in Figure 2.13. Finding the elevation z of a point (x, y) is equivalent to computing the intersection of the surface observed by the sensor with a vertical line passing through (x, y). Knowing the geometry of the sensor, the line can be represented in image space by an analytical equation of the form $range = f(r, c)$, where r and c are the row and column coordinates in the image. (The projection of this line into the image defines a locus of points that gives the algorithm its name.) The intersection between the line and the observed surface is found between two adjacent pixels (r_1, c_1) and (r_2, c_2) such that $range_1 < f(r_1, c_1)$ and $range_2 > f(r_2, c_2)$, where $range_1$ and $range_2$ are the values in the image. The final value of z is obtained by interpolating the range between (r_1, c_1) and (r_2, c_2).

The key point of the locus algorithm is that the interpolation is taking place in the *image* instead of in the map. This allows us to explicitly take into account the sensor model: the uncertainty on z is computed by combining the known uncertainties at (r_1, c_1) and (r_2, c_2). The unknown regions in the map can be detected by observing that (x, y) belongs to an unknown region of the map if (r_1, c_1) or (r_2, c_2) is on a range discontinuity in the image. Another important consequence is that the elevation can be computed at any point of the map without having to recompute the entire map, whereas standard map interpolation would have to compute the entire sparse map before interpolating the dense map. Finally, there is no constraint on where the map coordinate system is located with respect to the image. In particular, we can generalize the algorithm to compute the intersection of any line in space with the image.

This technique has been used to build terrain maps, with resolution as fine as 10 cm, that include uncertainty and explicit representations of unknown regions. The locus algorithm can also be used to build large maps by matching maps from individual images. Two images of the same area taken from two different locations are related by a transformation T (rotation and translation) between the two locations. The matching problem is essentially to compute T as accurately as possible. Once this is done, the maps can easily be merged into a larger composite map. Given some value of T, we can compute from the images two elevations z_1 and z_2 for each point (x, y) in the map. The squared difference $(z_1 - z_2)^2$ is a measure of how good our knowledge of T is. Since one point in the map is not sufficient because of the uncertainty, we can use the sum of the squared differences, $E(T)$, over the part of the map that is visible in both images. $E(T)$ is minimum when T is the exact transformation between the locations at which the two images were taken. Starting with an initial estimation of T, we can therefore apply a minimization algorithm (gradient descent) to $E(T)$ to compute the best possible value of T. In practice, the initial value of T is computed from feature matching or from the positioning system of the vehicle. $E(T)$ is not computed over smooth areas of the map, which would provide little variation as a function of T.

This matching technique has been applied to the building of large maps (several hundred meters) using many range images collected as the vehicle travels. Experimental results show that the locus algorithm can be used to build accurate maps over long distances of travel.

2.4.5 Discussion

The terrain representations that we have developed have proved to be critical in building a successful mobile robot that includes capabilities of obstacle avoidance, open terrain navigation, and map building. We have also demonstrated that laser range finding is a sensor modality that should be used, being superior to passive techniques at least at this stage of the research. There are, however, a number of additional problems, including the following.

Sensor fusion: Using geometric information is sufficient for most navigation tasks. In other areas it would be beneficial to explicitly merge geometric information with appearance information. For example, shape and color are equally important in object recognition. In road following, the use of geometric information would help distinguish between shadows and other illumination effects and the presence of real objects on the road. This would greatly improve the performance of color classification, but requires that color and range information be merged. We have done some limited experiments with sensor fusion both at the level of the images, constructing a "colored-range" image, and at the level of the features extracted from color and range images. The main issues are the choice of the level at which information should be merged (images, features, or interpretation) and the difficulty of accurately registering range sensor and color cameras. Much more work remains to be done in this area.

Use of reflection: In addition to range, a laser scanner can also measure an image of the energy of the reflected laser beam. This image, usually called the *reflectance* image, is similar to an intensity image except that it is largely insensitive to outside illumination. Therefore, it does not exhibit effects such as shadows, highlights, or interreflection, all of which are hard to model. We have used reflectance data for road following and object recognition with some success. The main limitation was poor performance of the reflectance measuring, which we believe is not inherent to the technology but is due to the particular sensor that we were using. Those preliminary experiments have shown that active reflectance images are an attractive alternative to intensity images and that research in this direction should be pursued further.

Uncertainty: We have proposed ways of representing uncertainty for various terrain representations. However, the way we deal with uncertainty is still somewhat ad hoc in that it uses heavily the characteristics of our sensor and the particular representations that we have developed. A more systematic approach to modeling uncertainty in 3-D terrain is needed.

2.5 Planning

Intelligent action requires perception, planning, and control. While our main emphasis has been on perception, we have also developed the planning and control needed for smoothly following roads and for traversing rugged off-road terrain that challenges the limits of the vehicle hardware.

Early Navlab trajectory planning used a variety of heuristics for steering and obstacle avoidance. We eventually settled on a simple "pure pursuit" trajectory planner for road following. Pure pursuit steers the vehicle to point toward the desired path, a given "look-ahead" distance in front of the vehicle. It is stable, simple, and fast.

Obstacle avoidance and cross-country navigation in rough terrain require more complex planning. Typical cross-country planners consider vehicle traversability constraints, such as finding areas that are too steep, have too large a vertical step, or have vertical spikes that would hang up the vehicle's undercarriage. Our cross-country planner, in addition, considers accuracy constraints (for both perception and vehicle motion) and sensor positioning. Two distant objects may appear to be far enough apart to allow safe passage, but the planner may not be able to guarantee that the vehicle can move to and past the objects accurately enough to miss them. In that case, the planner must generate a path that moves closer to the objects, orient the vehicle so that sensors can see the objects and update their relative positions, and replan to go through the space between the two obstacles. The planner also reasons about nonholonomic motion constraints (limited turning radius) and variation of traversability with vehicle orientation (both because of wheel orientation and vehicle shape). So, for instance, a particular ditch may be traversable if the vehicle approaches perpendicularly, but would entrap a wheel if the vehicle's path were nearly parallel to the ditch.

Obstacles are represented in the three dimensions of x, y, and z in the vehicle's coordinate frame. The search starts at the current vehicle configuration and expands, following the constraints of turning radius, to reach the goal. Each path is "fattened" by maximum expected error, which turns a planned "ray" in configuration space into a "cone" of possible trajectories. In order for a path to succeed, all paths within that cone must arrive within the goal configuration envelope without encountering obstacles. If that is impossible, the planner must select an intermediate goal and replan. The search is made efficient by an oct-tree representation of obstacles and free space, and by considering various pruning strategies. Figure 2.14 shows the planned

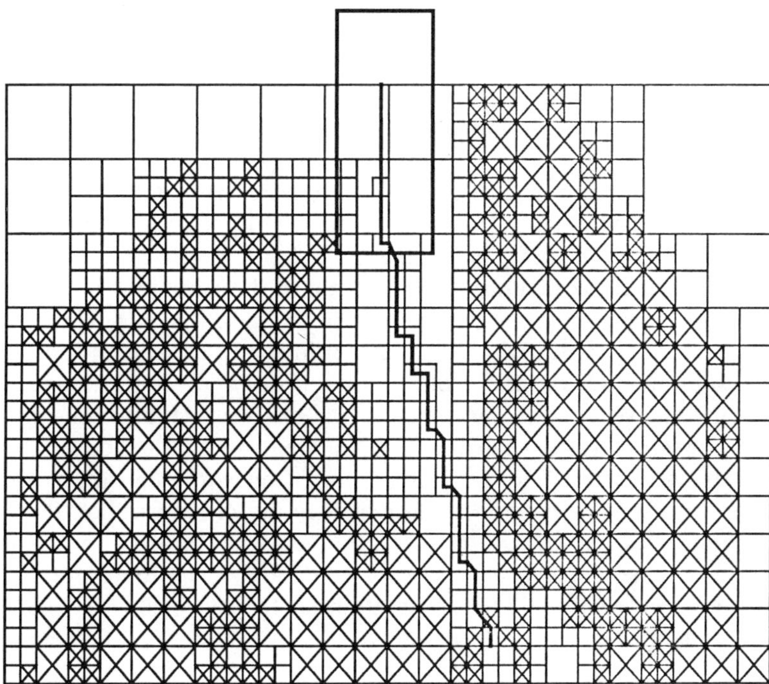

FIGURE 2.14. Planned trajectory, showing one slice of the oct-tree, with obstacles as x's and passable areas as empty squares.

trajectory through a series of 3-D scans. The ellipses indicate the bounds of expected vehicle error along the planned path.

2.6 Architectures

The first architecture used for robot vehicle research at CMU was hard-coded software to invoke each module in turn—vision, path planning, motion. With this architecture, we wrote a number of "pure pursuit" programs in which a vehicle would take a picture, calculate a single number, such as a steering angle, and move according to that direction, then repeat the cycle. The control paradigm was that of a simple servo loop, and the perception consisted of a simple estimation of a single parameter, such as the angle of the centerline of the road with no semantic interpretation. The mission of the robot was to follow a road, and it required no concern for conflicting goals or resource allocation. As long as these systems were written by one person and only had to do one thing at a time, there was no need for any "architecture" other than simple subroutine calls to the modules. But, as the complexity of the system grew, a more organized architecture was needed. The CODGER blackboard system was built to fulfill this need. Having used CODGER for several years, we have now implemented EDDIE, a new system design based on what we have learned from CODGER.

2.6.1 CODGER

When we started to design the NAVLAB, we envisioned a system with several different modules running simultaneously: color vision to find roads, laser range data analysis for obstacle detection, a map to predict road and landmark locations and appearances, and a route planner to decide which routes to follow cross-country or at road intersections [31]. Rather than build up a system a piece at a time using existing approaches for these tasks, we wanted to conduct research to advance each area and at the same time integrate the new software to demonstrate complete robot vehicle systems. This agenda required a very sophisticated architectural framework capable of supporting multiple simultaneous processes on multiple processors, coordinating planning at various levels, and integrating information from multiple sensors.

Our architecture was designed in three different facets: the module decomposition that separated functions and described the communications among them; the "driving pipeline" that discussed timing of information flow and data dependencies; and CODGER (COmmunications Database for GEometric Reasoning), the central tool that provided database, communications, synchronization, and geometric transformation facilities [15, 28, 30]. The modules consisted of

color vision, to see the road;
3-D vision, looking for obstacles;

the pilot, predicting road locations for color vision and planning vehicle
  trajectories;
the helm, which managed trajectory execution;
the controller, which handled the real-time tasks of vehicle driving; and
the map navigator, responsible for route planning.

The driving pipeline was managed by the pilot. Pieces of the road about
4 m long, called driving units, were first predicted, then detected by color
vision, then swept for obstacles by 3-D vision, then used for planning by the
pilot, then traversed by the helm and controller. By looking ahead different
distances, each module could be working on a different driving unit, keeping
the pipeline always full and maximizing throughput [16].

All communication between modules went through CODGER. But, the
various modules were looking at the same objects at different times, from
different vehicle positions, and in different coordinate frames corresponding
to different vehicle and sensor positions. To manage this complexity, COD-
GER automatically translated geometry into any requested coordinate frame.
By keeping a "history list" of vehicle positions, CODGER could take a location specified relative to the vehicle at a particular time and return the coordinates of that point, either in the world frame or relative to the vehicle, at a
different time. CODGER also had a rudimentary mechanism for including
uncertainty in locations and transforms and was designed to eventually use
filtering and multiple observations to adjust object positions and reduce
uncertainty.

The communications part of CODGER included features such as anonymous data flow (a module could request or generate data of a particular type
with no knowledge of other modules listening or sending that data), message
formats specified by the user at run time, and a variety of module synchronization and interrupt generation methods.

CODGER CRITIQUE

CODGER and its associated architecture were designed to be a general system, to support whatever missions, sensors, vehicles, processing, and representations we would build. At the time, this was a reasonable decision, since
we did not have well-formed ideas of future systems. This is a natural approach to take in designing an integrated system whose components are not
well understood. However, from a vantage point of four years later, we now
see some shortcomings in the design.

The entire architecture runs on a single level, the level of module synchronization in the driving pipeline. The support built in to CODGER for geometry
and transformations is ideal for that level. But, CODGER had inadequate
support for lower-level processing, such as high-performance real-time control. It was difficult to add reflex-level behavior, such as stopping the vehicle
when an obstacle was detected with short-range sonars. The pilot was expected to manage all vehicle control; and CODGER's internal vehicle history

mechanism depended on the helm knowing what the vehicle was doing. So, any lower level of interaction would greatly disturb the system.

Not only was communication on a single level, but map representations were also basically flat. Map objects had several descriptors, from bare geometry to symbolic roads. But, each map object was described at all levels, so there was no concept of data abstraction or differentiation between local and global maps. All the geometric objects ever used were retained in CODGER's database. All the geometric transforms from a particular object back to the world coordinates had to be calculated as a string of transforms representing estimated vehicle motion from the last landmark sighting, which tied vehicle coordinates to the world. This started to affect system performance after several hundred meters of vehicle travel, when transform chains became longer and longer and searches for geometric objects had to look at more and more database entries.

CODGER also suffered from "architectural overkill" and had little-used features that added to the difficulty of system maintenance. Anonymous messages and run-time specification of message formats turned out to be unnecessary. In retrospect, message communication would have been far simpler if message formats were compiled into the modules. As for anonymity, the only real use of anonymous communications was graphic programs that could be run optionally, to listen to message traffic and display current vehicle position and map objects. In all other cases, the sender and receiver of the message knew each other's identity, and in fact, they had to somewhat distort CODGER's message mechanisms to accomplish some types of message sequencing. Moreover, the design for anonymous communications forced all messages through a central process, which created the potential for a bottleneck and limited the speed of low-level communications.

Finally, CODGER was complicated. The helm module, for example, had 4000 lines of code; CODGER itself was 14,000 lines long; and the CODGER libraries that were loaded with each module added an additional 25,000 lines. This complexity, caused in great measure by providing features we did not really need, was one of the major reasons for our new system design, EDDIE.

2.6.2 EDDIE

Our new EDDIE system addresses the concerns we have with CODGER. It is much more tightly focused on the real issues of local vehicle navigation, and it provides a high-speed lower level. Vehicle positions are maintained by the lowest level controller, which has the closest access to the vehicle and therefore the most accurate information. Communications are greatly simplified and are now point to point, increasing their efficiency. The map is divided into local and global representations. By splitting CODGER's functions into separate pieces for local communications, vehicle history, and map handling, the individual modules are much smaller and easier to maintain.

The first part of EDDIE is the new real-time controller. This module subsumes the functions of the old controller and the helm, and in addition, it maintains the current vehicle position, which supplants CODGER's history mechanism. Vehicle motion commands arrive at the controller labeled as either "immediate" or "queued." The controller parses incoming commands, handles the queue, and talks to the hardware motion controller at the appropriate times to set new steering wheel positions and vehicle velocities. By querying the vehicle's encoders at frequent intervals, the controller is able to maintain an accurate dead-reckoned position estimate. In EDDIE, no vehicle position history is kept. The only times when it is necessary to know vehicle position are when new data are required or during trajectory planning. It is easier, and more accurate, to dispense with history mechanisms and instead to query the controller for the current vehicle position each time an image is digitized and whenever a planner needs to know the vehicle's location.

The vehicle controller uses different tracking strategies to keep the vehicle on the desired path. It can also be called upon to follow a previously recorded map if the perception clients are temporarily unable to navigate the vehicle. This keeps the vehicle on a safe path while the vehicle turns sharp corners outside the camera's field of view or travels through featureless or confusing visual scenes. Figure 2.15 shows the accuracy of the executed path (solid line)

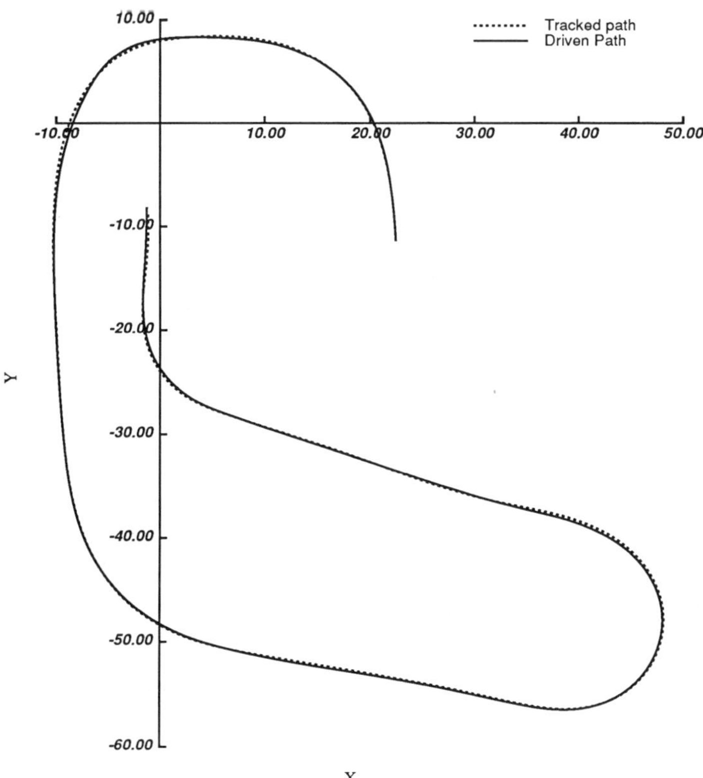

FIGURE 2.15. Path following: recorded (solid line) and executed (dotted).

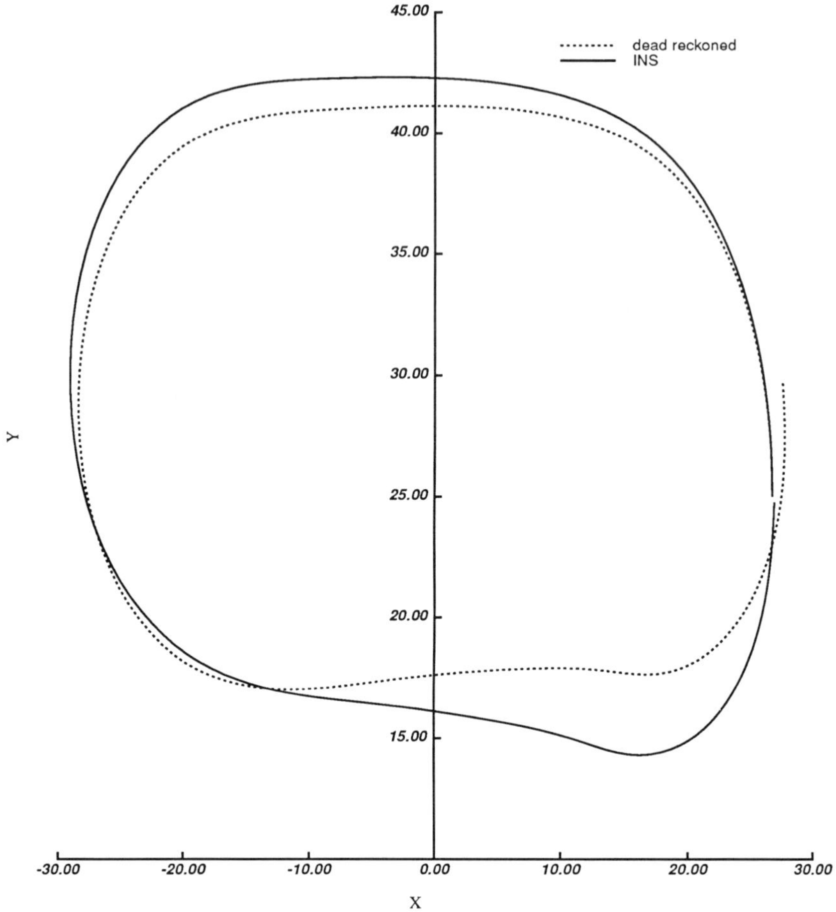

FIGURE 2.16. Position estimation: highly accurate with INS (solid line) and less accurate with dead reckoning (dotted line).

following a stored path (dotted line). Another safety consideration is smoothly regulating velocity, trading some reduction in accuracy of velocity for smooth accelerations and reduced vehicle roll around sharp curves. The controller warns against system failures and records a log of events for future reference. This is extremely valuable in system configuration and debugging. The low-level controller is also responsible for utilizing INS Inertial Navigation System and encoder data to find the best estimate of current position and relaying it to external clients through the ethernet. Figure 2.16 shows accurate vehicle position estimation using the INS (solid line) and the less accurate, but still stable, position estimation using only dead reckoning (dotted line). The clients are managed by a software server that prioritizes the connections in order to meet the needs of many clients without degrading the level of performance required by critical components of the system [1].

Closing all position-estimation loops through the controller has a powerful fringe benefit: transparent path modifications. We have implemented a joystick interface that allows a user to modify commanded trajectories. Joystick input is simply summed with computer input, so the user has the sensation of "nudging" the vehicle away from its planned path. The Navlab is also being equipped with a "soft bumper," a ring of ultrasonic range sensors to detect nearby objects before collision. When completed, the soft bumper will interact with the controller in the same manner as the joystick, by adding its control input to the input from planning, but it will have progressively higher gains as the time to collision decreases. Previous systems would have been destroyed by this subversion of planned paths, since CODGER kept vehicle position history by an open-loop expectation of perfect path tracking. In the EDDIE system, all position queries are handled directly by the controller and are therefore answered correctly even if the path has been modified.

Communications in EDDIE are unexotic and uninteresting, but fast, with point-to-point connections. We still use TCP/IP communications protocols over the ethernet, but we can now go to shared memory or other protocols for particular connections, as needed.

2.6.3 Discussion

The main philosophical differences between CODGER and EDDIE reflect our current thinking on architectures:

Task specific models: EDDIE does not impose particular connectivity or map structure, but instead it provides tools to let users build their own. In particular, much of the data that CODGER put into a central database properly belongs within a single module or pair of communicating processes, as encouraged by EDDIE.

Explicitness: The most important models maintained by an architecture are vehicle positions. Where CODGER implicitly assumed open-loop perfection, EDDIE explicitly queries the lowest level hardware for current position.

Architectural support: EDDIE uses annotated maps as a better form of CODGER's geometric tools at an intermediate level of abstraction, while adding support for soft bumpers, joysticks, and other physical-level control at lower levels. In both systems, we have left higher, AI-level support to be provided by other modules as needed.

The evolution from CODGER to EDDIE is a natural one in the evolution of the Navlab project. In the early days, the nature of the modules and their interactions were not known, and our major concern was to not preclude any conceivable system design. Thus, we built CODGER, which was very general and provided easy reconfiguration through anonymity of data storing and access. Now that we know the specific configuration of low-level modules that we need to run the NAVLAB and how they communicate and synchronize with each other, we seek the simplicity and higher performance that can be achieved by a more specialized architecture. EDDIE is that new design.

2.7 Maps and Missions

Much of the information that mobile robots need is tied directly to particular objects or locations. Maps, object models, and other data structures store useful information, but they do not organize it in efficient and useful ways. We have built a new map-based knowledge representation, the "annotated map," to index information to the relevant objects and locations. The annotations are used for a wide variety of purposes: describing objects, providing hints for perception or control, and specifying particular actions to be taken. We have provided a query mechanism to retrieve annotations based on their map locations. We have also built "triggers," which cause a specified message to be delivered to a particular process when the vehicle reaches a given location in the map.

These annotated maps serve a crucial role in enabling missions that are otherwise beyond the reach of autonomous systems. Control descriptors allow mission planners to specify what the vehicle is to do at particular locations, reducing the need for on-board planning. Object descriptors contain detailed instructions of how to recognize a particular object, or they contain the appearance of this object as seen by a particular sensor on a pevious vehicle run. Such information greatly simplifies the problem of seeing and recognizing objects. Geometric queries enable the vehicle to focus its attention on objects in its vicinity, reducing database access and matching time. The trigger mechanism frees individual modules from having to track vehicle position, allowing them to devote their processing to the task at hand or to lie dormant until they receive their trigger message.

Annotated maps do not by themselves solve difficult problems of sensing, thinking, or control for autonomous vehicles. Their contribution is to provide a framework that makes it easy for other modules to cooperate in planning and executing a mission. Annotated maps thus fill a need that is common to many different vehicles, missions, and architectures.

Many analogous annotated maps exist for human use. Aeronautial navigation charts contain symbolic descriptions of routes (airways) and landmarks and include annotations such as the Morse code call letters of radio navigation beacons. The AAA produces "Triptiks,"[2] which include annotations for route, current conditions (construction, speed check), road type (interstate, two lane, etc.), general conditions ("winds through rolling hills"), points of interest (rest areas, gas, food, and lodging), etc. An intelligent person can usually drive a route without such aids, but they do provide a convenient framework for preplanning and make "mission execution" easier. Furthermore, as we drive a route, we build our own mental representations of landmark appearance, curves in the road, and so forth, which we use to follow the same route more easily at a later time. Our annotated maps provide the same kind of functionality for autonomous mobile vehicles.

----

[2] Triptik is a registered trademark of the American Automobile Association.

2.7.1 Related Work

Many other groups are working on related problems of mobile robots and knowledge representation. Rather than competing with the ideas of annotated maps, most of this research is providing useful tools and ideas that could use or help generate the annotated maps.

Fennema, Hanson, and Riseman at the University of Massachusetts are building world models and maps for their mobile robot, Harvey [14]. They have defined the concepts of neighborhoods (topological regions), locales (information to decide whether the robot is within a neighborhood), milestones (perception for verification), and actions. The UMass map and plan representations are similar to some of the uses of annotations, but have simple, fixed formats, are focused on declarative representations of 3-D object models, and do not provide map-based triggers.

Rod Brooks at MIT has long argued for simple robots with simple control schemes and simple world maps [7]. We concur that simple, sensor-based maps of particular locations are often useful. The lowest levels of our descriptor annotations are designed to contain precisely the sort of information that Brooks's robots use to calculate their position, or to cause a particular action, in a small local area. We disagree with Brooks's contention that this is the only sort of information that a robot should remember. Robots often work in open, featureless environments and need precise maps and accurate navigation even where no landmarks may be nearby. Annotated maps are designed to keep precise metric information in the geometric levels of annotations, as well as the lower level cues advocated by Brooks.

Kender gives a much more abstract view of planning for sensor-based navigation [19]. He describes the combinatorial problem of deciding which sensors to use and which landmarks should be recognized in order to reach a given goal. The results of analyses such as Kender's should be entered into triggers to tell the vehicle what to look for and into object descriptors to say how to look for those objects.

Blidberg and his associates at the University of New Hampshire's Marine Systems Engineering Laboratory have implemented world models for underwater mobile robots [8]. Most of their work has concentrated on efficient descriptions of space, such as quad-trees. These spatial descriptions are important, but do not include many of the other forms of knowledge (actions, descriptions) for which annotated maps are useful.

2.7.2 Scenario

A typical mission for the Navlab is a delivery task on suburban streets. In order to accomplish its mission, the Navlab must use several of its navigation modules: SCARF, YARF, ALVINN, 3-D perception, and inertial navigation. Road following using color vision will follow streets, but it will not be able to recognize intersections. Inertial navigation will drive through intersections, but it must have an accurate starting position. Landmark recognition will update vehicle position before intersections, but it is too slow to be run continuously. Only a combination of all those modules, each running at the appropriate locations, will produce an accurate and efficient mission.

KNOWLEDGE AND ORGANIZATION

In general, planning and executing such a mission requires several types of knowledge: what to look for and how to see it, what to do and how to accomplish it, and where to go and how to get there. The knowledge may range from high-level symbols to low-level raw data. Knowledge is both internal to a single module and used by controlling modules to switch between knowledge sources. Approaching the intersection, for instance, the perceptual knowledge includes

symbolic: intersection;
geometric: size and shapes of intersecting roads;
sensor-specific: use laser range finder to pinpoint the position by landmark
    identification; and
raw data: landmark 2 m tall, 0.4 m wide at position (x, y).

Control knowledge can also span a range of levels:

symbolic: turn left at intersection,
geometric: intersection angle $45°$,
vehicle-specific: turn with a circular arc of radius 15 m, and
raw data: steering wheel position left 1200 clicks.

This knowledge must be carefully organized if it is to be useful. If the vehicle has to sort through all bits of information it has about every possible object, it will overshoot the intersection long before it has decided how to recognize it or deduced that it was supposed to turn. It is far better to have information tied directly to the map or automatically retrieved as needed. The landmark recognition module, for instance, must be able to ask for a description of objects within its field of view and retrieve the knowledge it needs to recognize them.

ANNOTATED MAPS

Annotated maps provide the mechanism for organizing this knowledge by tying information to a map. The annotations contain knowledge about particular objects, locations, or actions. Annotations come in one of two classes: descriptors or triggers. Descriptors are passive and are retrieved by queries based on geometry and object type. A query for "all objects of type 'intersection' in this polygon" would return the annotation for the requested intersection if it were in range. Triggers are active, firing when the vehicle reaches a particular location or crosses a certain line. A trigger will send a message to a particular module, such as "controller, start turning hard left in five more feet."

The knowledge in these annotations comes from many sources, including human experts, mission planning software, and even the vehicle's own observations and experiences on previous missions. It is both declarative (data) and procedural (methods and procedures). The level of the annotations depends partly on the vehicle's computational capabilities. Simple vehicles in known environments are able to execute simple, preplanned missions by having every object and action completely annotated at low levels. A more challenging environment with more variation over time may require higher level symbolic descriptors in the map and more reasoning at run time. Practical missions will probably require a mix of levels of detail. Even a sophisticated vehicle may, for instance, decide to record the locations of specular reflections from a mailbox and use those specularities as recognition cues. It may be much more difficult to reconstruct a 3-D model from the observed data and later predict the appearance from the model.

EXAMPLE RUNS

Figures 2.17 and 2.18 show a typical annotated map. Figure 2.17 shows a map of a suburban area, including about 0.7 km of road with two T intersections and a variety of 3-D objects. Object information was collected using the ERIM laser range finder, and the road information was collected by using the inertial navigation system to provide accurate vehicle positions while we tra-

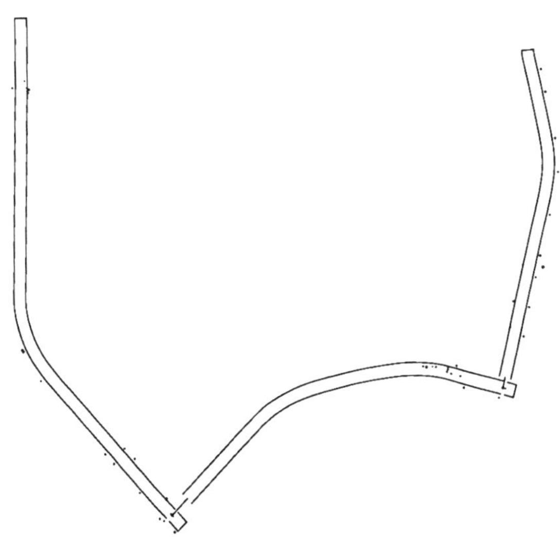

FIGURE 2.17. Map built of suburban streets and 3-D objects.

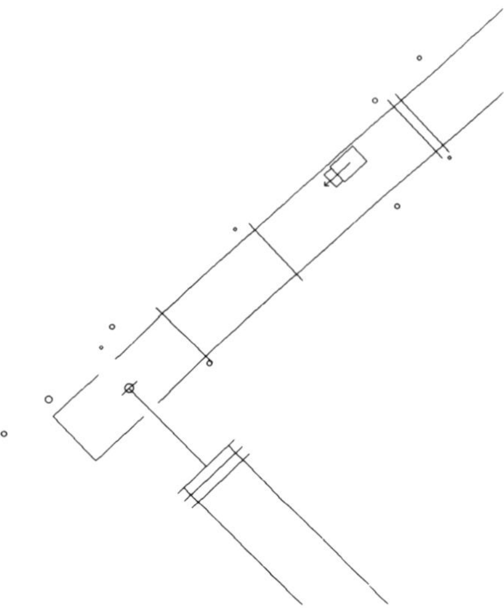

FIGURE 2.18. Trigger annotations for sensing and vehicle control.

versed the route. Figure 2.18 shows a detail of the first intersection, including the Navlab's position during a run, and several triggers.

The goal of this run was to drive from a house near the beginning of the map to a specified house near the end. Annotations were added to the map to enable the Navlab to carry out this mission. There were annotations to set the speed appropriately: up to 3.0 m/s in straightaways and down to 0.5 m/s in intersections. Other annotations activated and deactivated the module that uses the laser range finder to correct vehicle position based on detected landmarks. Before every intersection, there was an annotation that switched driving control from the ALVINN vision program to a module that used knowledge from the map of the intersection structure and dead reckoning to traverse the intersection. Finally, there was an annotation at the end of the route that caused the vehicle to stop at the appropriate object. The route was successfully traversed autonomously.

In this run, and a variety of other runs, we have successfully used nine different types of trigger annotations:

set speed,
dead reckon through intersection,
resume vision after intersection,
start landmark matching,
stop landmark matching,
stop at objects,
stop and start fast obstacle detection,
use vision through intersection, and
switch perception modules.

2.7.3 Tenets of Map Construction and Use

Several key ideas underly our design for annotated maps, reflecting our experience in building perception and navigation systems for a variety of robots.

MINIMIZE SEMANTIC INTERPRETATION

No one can predict all the kinds of knowledge that will be placed in annotations. Moreover, the map module need not understand the annotations. The only common knowledge in annotations should be enough header information to store and retrieve the annotation. All the rest of the annotation belongs to the modules that create it and interpret it, with the format to be decided upon by the module creators. The annotated map serves only as a scratch pad.

NO SPECIALIZED QUERY LANGUAGE IS NEEDED

The standard queries ask for all objects of type X within polygon Y. Any query more ambitious than that need not be supported. Any more detailed query would require that the map module know the internal details of each type of annotation. It is more efficient, and a better abstraction, to let the querying module sort through the returned objects.

SEPARATE GLOBAL POSITION TRACKING FROM LOCAL SERVOING

Maintaining the current position estimate in local coordinates is a real-time job and is best done by the low-level real-time controller. In order that locations stored in local coordinates will always be consistent, the controller's local coordinates should never be updated. Commanded trajectories, current positions of obstacles to be avoided, and other phenomena that are used once and then discarded should be kept in local coordinates and never entered into the map. Map-based calculations, such as matching landmarks against a map or interpreting a position fix, are aperiodic events best done by a separate navigator module. The navigator maintains the transform from local to world coordinates. Any module that needs to know current vehicle position in world coordinates must acquire the navigator's transform and then apply that to the running position reports of the controller. In practice, acquiring the Navlab's current transform is done in one of two ways, specified at start-up.

The navigator can send its transform every time it is updated. This is used by
  fast-running modules that always need the latest update.
Slower modules, that have a longer cycle time, may not need every updated
  transform. Worse, receiving too many updates before the module is ready
  to read them may cause the input queue to overflow. Instead, these modules are notified that a new transform is ready, but they do not receive the
  update until they request it. The navigator remembers which modules have

been notified and have not yet requested updates to avoid sending repeated notifications.

CENTRALIZE POSITION TRACKING

Modules often want to perform specific actions when the vehicle arrives at particular locations in the map. If each module were to continuously poll the navigator and controller for current position, the controller could become overloaded. Active polling also means that those modules are using computer cycles. Moreover, a navigator position update may skip the vehicle position estimate past the point for which a module is waiting. For each update, each module would have to decide if any of its target positions had been passed. We prefer to have a single module, the map manager, doing position tracking for all modules. On reaching the points of interest, it awakens or signals the appropriate module. This is the function of "trigger" annotations.

NO MASTER CONTROL

The map module is best thought of as an alarm clock (for the triggers) and a scratch pad (for the descriptors and trigger messages). It is not some "master" module that controls all thinking and that therefore can become a major bottleneck. We prefer point-to-point communication between modules, with flow of data and control decided on module by module, rather than forcing all information through a single controller.

PLAN INCREMENTALLY

The map module is designed to be used by many programs, for many purposes, at many times. Some information may be permanent; other annotations may be added to provide directions for only a single mission. It is an advantage to be able to update, add, and delete at various times. In particular, display and user interface modules may read the annotated map from a file, look at it, display the annotations, change things, and write the updated file to disk.

2.7.4 Implementation of Annotations

The annotated map needs to provide efficient access, indexed by position. The annotations need to contain an arbitrary amount of data with a minimum of externally imposed organization on the contents. We have designed and implemented a two-part representation consisting of a map grid and an annotation database. Each square of the grid contains a list of any annotations that are included in that square's area.

Adding an annotation to the map is a two-step process. First, the actual annotation is added to the annotation database. Second, the map grid must be updated. The location of the annotation is a point, line, or polygon. This location can either be specified directly, for those annotations tied to a location, or retrieved from an object description, for those annotations that describe an object. The location is then scan-converted (converted to a list of cells) into the grid, and a pointer to the appropriate entry in the database is written into each of the corresponding grid cells.

Retrieval of annotations in response to a query is also a two-step process. Queries can specify a polygon and an annotation type. The query polygon is scan-converted into grid cells. The annotations pointed to by each of those cells are collected, checked to see if they match the specified type, and returned.

Triggers work similarly. At each cycle, the map module calculates the current vehicle position. It calculates the line on which the vehicle has moved since the last cycle and scan converts that line into the grid. Each cell through which the vehicle has moved is checked for trigger annotations. If any are found that have not already been fired, their messages are sent to their destination modules. Since the location of a trigger can be a point, line, or set of lines, a trigger can be fired when the vehicle reaches a certain location or when it enters a given polygon.

REPRESENTING ANNOTATIONS

Annotations are represented with a uniform header format plus a free-format data field. Typical header fields include the following.

header: type, destination module, used flag, text description, location, next
    object, previous object, data size;
data: pointer to data.

The header portion contains all the information that the map module needs to understand. "Type" and "location" are sufficient for answering queries; "destination module" is required for sending trigger messages. The "used flag" is set when a trigger is fired, to avoid firing the same trigger repeatedly if the vehicle stays in the area covered by the trigger for more than one cycle. "Text description" is used by graphics display modules. This information is also sent as part of messages to make it easier to debug receiving modules. The "location" of the annotation is used both in initially setting up the grid pointers and for the use of the receiving module. "Next object" and "previous object" are used to describe extended linear objects. Extended objects may also have branches, which meet at intersections. Intersections have a center point and any number of vertices, each of which points to the beginning of an extended object. The most common extended objects are roads, which are represented as short segments pointing to their preceding or following segments or pointing to intersections.

The data portion of the annotation is, in the view of the map, an undifferentiated field of bytes. Any internal structure need only be understood by the modules that create and read the annotation. Since the headers have a known, fixed size, they can be stored in a random-access file. The data may be stored as a stream of bytes, with the header containing only a pointer to the beginning of the data and the number of bytes.

IMPLEMENTATION DETAILS

Our prototype implementation has tested some of our design decisions, while other details will be decided after further data collection and analysis.

Grid Cell Size

If grid cells are too small, queries will have to look at large numbers of cells, and map storage will become a problem. But, the querying becomes simpler, because any object found in any of the cells can be returned. Larger cells give faster lookups, but are no longer selective enough to answer queries on their own. Instead, objects within grid cells must still be checked to make sure they are within the query polygon. For autonomous land vehicles with sensor ranges of 2 to 30 m, a grid with 0.5- to 1.0-m cell spacing probably provides the right trade-off; our current implementation uses 0.5-m cells.

Handling Large Maps

For a grid with 1.0-m cells, each square kilometer will contain a million cells. Each cell can be represented with at most a few bytes of data, depending on annotation density. The amount of memory required by a grid this size is easily within the capability of today's computer systems, but for missions spanning several kilometers, we will not be able to keep the whole grid in main memory at once. One possible solution is implementing quad-trees to take advantage of sparse data requirements over most of the grid. A more likely strategy is to keep the grid on secondary storage and only keep a window around the current vehicle position in main memory. The annotation databases themselves may also need to be kept on backing store, and only read in as needed.

Distributed Databases

Object descriptions might be most easily implemented in separate databases internal to the modules that use them. Then, the annotations need only return the index of the database entry. The problem with this method is ensuring consistency between databases in the modules and indices in the grid. At the opposite extreme, the map annotations could contain all the data. The disadvantage of this approach is requiring more traffic between maps and objects. An intermediate approach is to start with all the knowledge in the map annotations, but have it automatically replicated in the appropriate modules at system initialization time. This ensures consistency while reducing run-time overhead at the expense of start-up costs. The design of distributed databases interacts with the design for handling large maps. Keeping annotations in individual modules would decrease the amount of information needed by the map module and thus make building large maps somewhat easier.

In the current implementation, the annotation database is static during a run. When the system is initialized, the user adds stop points, turn points, or other triggers to specify the current mission. When the user is ready, the interface module saves the current annotation database and sends the name of the file to the map module. At start up, each module that needs a copy of the annotation database requests the name of the file from the map module. So, modules contain a complete, consistent copy of the annotation database. The map module builds the grid, so it can handle geometric queries. It communicates with the other modules by specifying the index in the annotation database of the objects that match the current query. The map module also watches the grid for triggers.

Map Update

Changing an annotation during a run is conceptually easy. Moving objects and annotations is more difficult. If a single object moves, it is easy to erase it from one part of the map and write it into another location. But, if an entire portion of the map moves, such as discovering that a portion of the road is really longer than previously thought, the changes can be very hard to handle. Many objects would have to move: the road, all objects attached to it, all landmarks that were seen on previous inaccurate runs and indexed to the road, planned mission steps based on following the road or on seeing those landmarks, etc. It is probably better to note the new information, keep running with the flawed map, and build a new map at the end of this run, rather than try to do updates on the fly. Map update strategy is also influenced by the "large map" and "distributed database" design issues. If an individual module updates its copy of an object description annotation, it will need to make sure any permanent information is written when the run is terminated or when that portion of the map is overwritten by a new data window.

Since, in the current implementation, each module keeps its own internal copy of the annotation database, map updates must be specially handled while building a new map. Under most circumstances, the map updates refer to objects that the vehicle will not see again on this run, and therefore the updates need not be propagated to all the modules. At the end of a run, all the new objects can be written to a new map file to be used on succeeding runs. The exception is for building maps of intersections. Our procedure is to drive through the intersection following one branch and build a map and then reposition the vehicle before the intersection and follow the second branch. In order to register the two branches correctly, the perception and matching systems need to find newly mapped landmarks. The map manager writes the annotation database to a file and notifies the relevant modules, which read in the updated database.

Interfaces

Conceptually, it is easy to add annotations to the map. A program reads in the annotation database, adds new annotations, and writes the updated files.

Machine-generated annotations, such as object descriptions, use interface routines to read and write the map and to insert annotations into the annotation database. Annotations added by hand require, besides the basic map interface routines, a user interface to point to locations or objects on the map, type or read the annotation data, display the resulting map, and ask for verification. While the format and contents of the annotations will vary, there is still a large body of common functions that use standard modules. We have built an interface, using X windows, that allows a user to add new objects and triggers to the map. The same interface is also used to display the vehicle and map during a run.

TRIGGER DETAILS

In order for the map module to track vehicle position, it must know both the controller's current local position estimate and the navigator's transform that relates local to global coordinates. Position queries to our vehicle controllers are efficient, returning in less than 10 ms. Our current implementation uses an efficient process for getting transforms from the navigator, by having the navigator send the transform each time it is updated. Sine landmark sightings or position fixes are relatively infrequent, an event-driven transform update is much more efficient than polling.

When the navigator updates position, the map module has to pay special attention to triggers. It may be that the vehicle position estimate will jump forward, skipping some triggers; or it may be that it will move backward, creating the potential for firing triggers that have already been fired. If the position update is relatively small, it makes sense to use the line of vehicle travel, plus the used flag, to make sure that all appropriate triggers get fired once. If the update is large, it may no longer make sense to fire triggers that should have been fired long ago; and it may make sense to refire triggers that were fired very prematurely. Details of these design decisions are yet to be determined.

The mechanism of notifying a module of a trigger is by sending a message over a port. In the Unix operating system, ports can be set up by broadcasting their address and listening to the net to find out who would like to talk to them. Once connected, ports appear as files and can be read and written easily. A module can easily check if there are any bytes waiting on its trigger port. If not, it has not received a message and can continue running. If so, it can read the message header, allocate the memory structure for the message, and read the appropriate number of bytes into its memory. A running module can periodically check to see if a message is waiting. A sleeping module can simply block on read, which will cause it to pause until data arrives. It is possible to set timers, so a module can wait until either a timer expires or a message arrives, whichever occurs first. It is also possible to have an incoming message generate an interrupt, so the module can be notified while running even without checking for incoming data.

2.7.5 Discussion

Annotated maps provide a framework to organize knowledge storage and retrieval for autonomous mobile robots. Our group at CMU and other groups around the world have many of the individual pieces of a complete system: sensing, sensor understanding, local trajectory planning, control, and vehicles. These pieces in themselves are only sufficient to perform limited tasks. Integrating these components into an efficient system is one of the difficult remaining gaps. The annotated map helps fill that gap. By providing generic data handling, it allows diverse modules to communicate their specialized knowledge. By tying this knowledge to specific locations and objects, the annotated map provides a focus of attention, using an efficient grid structure to answer queries about specific parts of the map. Through the automatic triggers, the annotated map eliminates the need for individual modules to attend to vehicle position and map location. We have built our first prototype annotated map, interfaced several modules to it, and used it to store and retrieve data during real Navlab runs. We are currently addressing the issues of large maps, and we continue to interface more modules and use annotated maps to manage a wider variety of knowledge.

2.8 Contributions, Lessons, and Conclusions

We began the Navlab project six years ago with the firm conviction that the best way to make real progress on outdoor mobile robots was to build complete systems and to concentrate our efforts on eliminating the bottleneck of inadequate perception. We continue to agree with, and to follow, those convictions. Following those general guidelines, we have built a number of successful perception, planning, and control modules, and we have integrated them into systems that drive the Navlab on a wide variety of test sites. During the course of our work, we have also been surprised (usually unpleasantly) by several other aspects of building mobile robots: problems with sensors, difficulty of using experimental computers, questions of how to evaluate our work and how to compare it with results from other groups, and the critical importance of simplicity and defining the environment in which the vehicle must operate.

2.8.1 Contributions

Navlab experiments have validated and demonstrated several new ideas.

1. Color classification works for tracking unstructured roads. It is important to use full RGB color to handle difficult illumination, and it is important to use multiple classes, and Gaussian probabilities, to handle degraded roads.
2. Fitting complex road models to noisy data can generate gross errors; the simple straight road approximations used by SCARF are the right approach for the low vehicle speeds and difficult scenes of unstructured roads.

3. Specialized operators for tracking individual features can be combined into a reliable road follower for structured roads.
4. Neural nets can learn to track roads. A single algorithm can learn many different roads, with only a few minutes training time for each new road.
5. Accurate descriptions of unstructured terrain can be built from 3-D data. Different levels of descriptions are available, depending on the task requirements and available processing power. This information is directly useful for cross-country navigation.
6. Map building is possible, combining many noisy 3-D range images to form large-scale maps. Our approach uses a combination of iconic matching, feature matching, and vehicle position sensing. This has been shown before for simple indoor environments, but we invented new techniques and representations for outdoor unstructured terrain.
7. Cross-country trajectory planning requires not only a representation of obstacles, but also reasoning about vehicle capabilities, limits, and accuracies. These constraints can be combined efficiently and powerfully to guide vehicles up to the limits of their sensing and mechanisms.
8. Simple architectures work best. Dictating the structure of the data and control flow is not needed. It is better to build a too kit that provides communication, synchronization, map data handling, and clean interfaces to the low-level control and let individual system builders tailor the system structure to their own needs.

2.8.2 Perception Lessons

PERCEPTION

Perception is always the bottleneck. That is not to say that the other aspects of mobile robots are solved problems (path planning, map representation, etc.), but rather that they cannot be properly explored until robust perception components are built. The performance of a mobile robot system depends on the performance of the perception components. It is often assumed that robots are control systems, and that perception will provide clean numerical imput; or that robots are cognitive problem solvers, and that perception will provide clean symbolic scene descriptions. Neither of these assumptions are justified by the current state of the art in perception. Robots will not fulfill their potential unless we continue to improve perception capability.

SENSORS

While the most important scientific bottlenecks to perception involve inadequate algorithms, the current state of the art of sensor design is also a stumbling block. Too much of our effort has been spent in overcoming sensor limitations, which is necessary to do real experiments but makes no lasting scientific contribution. A few examples are given here. Laser scanning technology is a great advent in 3-D sensing. It still has considerable limitations, however, such as slow image acquisition, which puts a severe limit on the speed of the vehicle; ambiguity intervals; and bad behavior on certain material types. Color cameras also have problems, including limited field of view, inadequate dynamic range for mixed sun and shadow conditions, and unpredictable response from automatic irises and gains. We do not believe that any one magic sensor will solve the outdoor robot problem, but advances in sensors will certainly enable and encourage advances in the image understanding algorithms. We continue to build better algorithms, but their full power will not become useful until we have adequate sensors.

2.8.3 Systems Lessons

DESIGN FOR TASK AND ENVIRONMENT

Mobile robots operate in a certain environment to carry out a certain task. In the current state of the art, there are no such things as a completely general-purpose robot, universal vision system, and generic architecture. Tracking highways requires substantially different processing from driving cross-country. Some of the concepts are common (local map building, control); and learning algorithms, such as neural nets, are beginning to try to use a single algorithm to automatically extract the specialized knowledge needed for each situation. But, currently the right way to build mobile robot systems is to incorporate in the design, from the beginning, knowledge of the task and the environment. Too often, neat ideas are investigated in perception or planning and then artificially matched to an environment and a task. While this is great to demonstrate some new research results, it usually does not contribute much to mobile robots.

SIMPLICITY

The simplest approach is always the best. Designing a complex system does not solve any problems, especially if the components of the system (e.g., perception components) have not even been considered yet. The research community is full of proposed architectural standards that needlessly complicate mobile robots and that are not based on experience with working perception systems. Simpler is better. For example, the following is the approach that we have followed in our AMV system.

1. Define the task, for example, track roads with the help of a map, and perform actions at specific locations.
2. Develop and analyze the necessary components, such as road following, object detection, and map building.
3. Build and evaluate the components separately to understand their limitations. For example, we first built a smaller system that tracks a road map and stops at specific objects, and then we expanded to annotated maps and the AMV.

4. Define representations that are matched with the task, such as the annotated maps.
5. Put together components and representations in a system that is configured for the task. The system is simple in the sense that it includes only the functionality that is needed for the task using the selected components.
6. Experiment. The important point is that the experimental phase is used to evaluate how well the mission is carried out and maybe to add new perception components or modify the representations. It is *not* used for debugging a giant complex system.

Our early CODGER system started in the right direction, because it provided the specialized tools mobile robots need for representing geometry and motion. But, since we did not know exactly what interactions would be needed, the CODGER architecture provided for all possibilities and was large and complex. At a certain stage in our research, we found that the complexity of CODGER outweighed its advantages. Matching our perception components to CODGER's framework, and debugging the whole system at once, became more difficult. The design of EDDIE benefited from our experience with CODGER and from our increasing understanding of the needs and capabilities of our modules to build a much more simple, streamlined, specialized system.

COMPUTATION

Fast computation is, of course, of great help in building a mobile robot system. Not only does it improve the performance of the final system, it also holds the promise for more images processed, faster runs, and more experiments, and thus faster progress in the basic research. We have found, however, that faster computation should not be the highest priority. In the early stages of a mobile robot project, especially, the researchers need to try many different possible approaches to perception. It is more important to have easy-to-use computers, with well-supported and efficient compilers, than to have the ultimate in running speed. It is also crucial that input/output (I/O) be well supported, both for image digitization and for communicating results. Using experimental computers, especially as a demonstration platform for the computers, can greatly impede robotics research. Now, after six years of the project, our algorithms are stable enough that we can properly take advantage of nonstandard high-speed machines; but those machines should be stable and well supported. It is very difficult to do robotics research simultaneously with hardware or operating systems research.

VEHICLE

The vehicle itself must be considered an integral part of a mobile robot system, not just a platform on which experiments are conducted. The Navlab was specialized for our early systems and provides the high-accuracy motion and slow speeds we needed [22]. It was not designed for rough terrain motion, nor for highway speeds. We are currently rebuilding the drive train for the higher speeds that our perception and control can now handle. We are exploring other vehicles to host our off-road work, which will be selected and modified to complement the capabilities of our sensors, perception algorithms, and planners.

CONTROLLER

Real-time mobile robot controllers need to integrate a wide range of capabilities beyond just control theory: position estimation, mapping and tracking of paths, human interfaces, fast communication, multiple client support, and monitoring vehicle status for safety and debugging. Most mobile robots do not push the limits of current control theory. Control theory is not the major issue in controller design, design for system integration is.

DEBUGGING AND MONITORING

At slow speeds, it is relatively easy to watch the performance of a system. Our first color road trackers, for instance, ran in tens of seconds, which gave ample opportunity for watching graphics, saving the files to disk, noting the response of the vehicle, and so forth. It is much more difficult to debug a system running at higher speeds. YARF now runs in less than a second, which is faster than we can write an image to disk (for later examination), faster than we can examine the debugging graphics, and even too quick to read text output. As a corollary, YARF can now process hundreds of images in a typical run or thousands of images during a day's experiments, which makes examining the output by hand tedious at best. We need both better technology (faster disks, better video recorders, etc.) and better ideas for debugging complex real-time systems.

EXPERIMENTAL EVALUATION

Even with proper tools to monitor a particular system, it is difficult to measure progress. The basic problem is how to answer the questions "Does it work?," and, "Does it work better?" Some systems are easy to measure: did an obstacle avoidance system run over a tree or not? Others are more difficult: did the vehicle clip a corner because of bad calibration, bad trajectory planning, bad image processing, or bad control? The problems become worse when comparing work from different research groups. All papers on road following claim success. Most are missing crucial details that would enable evaluating competing algorithms. Even where all the details of the software are spelled out, crucial differences in hardware (processing rates, camera capabilities, vehicle and camera control, etc.) make head to head comparisons difficult. Common image databases provide only a small part of the solution, since different algorithms and vehicles may need different sensor vantage points, image collection frequency, auxiliary data, and so forth.

2.8.4 Conclusions

One of the greatest lessons of our research is the definition of what mobile robots research is *not*. Mobile robot research is not just research in perception algorithms, sensors, architectures, computers, vehicles, or controllers; nor should it be driven by the desire to demonstrate a particular neat result in one of those areas, with the risk of warping the entire project to showcase one component. At the same time, a mobile robot project cannot be driven strictly top-down from requirements; it is no help at all to draw block diagrams that specify what vision should do without a deep understanding of what vision might or might not be able to provide. The true nature of mobile robot research, rather, is a give and take, in which existing modules are used to drive system design, and existing system capabilities are used to motivate research.

The only way to continue to make progress on mobile robots is to continue to build more systems. We have far too few mobile robots, especially operating in unconstrained outdoor environments. The best of our robots to date are still woefully inadequate at most tasks. We need to persist in designing robots for particular domains, such as following roads or traveling across rough terrain. We need better technology in sensors, computers, and vehicles. We need to continue to work on the bottlenecks of perception, and we need to remain focused on building complete, integrated systems that solve real-world problems.

Acknowledgments. Navlab work is the product of many people. Takeo Kanade, William Whittaker, and Steve Shafer have all shared in principal investigator responsibilities. Navlab planning and systems have been done by Tony Stentz and Eddie Wyatt. The new controller is the work of Omead Amidi. Martial Hebert is the CMU expert on 3-D perception, including the Navlab's medium resolution mapping. Dave Simon built the first AMV prototype, and Jay Gowdy continues development. Karl Kluge is following structured roads with explicit models, while Jill Crisman and Didier Aubert work on unstructured roads with simple appearance models. Dirk Langer is working on the sonar "soft bumper." Ken Rosenblatt is developing new system integration approaches. Dean Pomerleau, a student of Dave Touretzky, studies neural nets on the Navlab. Thanks to those who keep the Navlab alive and productive, especially Jim Frazier, Bill Ross, Jim Moody, and Eric Hoffman. This paper benefitted from comments and contributions of figures from many people, especially Dirk Langer, Didier Aubert, Karl Kluge, Omead Amidi, Jill Crisman, Dean Pomerleau, and Jay Gowdy. This research is sponsored in part by contracts from DARPA (titled "Perception for Outdoor Navigation" and "Development of an Integrated ALV System"), by NASA under contract NAGW-1175, by the National Science Foundation under contract DCR-8604199, and by the Digital Equipment Corporation External Research Program.

References

[1] Amidi, Omead (1990). *"Integrated Mobile Robot Control."* Technical Report, Robotics Institute, Carnegie–Mellon University.

[2] Anandan, P. (1989). "A Computational Framework and an Algorithm for the Measurement of Visual Motion." *IJCV* **2**(3), 283–310.

[3] Aubert, Didier, and Thorpe, Charles (1990). *"Color Image Processing for Navigation: Two Road Trackers."* Technical Report CMU-RI-TR-90-09, Robotics Institute, Carnegie–Mellon University.

[4] Bergman, A., and Cowan, C. K. (1986). "Noise-Tolerant Range Analysis for Autonomous Navigation." In *Proc. IEEE Conf. on Robotics and Automation. San Francisco.*

[5] Bhanu, Bir, Symosek, Peter, Ming, John, Burger, Wilheml, Nasr, Hatem, and Kim, Jon (1989). "Qualitative Target Motion Detection and Tracking." In *Proc. Image Understanding Workshop.* Morgan Kaufmann Publishers.

[6] Bobick, Aaron, and Bolles, Robert (1989). "Representation Space: An Approach to the Integration of Visual Information." In *Proc. Image Understanding Workshop.* Morgan Kaufmann Publishers.

[7] Brooks, R. (1986). "A Robust Layered Control System for a Mobile Robot." *IEEE Journal of Robotics and Automation* **RA-2**(1), 14–23.

[8] Chappell, Steven G. (1989). "A Simple World Model for an Autonomous Vehicle." In *Sixth International Symposium on Unmanned Untethered Submersible Technology.* Marine Systems Engineering Laboratory, University of New Hampshire, June 1989.

[9] Crisman, Jill D., and Thorpe, Charles E. (1990). "Color Vision for Road Following." *Vision and Navigation:The Carnegie Mellon Navlab.* Kluwer Academic Publishers, Chapter 2.

[10] Daily, M. J., Harris, J. G., and Reiser, K. (1987). "Detecting Obstacles in Range Imagery." In *Proc. Image Understanding Workshop. Los Angeles, 1987.*

[11] Daily, M. J., Harris, J. G., and Reiser, K. (1988). "An Operational Perception System for Cross-Country Navigation." In *Proc. Image Understanding Workshop. Cambridge, Massachusetts, 1988.*

[12] Dunlay, R. T., and Morgenthaler, D. G. (1986). "Obstacle Detection and Avoidance from Range Data." In *Proc. SPIE Mobile Robots Conference. Cambridge, Massachusetts, 1986.*

[13] T. Dunlay (1988). "Obstacle Avoidance Perception Processing for the Autonomous Land Vehicle." In *Proc. IEEE Robotics and Automation. Philadelphia, 1988.*

[14] Fennema, Claude, Hanson, Allen, Riseman, Edward (1989). "Toward Autonomous Mobile Robot Navigation." In *DARPA Image Understanding Workshop.* Morgan Kaufmann.

[15] Goto, Y., and Stentz, A. (1987). "Mobile Robot Navigation: The CMU System." *IEEE Expert.*

[16] Goto, Yoshimasa, Shafer, Steven A., and Stentz, Anthony (1990). "The Driving Pipeline: A Driving Control Scheme for Mobile Robots." *Vision and Navigation: The Carnegie–Mellon Navlab.* Kluwer Academic Publishers, Chapter 10.

[17] Kehtarnavaz, N., and Griswold, N. (1989). "Establishing Collision-Zones under Uncertainty." In *Proc. Mobile Robots IV*, Society of Photo-optical Instrumentation Engineers, Bellingham, Washington, pp. 66–76.

[18] Keirsey, D., Payton, D., and Rosenblatt, J. K. (1988). "Autonomous Navigation in Cross Country Terrain." In *Proc. Image Understanding Workshop.* Morgan Kaufman, San Mateo, California.

[19] Kender, J. R., and Leff, A. (1989). "Why Direction-Giving is Hard: The Complexity of Linear Navigation by Landmarks in One-Dimensional Navigation." *IEEE Transactions on Systems, Man, and Cybernetics* **19**(6), 1656–1659.

[20] Kenue, Surender K. (1989). "Lanelok: Detection of Lane Boundaries and Vehicle Tracking Using Image-Processing Techniques. Part I: Hough-Transform, Region-Tracing, and Correlation Algorithms." In *Proc. Mobile Robots IV*, Society of Photo-optical Instrumentation Engineers, Bellingham, Washington, pp. 221–233.

[21] Kenue, Surender K. (1989)."Lanelok: Detection of Lane Boundaries and Vehicle Tracking Using Image-Processing Techniques. Part II: Template Matching Algorithms." In *Proc. Mobile Robots IV*, Society of Photo-optical Instrumentation Engineers, Bellingham, Washington, pp. 234–245.

[22] Dowling, Kevin, Guzikowski, Robert, Ladd, Jim, Pangels, Henning, Singh, Sanjiv, and Whittaker, William (1990). "Navlab: An Autonomous Navigation Testbed." *Vision and Navigation: The Carnegie–Mellon Navlab.* Kluwer Academic Publishers, Norwell, Massachusetts, Chapter 12.

[23] Kluge, Karl, and Thorpe, Charles E. (1990). "Explicit Models for Robot Road Following." *Vision and Navigation: The Carnegie–Mellon Navlab.* Kluwer Academic Publishers, Norwell, Massachusetts, Chapter 3.

[24] Levitt, T., Lawton, D., Chelberg, D., and Nelson, P. (1987). "Qualitative Navigation." In *Proc. Image Understanding Workshop.* Morgan Kaufmann, San Mateo, California, pp. 447–465.

[25] Mysliwetz, Birger D., and Dickmanns, E. D. (1987). "Distributed Scene Analysis for Autonomous Road Vehicle Guidance." In *Proc. SPIE Conference on Mobile Robots. November, 1987.*

[26] Ozaki, T., Ohzora, M., and Kurahashi, K. (1989). "Image Processing System for Autonomous Vehicle." In *Proc. Mobile Robots IV*, Society of Photo-optical Instrumentation Engineers, Bellingham, Washington.

[27] Pomerleau, Dean A. (1990). "Neural Network-Based Autonomous Navigation." *Vision and Navigation: The Carnegie–Mellon Navlab.* Kluwer Academic Publishers, Norwell, Massachusetts, Chapter 5.

[28] Shafer, S., Stentz, A., and Thorpe, C. (1986). "*An Architecture for Sensor Fusion in a Mobile Robot.*" Technical Report CMU-RI-TR-86-9, Robotics Institute, Carnegie–Mellon University.

[29] Stentz, Anthony (1990). "Multi-Resolution Constraint Modeling for Mobile Robot Planning." *Vision and Navigation: The Carnegie–Mellon Navlab.* Kluwer Academic Publishers, Norwell, Massachusetts, Chapter 11.

[30] Stentz, Anthony (1990). "The CODGER System for Mobile Robot Navigation." *Vision and Navigation: The Carnegie–Mellon Navlab.* Kluwer Academic Publishers, Norwell, Massachusetts, Chapter 9.

[31] Thorpe, C., Hebert, M., Kanade, T., and Shafer, S. (1988). "Vision and Navigation for the Carnegie-Mellon Navlab." *IEEE PAMI* **10**(3), 361–372.

[32] Turk, M., Morgenthaler, D., Gremban, K., and Marra, M. (1988). "VITS—A Vision System for Autonomous Land Vehicle Navigation." *IEEE PAMI* **10**(3), 342–360.

[33] Waxman, A. M., LeMoigne, J. J., Davis, L. S., and Siddalingalah, T. (1987). "A Visual Navigation System for Autonomous Land Vehicle." *IEEE J. Robotics and Automation* **RA-3**, 124–141, April.

3
Algorithms for Road Navigation

LARRY S. DAVIS, DANIEL DEMENTHON, SVEN DICKINSON, AND PHILIP VEATCH

3.1 Introduction

This paper provides a summary of the research conducted at the University of Maryland during the past five years on problems associated with visual navigation of ground vehicles. This research has been driven by a variety of scientific and engineering goals, including

1. the identification of principles of organization for autonomous navigation systems,
2. the identification of fundamental scientific problems that must be addressed in the course of designing and developing visual navigation systems, and
3. the implementation of prototype visual navigation systems that operate in the real world (ideally in real time) and demonstrate progress toward the solution of a specific problem in visual navigation.

Our research has focused primarily on the development of a complete system for visual navigation of roads and road networks (described in detail in Refs. [22] and [10]). We call this system **MARF** for **MA**ryland **R**oad **F**ollower. MARF was developed as part of the Autonomous Land Vehicle (ALV) program sponsored between 1984 and 1988 by the Defense Advanced Research Projects Agency. It was able to visually navigate a ground vehicle over unfamiliar roads. If *a priori* information was available in the form of a map, then MARF was able to use that information to navigate through intersections whose geometries were coarsely specified in the map. More important than the actual navigation problem that MARF attempted to solve was the organization of the system. That organization emphasized the importance of **focus of attention** (both to minimize computation time and to make maximal use of accumulated expectations), **explicit representation of visual search strategies**

The support of the Defense Advanced Research Projects Agency (ARPA Order No. 6350) and the U.S. Army Engineer Topographic Laboratories under Contract DACA76-88-C-0008 is gratefully acknowledged.

(to support extensibility and ease of modification), and **sensor integration at the symbolic level**.

An important dimension in the design of autonomous vehicles is the decomposition of the system into modules. The most prevalent decomposition is a functional one (referred to by Brooks [4] as a **horizontal** decomposition). So, for example, our road following system has modules for image processing, sensor control, inverse perspective, etc. Brooks [4] argues that such a functional decomposition will lead to systems with very complex interfaces between the modules that are, additionally, very difficult to modify or extend. Our experience with the initial implementation of our road following system certainly supported Brooks contentions; however, our last implementation, based on a set of interacting blackboards with explicit representations for objects and processes (described in detail in Dickinson and Davis [10]), allowed us to overcome many of these difficulties and, perhaps, could form the basis for the design of even more general systems. We describe this system in some detail in Section 2.

The problem of recovering an explicit model for the three-dimensional geometry of the road is especially important when the sampling interval for the vehicle (time taken to acquire and process a frame of data) is large compared to the speed of the vehicle. The most straightforward approach to reconstructing the geometry of a road once its boundaries or lane markings are detected is to assume that the road is flat and that one can measure (using either an inertial sensor or a range sensor) the orientation and height of the camera with respect to the road plane. However, small errors in the estimation of these parameters, or small deviations of the road from flatness, can result in extremely large errors in the recovered three-dimensional road coordinates. If the road contains a curve, and the distance to the curve is either over- or underestimated because of such errors, then the vehicle might be driven off the road. In Section 3, we describe a monocular inverse perspective algorithm developed by DeMenthon [8]. This so-called "zero-bank inverse perspective algorithm" was tested extensively using reference road reconstructions obtained by data fusion between range images and video images from the Martin Marietta ALV [17].

While the ultimate goal of autonomous road following systems is to navigate in the presence of other moving objects, it is also important to be able to navigate around stationary obstacles on the road. The most straightforward approach for identifying road obstacles is based on range images and involves comparing the observed height of points from the range image against the predicted height of road points. This approach suffers from two serious problems:

1. Small errors in estimating the geometry of the road can lead to very large errors in estimating the heights of road points.
2. Even if the road geometry is known, small errors in estimating the orientation of the range sensor with respect to the road can lead to very large errors in height estimation.

In Veatch and Davis [21], we proposed an obstacle detection algorithm based on comparing the observed range derivatives of road pixels against the predicted derivatives. We showed that such an approach is much less sensitive to errors than algorithms based on height comparison or other similar approaches. Section 4 contains a description of the obstacle detection algorithm and experimental results.

3.2 Maryland Road Follower

While the development of MARF involved the solution of many difficult technical problems (image analysis algorithms for detecting road boundaries and markings in color images of roads, robust inverse perspective algorithms for road geometry reconstruction, algorithms for obstacle detection in range imagery, etc.), we would like to focus here on the organization of the system at a high level and describe the representations adopted that allow MARF to reason about what to look for, which sensors to use, and how to recover from failure.

Figure 3.1 contains a road image taken from the Martin Marietta ALV at their Denver, Colorado, test site. From images such as these, MARF constructs a partial model of its three-dimensional environment. This *scene model* contains the objects visually identified by MARF and forms the basis for planning a path through the environment.

For road following, the scene model contains objects from which the location of the road can be determined. Most simply, the scene model might contain the locations of the left and right boundaries of the road. However, it would also be possible to navigate the vehicle based on locating the right boundary and the centerline, the centerline only, the location of a ditch that is known to run parallel to the road at a given distance from the road, etc.

Generally, there are many cues in the environment that contain direct or

FIGURE 3.1. Typical ALV road image. Reprinted with permission from *IEEE Transactions on Robotics and Automation* ("A Flexible Tool for Prototyping ALV Road Following Algorithms" by Sven Dickinson and Larry Davis, **6**(2), 1990). © 1990 IEEE.

indirect evidence concerning the location of the road. MARF must have some basis for deciding which objects in the environment to search for and how to search for them. These decisions can be based on information including the recent history of object detection (i.e., the right road boundary might have been robustly detected recently because of high color contrast and so is a good candidate for the current road tracking method), the current contents of the scene model (i.e., what has already been successfully detected in the current frame and where), and information from a road map.

Detection, verification, and accurate delineation of these objects are themselves complex tasks involving sensor control, computationally demanding sensor data processing algorithms, and fusion of either raw data or analyses from multiple sensors. Methods for performing such tasks evolve as the road following problem becomes better understood. We would like to increase the class of objects that the vehicle can use to determine its motion, include new sensors on the vehicle, embed new sensor data processing algorithms into the system, and modify and extend the navigation strategies employed by the vehicle. The successful evolution of such a system depends on the ability of its control structure and knowledge representations to accommodate such changes.

In the system to be described, the scene model is represented as a network of frames, with each frame corresponding to a class of objects and encapsulating the relevant information pertaining to that class. The control structure used to construct the network is based on a system of communicating production systems that implement a structured blackboard. The blackboard is partitioned into regions, each of which corresponds to a specific class of frames and contains the rules that define the attributes of the class. The system promotes modularity and maintainability through a structured object representation and a structured control scheme.

3.2.1 System Overview

A scene model is constructed in MARF through the cooperation of two modules:

1. the scene model planner (Planner) is responsible for deciding what objects to search for and where in the image to search for them, and
2. the scene model verifier (Verifier) is responsible for sensor control and sensor data processing required to verify the scene predictions made by the Planner.

The Planner determines its sequence of object predictions based on knowledge about the current scene, the history of processing of recent frames, and the goals associated with the current navigation task. The Verifier controls the acquisition and analysis of data acquired from the sensors. For example, if requested to locate the image of the road's left boundary at a distance of 10–15 m in front of the vehicle, it would choose an appropriate sensor based on map information and its current history of road boundary detection, choose a sensor data processing algorithm, determine the appropriate pointing direction for the sensor based on estimates of the vehicle's position in the world and its motion, acquire the sensor data, and process it. All of these decisions are arrived at by the application of a rule-based system to structured databases of facts and conjectures about the world, the vehicle, and its sensors.

Here, we would like to present a detailed description of the Planner, and we would specifically like to illustrate how the explicit representation of visual search strategies in the Planner makes it relatively straightforward to experiment with competing strategies.

The Planner is implemented as a frame whose slots point to the modules with which the Planner communicates. There are currently slots for

1. the scene model,
2. the a priori road map,
3. the local navigation task, and
4. the Verifier.

The unique feature of this framework is that it inherits the capabilities of a YAPS [1], a production system providing a rule database, a factual database, and a conflict resolution strategy. The Planner's principal goals are to choose what objects in the world to search for, deduce the location of the road from the detection of some subset of these objects (the Verifier may fail to detect some of the objects sought by the Planner), and decide when to terminate processing of a frame.

In our current implementation, the only local navigation task is to follow the road. A road map is available that specifies the geometry of the road network at a coarse level (i.e., where road intersections occur and how many roads meet at those intersections). Detailed information about the road geometry between intersections is not provided.

In our original road following system [22], a fixed search strategy was employed to detect the road. It operated by predicting where in a video image the left (and right) boundaries at a fixed distance in front of the vehicle would appear. This prediction was based on its accumulated three-dimensional model of the road geometry and an estimate of the vehicle's position in that model. The system (1) placed windows in the image surrounding those points, with the size of the window heuristically determined by estimates of the prediction error; (2) applied simple image processing algorithms to those windows (based on edge detection and line extraction) to locate the road boundary; and (3) placed subsequent windows in the image that overlapped the previous windows by a fixed percentage, oriented along the extrapolated direction of the road. Because the strategy was implemented directly in code, it was very difficult to modify or extend.

In MARF, these strategies are represented explicitly in the Planner as rules, making it straightforward to specify the conditions under which a strategy should be employed, what new strategy to invoke should the current one (or ones, since several could be pursued in parallel) fail, etc. We illustrate this with a simple example.

Since the road patches seen in successive images taken from the vehicle have a large intersection, one might imagine that for straight roads, at least, it is not necessary to reprocess all those parts of the current image that correspond to road segments identified in previous frames. For example, we might want to experiment with a search strategy that, after having detected 10 m of straight road, would *skip over* the image of the next 10 m of road, thus saving the time required to process those parts of the image containing the skipped 10 m (of course, in order to do this, we must either have determined the three-dimensional geometry of that 10-m patch from the analysis of previous frames or we must make some simplifying assumptions, such as the road is straight and flat over that 10-m patch).

Among the rules that collectively define the road patch search strategy, the following rule defines the search strategy as disconnected:

```
(defp define-disconnected-search-strategy
     (hypothesized object -object)
     (goal (define search strategy for road patch -object))
test(null ( <- object 'search-strategy))
     (cond ((not (null ( <- object 'priori-road-straightness)))
     ( >= ( <- -object 'priori-road-straightness)
     MIN-ROAD-STRAIGHTNESS)))
—>
     ( <- -object 'set-search-strategy 'disconnected)).
```

The antecedent of the rule (the conditions preceding the —>) is a conjunction of conditions that must be satisfied in order for the consequent (the expression following the —>) to be executed. The first antecedent expression matches a road patch hypothesis created by the Planner. The second expression represents the current goal of the Planner; in this case, the Planner is attempting to define the search strategy of the road patch hypothesis. The next two expressions, called test clauses, specify further conditions that must be met; the search strategy must be previously undefined and the prior road straightness must exceed the value of MIN-ROAD-STRAIGHTNESS. The symbol <- indicates message passing between objects; for example, in the first test condition, a message is passed to the road patch hypothesis, bound to the variable-object, requesting the value of the search strategy attribute. If both of these conditions are met, then the search strategy is defined as disconnected.

A second rule defines the search location of the road patch hypothesis:

```
(defp define-disconnected-search-location
     (hypothesized object -object)
     (goal (define search location for road patch -object))
test(null ( <- -object 'search location))
```

```
(eq ( <- -object 'search strategy) 'disconnected)
—>
  ( <- -object 'set-search-location
  (list ( <- ( <- * yaps-db * 'scene-model)
  ':predict-extended-left-feature-seed
  MAX-EXTENDED-SEARCH-DISTANCE)
  ( <- ( <- * yaps-db * 'scene-model)
  ':predict-extended-right-feature-seed
  MAX-EXTENDED-SEARCH-DISTANCE)))).
```

In this rule, the consequent defines the search location as a value resulting from sending two queries to the scene model, requesting points extrapolated from the left and right road patch segments, respectively, of the last patch in the scene model.

Suppose, now, that we wanted to change this strategy by having the vehicle look further and further ahead as it identifies more and more road patches. So, if the strategy initially attempts to skip 5 m of road, then if it were successful, it might next try to skip 10 m of road. To accommodate this new "dynamic search strategy," we add the following rule to the Planner:

```
(defp define-dynamically-disconnected-search-strategy
       hypothesized object -object)
       (goal (define search strategy for road patch -object))
test(null ( <- -object 'search strategy))
    (cond ((not (null ( <- -object 'prior-road-straightness)))
    ( >= ( <- -object 'prior-road-straightness)
    MIN-ROAD-STRAIGHTNESS)))
    (eq ( <- ( <- ( <- * yaps-db * 'scene model)
    ':retrieve-most-recent-road-patch) 'search-strategy)
    'disconnected)
—>
  ( <- -object 'set-search-strategy 'dynamically-disconnected)).
```

In the test clauses, we check that the search strategy of the road patch hypothesis is undefined, and we make sure that we have accumulated a sufficient amount of straight road in the scene model. In addition, we check that the previously verified road patch was verified using the disconnected search strategy. If all of these conditions hold, the search strategy is defined to be dynamically disconnected. To define the search location for this new strategy, we add the following rule:

```
(defp define-dynamically-disconnected-search-location
       (hypothesized object -object)
       (goal (define search location for road patch -object))
test(null ( <- -object 'search-location))
    (eq ( <- -object 'search-strategy) 'dynamically disconnected)
—>
  ( <- -object 'set-search-location
  (list ( <- ( <- * yaps-db * 'scene-model)
```

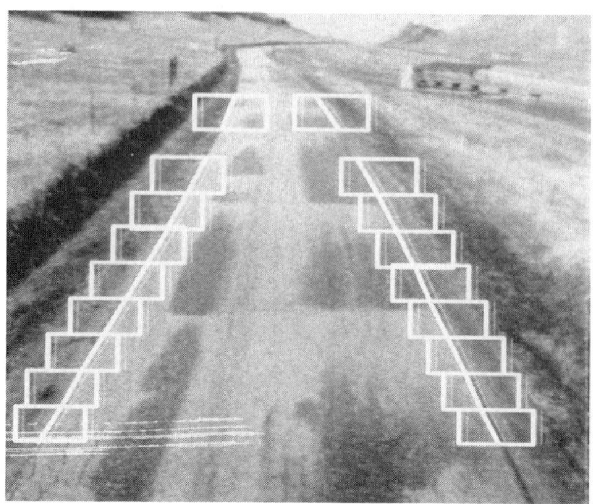

FIGURE 3.2. Tracking a straight road. Reprinted with permission from *IEEE Transactions on Robotics and Automation* ("A Flexible Tool for Prototyping ALV Road Following Algorithms" by Sven Dickinson and Larry Davis, **6**(2), 1990). © 1990 IEEE.

```
': predict-extended-left-feature-seed
MAX-EXTENDED-SEARCH-DISTANCE)
( <- ( <- yaps-db 'scene-model)
':predict-extended-right-feature-seed
MAX-EXTENDED-SEARCH-DISTANCE)))).
```

In this rule, the rule consequent defines the search location to be **MAX-EXTENDED-SEARCH-DISTANCE** from the last road patch in the scene model. By controlling the value of this parameter, we can control the extent of road skipped by the road detection strategy.

We illustrate the results of these search strategies with an image from the Martin Marietta test site. Figure 3.2 is an image of a straight segment of road. Initial search windows are placed near the bottom of the image based on the predicted location of the road in the image. Using these initial search windows, the connected search strategy is invoked until 10 m of road have been inserted into the scene model. At that time, the disconnected search strategy is invoked, and 10 m of road are skipped in the image. Since the road is straight, the search strategy is successful in identifying the road boundaries in the next pair of windows. With approximately 20 m of straight road in the scene model, the disconnected search strategy is again invoked; however, when the three-dimensional search location of the next road patch is mapped to the current image, the search windows are out of bounds (off the top of the image).

3.3 Recovery of Three-Dimensional Road Geometry

We now present an algorithm for reconstructing the road shape from a single image, providing the three-dimensional profile of the road in front of the vehicle, often up to the point where the road becomes hidden. Reconstructing the road over a large distance presents several advantages. The reconstructions from several video frames can be overlapped, and the evidence from each reconstruction can be combined for added reliability. The road reconstruction can be registered to a stored map of the road network and contribute to locating the position of the vehicle on the map. Finally, a system that makes estimations of turns well in advance can adjust its speed accordingly. This long-range observation of the road does not preclude the use of a shorter range road analysis in the control loop of the vehicle steering [11].

Once the images of the road boundaries are computed, road reconstruction can be posed as a "shape from contour" problem. The problem is, of course, underconstrained—an infinite number of world curves can correspond to each road image boundary unless additional assumptions are made about the three-dimensional geometry of the road.

The simplest assumption one can make is that the earth is flat, and that the vehicle is supported on the ground plane. The three-dimensional location of any image point is then simply the intersection of the line of sight through this point and the known ground plane. Reconstruction from the flat earth model is very fast, and it is not sensitive to slight misplacements of the road boundaries in the image. However, it is very sensitive to errors in estimating the camera tilt angle with respect to the ground plane. While the rocking of the vehicle on its suspension can be measured, changes of ground slopes of several degrees within the field of view are common. If the camera is situated at a height H above the ground and sees a point at a distance L from the vehicle (Figure 3.3), then if the ground plane is overestimated by an angle ε, the estimated value L' of L will be

$$L' = L/[1 - (L/H) \tan \varepsilon].$$

For a camera mounted on a vehicle with $H = 3.5$ m, a world point 30 m in front of the vehicle will be located 55 m in front of the vehicle if the ground plane angle is overestimated by $3°$ and at 21 m if it is underestimated by $3°$—an error range of over 100%.

In an attempt to overcome the limitations of the Flat Earth method, authors have added constraints to the road model, assuming that a road generally keeps an approximately constant width [20] or that directions of road edges may be parallel and may be deduced from their vanishing points

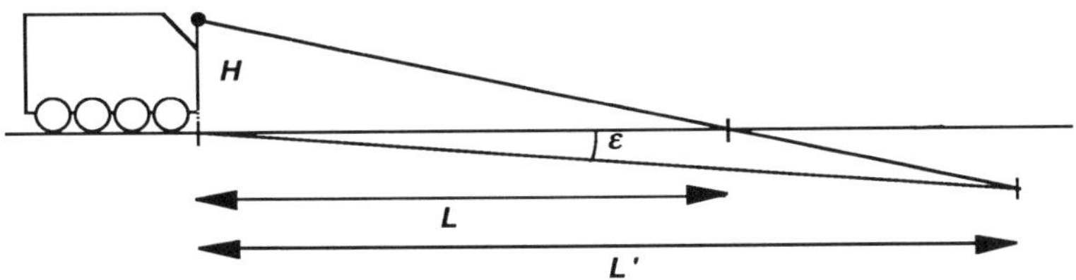

FIGURE 3.3. Error in distance estimates due to ground slope in the Flat Earth geometry model.

[16, 19]. A problem with applying these constraints is that one must find *which* pairs of points are separated by a distance equal to the road width and *which* pairs of points have parallel edge directions in straight or curved parts of the road. We call the problem of locating the correct pairs of points in the image the *matching point problem.*

The constant road width constraint is not sufficient. We find that another constraint must be added for the reconstruction to be possible. We have chosen the *zero-bank* constraint, specifying that the road does not tilt sideways. A road model combining constant width and zero bank was originally suggested in Ozawa and Rosenfeld [18].

In previous work, we had developed an incremental road reconstruction method based on these constraints [7] in which a new pair of edge points could be found if we had already found a neighboring pair of edge points; the road edges were reconstructed incrementally from edge points close to the vehicle to edge points in the distance. This method was fragile because any increment of construction depended on the previous elements in the chain.

This incremental method used a discrete approach. Incremental road reconstructions based on a differential approach can be found in Kanatani and DeMenthon [12] and Kanatani and Watanabe [14]. An interesting alternative to the global dynamic programming optimization proposed in the present paper can be found in Kanatani and Watanabe [13].

3.3.1 Summary

The proposed algorithm can be decomposed into the following steps:

1. In a preliminary step, not detailed here, appropriate image processing techniques have isolated the two curves of the edges in the image, and a polygonal approximation has been found for each edge curve.
2. Picking image points anywhere on one image edge curve, we are able to find the points that are candidates for being matching points on the other image edge curve. (Two image points are called matching points if they are images of the endpoints of cross segments of the 3-D road). This matching is made possible by making reasonable hypotheses about the shape of the road, which add enough constraints to make the problem solvable. Specifically, the road is modeled as a space ribbon defined by a centerline spine and horizontal cross segments of constant length cutting the spine at their midpoints at a normal to the spine. We further assume that tangents to the ribbon edges at endpoints of cross segments are approximately parallel (Section 3.3.2). We find an expression that must be satisfied by the two image points located on the facing image edge curves and the tangents to the edge images in order for the two points to be matching points (Section 3.3.3). If a_1 and a_2 are matching points and $\mathbf{a}'_1$ and $\mathbf{a}'_2$ are the tangent directions to the image edges in these points, the following relation holds:

$$[\mathbf{V} \times (\mathbf{a}_1 \times \mathbf{a}_2)] \cdot [(\mathbf{a}_1 \times \mathbf{a}'_1) \times (\mathbf{a}_2 \times \mathbf{a}'_2)] = 0,$$

where **V** is the vertical direction. For edge curves approximated by polygonal lines, the matching point a_2 can be on a line segment, and its position between the endpoints of the line segment can be expressed by a number between 0 and 1, whereas its tangent vector $\mathbf{a}'_2$ is constant; or the matching point a_2 can be at an endpoint of a line segment with a constant position but with a tangent angle that can be expressed by a number between 0 and 1 within the range of angles of the two adjacent line segments (Section 3.3.8). For each point chosen from one image edge, we check for each of the line segments of the other image edge to determine if a matching point belongs to that line segment, that is, if our expression gives a linear coordinate between 0 and 1 for this line segment. Then, we look for matching points at the nodes of the polygonal line by checking that the expression gives a number between 0 and 1 for the tangent angle.

3. For each point chosen from one edge image, the previous step may give several matching points on the other edge image. One of the reasons is that the images of the edges can be very rough and wiggly. Another reason is that the condition used is only a necessary condition for two points to be matching points in the image of the road. This condition is local and we must still choose the matching points pairs that are the most globally consistent and discard the other pairs. The criteria of optimization are three dimensional (Section 3.3.9); thus, at this step of the algorithm, from the pairs of matching points, the corresponding three-dimensional cross segments must be found. This correspondence is unique if the cross segments are assumed horizontal[1] and of known constant length (Section 3.3.2). The constant length is the width of the road, and it cannot be defined by this method. The assumed road width is a scaling factor in the reconstruction, whereas the optimization is based on angular considerations, which are independent of scaling. For driving a vehicle, the road width must eventually be obtained from other methods, such as stored data about the road, the Flat Earth method, or close-range methods, such as stereoscopy or time-of-flight ranging.

4. The group of matching point pairs corresponding to a single point chosen on one edge is the image of a group of world cross segments obtained at the previous step, and the world road can go through *at most one* of these cross segments (Section 3.3.9). If a sequence of points along one road edge is taken, a sequence of groups of cross segments is obtained, and the world road must go through at most one of the cross segments of each group, in the same order as the sequence of points chosen on the first road image edge. Each cross segment can be represented as a node in a graph. A path must be found in the graph that visits each group in the proper sequence and goes through at most one node of each group and that maximizes an

---

[1] The vehicle reconstructs horizontal cross segments on the basis of its knowledge of the vertical direction, and therefore, it must be equipped with a vertical direction sensor.

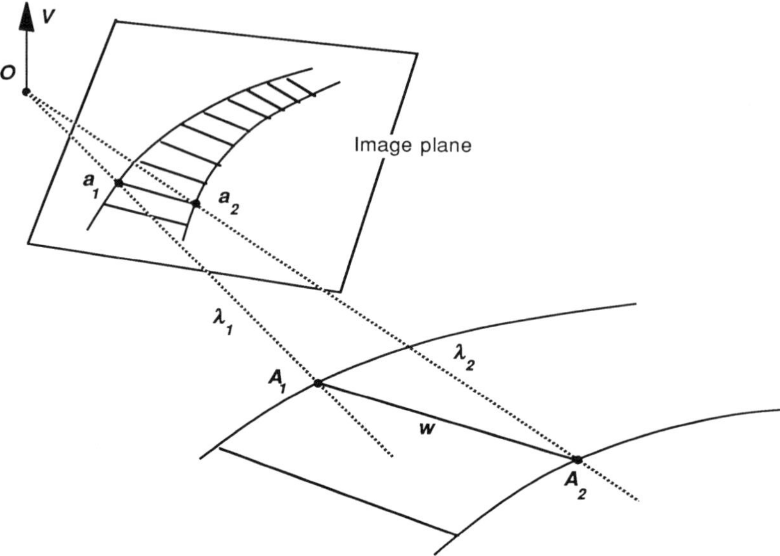

FIGURE 3.4. Reconstructing positions of world railroad ties from their images. Reprinted with permission from *IEEE Transactions on Robotics and Automation* ("Reconstruction of a Road by Local Image Matches and Global 3D Optimization" by Daniel DeMenthon and Larry Davis). © 1990 IEEE.

evaluation function that characterizes a "good road." The total evaluation function is the sum of the functions of each of the arcs of the graph. The evaluation function for an arc is the sum of weighted criteria, which grade the choices of individual cross segments and the neighborhood of consecutive cross segments based on angular considerations.

3.3.2 The Matching Point Problem

Consider the image of a railroad track and its railroad ties and assume that some appropriate image processing techniques have reduced the images of the rails to curves and the images of the ties to line segments between these curves (Figure 3.4). The positions of the endpoints of the tie segments on the curves of the rail are the matching points in the image. The reconstruction of the shape of the railroad track in 3-D space uses the matching points and is straightforward once three hypotheses are made.

1. The width w of the railroad track is constant and known.
2. The coordinates of the vertical unit vector $\mathbf{V}$ are known in the camera coordinate system.
3. The railroad ties are approximately horizontal.

Note that the last hypothesis does not mean that the railroad itself should be horizontal. Similarly, the stairs in a spiral staircase have horizontal step edges, but the ruled surface defined by these step edges is far from horizontal.

Consider two matching points a_1 and a_2, the endpoints of the image of a tie. The corresponding vectors from the viewpoint O to these image points will be denoted by $\mathbf{a}_1$ and $\mathbf{a}_2$. The corresponding world points A_1 and A_2 are defined by

$$\mathbf{A}_1 = \lambda_1\mathbf{a}_1, \qquad \mathbf{A}_2 = \lambda_2\mathbf{a}_2$$

since world points and their images are on the same line of sight.

The world line segment is assumed horizontal; the two parameters λ_1 and λ_2 are then related by

$$\lambda_2 = m\lambda_1,$$

with

$$m = \mathbf{a}_1 \cdot \mathbf{V}/\mathbf{a}_2 \cdot \mathbf{V}.$$

The requirement that the distance between $\mathbf{A}_1$ and $\mathbf{A}_2$ be equal to the width w completely constrains the parameters:

$$\lambda_1 = w/(\mathbf{a}_1^2 + m^2\mathbf{a}_2^2 - 2m\mathbf{a}_1 \cdot \mathbf{a}_2)^{1/2}. \tag{3.1}$$

Thus, the two curves of the rails in the scene can be, in general, uniquely reconstructed from their images up to a scale factor if the ties are assumed horizontal and of constant length. Problems occur only if the railroad image crosses the horizon, as noted in Kanatani and Watanabe [13]. In this case, the ties are horizontal on the horizon line, and their range cannot be determined, as can be seen from the previous equations.

Consider now the problem of reconstructing a *road* from its image, once some appropriate image processing techniques have isolated the curves corresponding to the road edges in the image. Now, of course, we do not have the images of railroad tie segments to help us. The method we propose involves first finding the endpoints of line segments that correspond to images of railroad tie segments and then doing the 3-D reconstruction of the endpoints of the images of these segments by the method just described for the railroad. We call these world segments corresponding to railroad ties cross segments, and their endpoints opposite points. The images of these points are the matching points. The main problem of road reconstruction from an image can then be stated: Given a point on one edge of the road image, where is the matching point on the other edge?

We choose a road model similar to the railroad model: the road is modeled as a space ribbon generated by a centerline spine and horizontal cross segments of constant length cutting the spine at their midpoints at a normal to the spine. This modeling gives cross segments the properties of railroad ties.

Cross segments are horizontal, that is, perpendicular to the vertical (on the ALV the vertical was detected by trim sensors).

Cross segments have constant length (the road width).

Cross segments are perpendicular to both road edges, that is, locally perpendicular to the centerline of the road.

Saying that cross segments are normal to both road edges means that they are normal to the tangents to the edges at their endpoints. Note that this does not generally mean that the tangents to opposite points are parallel. In DeMenthon [8], however, we show that assuming the tangents to opposite points to be approximately parallel is a reasonable assumption in most configurations. This assumption considerably simplifies the recovery of opposite points from the image. It is added to our road model and used in the following section.

3.3.3 Conditions for Two Image Points to Be Matching Points

Consider a world road defined by two 3-D curves E_1 and E_2 and the road image defined by two image curves e_1 and e_2. Assume that two opposite points A_1 and A_2 on road edges E_1 and E_2 have been found. Their images are a_1 and a_2 (Figure 3.5.), and the following properties are dictated by the world road model.

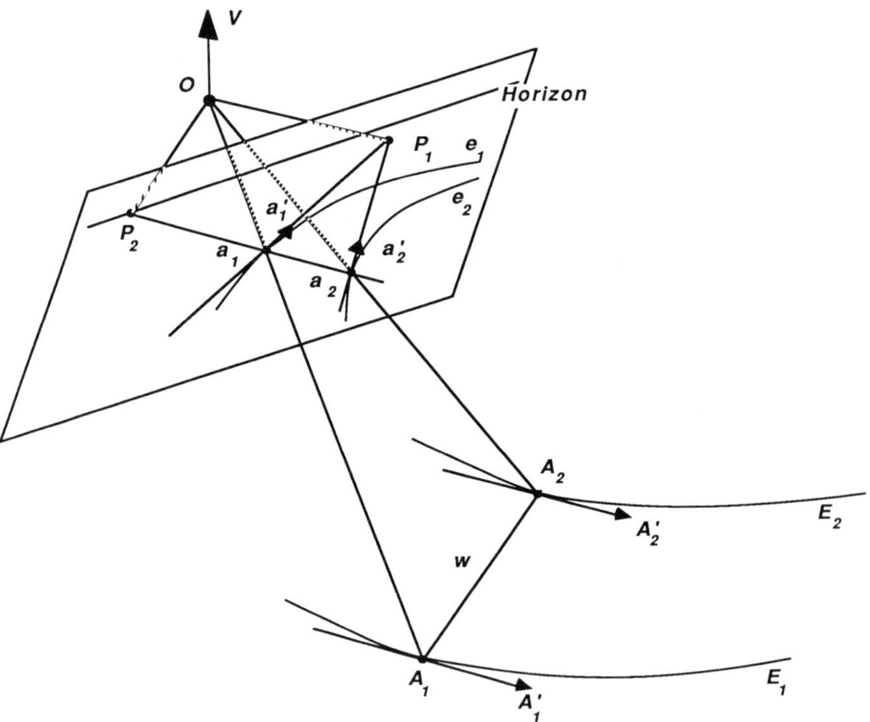

FIGURE 3.5. The cross segment of the world road is assumed horizontal and perpendicular to the tangents at its endpoints. The tangents are assumed parallel. A condition satisfied by the matching points in the image that also involves the image tangents and the vertical direction is deduced. Reprinted with permission from *IEEE Transactions on Robotics and Automation* ("Reconstruction of a Road by Local Image Matches and Global 3D Optimization" by Daniel DeMenthon and Larry Davis). © 1990 IEEE.

1. The segment $A_1 A_2$ is horizontal.
2. The tangents to the road edges at A_1 and A_2 are perpendicular to $A_1 A_2$.
3. The tangents to A_1 and A_2 are approximately parallel.
4. The tangent $\mathbf{a}'_1$ to the image edge e_1 at a_1 is the image of the tangent $\mathbf{A}'_1$ to the world edge E_1 at A_1; the tangent $\mathbf{a}'_2$ to the image edge e_2 at a_2 is the image of the tangent $\mathbf{A}'_2$ to E_2 in A_2. This is a general property of projected curves and tangents.

In deriving the following consequences, we make use of the property that the direction of the intersection of two planes is perpendicular to the normals of each plane and can be obtained by the cross product of the two normals.

3.3.4 Directions of Tangents to Opposite Points

If a_1 and a_2 are matching points and $\mathbf{a}'_1$ and $\mathbf{a}'_2$ are the tangents to the image edges at these points, the direction of the corresponding world tangents is

$$(\mathbf{a}_1 \times \mathbf{a}'_1) \times (\mathbf{a}_2 \times \mathbf{a}'_2).$$

Proof: If a_1 and a_2 are images of opposite points, the world tangents to the world edges are parallel. Since the images of the world tangents are $\mathbf{a}'_1$ and $\mathbf{a}'_2$, the world tangents lie on the planes $(O\mathbf{a}_1, \mathbf{a}'_1)$ and $(O\mathbf{a}_2, \mathbf{a}'_2)$, respectively. These planes are not parallel since they share the point O, and they do not coincide. Since the tangents are parallel, they must be parallel to the intersection of these planes. The direction of this intersection is given by the previous expression.

3.3.5 Direction of a Cross Segment

If a_1 and a_2 are matching points and $\mathbf{V}$ is the vertical vector, the direction of the world cross segment is $\mathbf{V} \times (\mathbf{a}_1 \times \mathbf{a}_2)$.

   Proof: $A_1 A_2$ belongs to a horizontal plane since it is horizontal. Since $a_1 a_2$ is the image of $A_1 A_2$, $A_1 A_2$ also belongs to the plane (O_1, Oa_2). This plane is generally not horizontal. Thus, the direction of the segment $A_1 A_2$ is given by the intersection of a horizontal plane with the plane (Oa_1, Oa_2). The normal to the horizontal plane is the vertical vector $\mathbf{V}$. The direction of the normal to the plane (Oa_1, Oa_2) is given by the cross product $(\mathbf{a}_1 \times \mathbf{a}_2)$. Thus, the direction of $A_1 A_2$ is given by $\mathbf{V} \times (\mathbf{a}_1 \times \mathbf{a}_2)$.

3.3.6 Matching Condition

If a_1 and a_2 are matching points and $\mathbf{a}'_1$ and $\mathbf{a}'_2$ are the tangent directions to the image edges in these points, the following relation holds:

$$[\mathbf{V} \times (\mathbf{a}_1 \times \mathbf{a}_2)] \cdot [(\mathbf{a}_1 \times \mathbf{a}'_1) \times (\mathbf{a}_2 \times \mathbf{a}'_2)] = 0. \qquad (3.2)$$

Proof: If a_1 and a_2 are images of opposite points, the direction of the cross segment $A_1 A_2$ is perpendicular to the direction of the parallel tangents.

3.3.7 Local Normal to the Road

If a_1 and a_2 are matching points and $\mathbf{a}'_1$ and $\mathbf{a}'_2$ are the tangents to the image edges at these points, the local normal to the world road has the direction given by

$$\mathbf{N} = [\mathbf{V} \times (\mathbf{a}_1 \times \mathbf{a}_2)] \times [(\mathbf{a}_1 \times \mathbf{a}'_1) \times (\mathbf{a}_2 \times \mathbf{a}'_2)]. \tag{3.3}$$

Proof: The local planar patch of the world road is defined by $A_1 A_2$ and by the parallel tangents at A_1 and A_2. The direction of the normal to this plane is the cross product of the directions of the cross segment and the tangents.

To summarize, when a point a_1 and the tangent $\mathbf{a}'_1$ to the road image are given, Equation (3.2) becomes an equation that must be satisfied by the coordinates of a_2 and the slope of the tangent to the edge in a_2 in order for a_2 to be a matching point to a_1. We can also find the direction of the normal along the corresponding world cross segment $A_1 A_2$.

3.3.8 Search for a Matching Point of a Given Image Point

If a point a_1 is chosen on one edge image and if the other edge image is a polygonal line, the matching point a_2 can be located on one of the line segments of the polygonal line or at one of the vertices between the segments. All the line segments and all the vertices are checked, because a single point a_1 can have several matching point candidates due, for example, to edge irregularities. Other reasons are considered in DeMenthon [8]. For each line segment and for each vertex, the equations developed in the next two subsections are applied.

Search for a Matching Point on a Line Segment

Assume that the segment being considered is the segment $p_2 q_2$. The matching point a_2 is on this segment if

$$\mathbf{a}_2 = p_2 + \lambda \mathbf{p}_2 \mathbf{q}_2, \tag{3.2}$$

with λ between 0 and 1. The point a_2 must also, with its tangent to the edge, satisfy Equation (3.2). The tangent $\mathbf{a}'_2$ to the edge image in a_2 is approximated by the vector $\mathbf{p}_2 \mathbf{q}_2$. We replace $\mathbf{a}'_2, \mathbf{a}_2$ by their values $\mathbf{p}_2 \mathbf{q}_2$, and $\mathbf{p}_2 + \lambda \mathbf{p}_2 \mathbf{q}_2$ in Equation (3.2), and transform cross product combinations into dot products by the well-known identity

$$\mathbf{a} \times (\mathbf{b} \times \mathbf{c}) = (\mathbf{a} \cdot \mathbf{c})\mathbf{b} - (\mathbf{a} \cdot \mathbf{b})\mathbf{c}.$$

The resulting value for λ is

$$\lambda = -\frac{(\mathbf{V} \cdot \mathbf{a}_1)(\mathbf{K} \cdot \mathbf{p}_2) - (\mathbf{K} \cdot \mathbf{a}_1)(\mathbf{V} \cdot \mathbf{p}_2)}{(\mathbf{V} \cdot \mathbf{a}_1)(\mathbf{K} \cdot \mathbf{p}_2 q_2) - (\mathbf{K} \cdot \mathbf{a}_1)(\mathbf{V} \cdot \mathbf{p}_2 q_2)}, \tag{3.4}$$

where $K = (\mathbf{a}_1 \times \mathbf{a}'_1) \times (\mathbf{p}_2 \times \mathbf{q}_2)$. If λ is between 0 and 1, the intersection is between the endpoints of line segment $p_2 q_2$, and the value of λ specifies the position of a_2 on $p_2 q_2$. The search also takes place among the vertices between the line segments.

SEARCH FOR A MATCHING POINT AT A VERTEX

We can think of a point q_2 linking two line segments $p_2 q_2$ and $q_2 r_2$ as a point at which the slope of the tangent to the edge changes from the slope of the segment $p_2 q_2$ to the slope of the segment $q_2 r_2$. An approach similar to the previous subsection is followed. A matching point a_2 is at the vertex q_2 if

$$\mathbf{a}_2' = \mathbf{p}_2 \mathbf{q}_2 + \mu(\mathbf{q}_2 \mathbf{r}_2 - \mathbf{p}_2 \mathbf{q}_2), \tag{3.2}$$

with μ between 0 and 1. For this point to be a matching point to a_1, it must also satisfy Equation (3.2). This produces the following value for μ.

$$\mu = -\frac{(\mathbf{M} \cdot \mathbf{q}_2)(\mathbf{n}_1 \cdot \mathbf{p}_2 \mathbf{r}_2) - (\mathbf{n}_1 \cdot \mathbf{q}_2)(\mathbf{M} \cdot \mathbf{p}_2 \mathbf{r}_2)}{(\mathbf{M} \cdot \mathbf{q}_2)[\mathbf{n}_1 \cdot (\mathbf{q}_2 \mathbf{r}_2 - \mathbf{p}_2 \mathbf{q}_2)] - (\mathbf{n}_1 \cdot \mathbf{q}_2)[\mathbf{M} \cdot (\mathbf{q}_2 \mathbf{r}_2 - \mathbf{p}_2 \mathbf{q}_2)]}, \tag{3.5}$$

where $\mathbf{n}_1 = \mathbf{a}_1 \times \mathbf{a}_1'$ and $\mathbf{M} = \mathbf{V} \times (\mathbf{a}_1 \times \mathbf{q}_2)$.

If the resulting value of μ is between 0 and 1, a matching point a_2 to the point a_1 is located at the vertex q_2.

In the following, we provide a detailed description of how the sets of all matching points for a point on one road boundary edge are determined. We then describe the dynamic programming optimization algorithm and illustrate its applications to some simple examples.

3.3.9 Dynamic Programming Road Reconstruction

Given a point from one road boundary edge, the shape-from-contour algorithm identifies several potential matching points from the other road boundary edge. This group of matching point pairs is the image of a group of world cross segments, and the world road can pass through at most one of these cross segments. If a sequence of points along one road edge is taken, then a sequence of groups of cross segments is obtained, and the world road must pass through at most one of the cross segments from each group. Once an optimization criteria is defined, it is then straightforward to find the "optimal" path through the set of groups of cross segments. Consider a directed acyclic graph (DAG) in which the nodes at level i correspond to the cross segments constructed from the matching pairs for the ith boundary point on one road edge, and arcs connect all pairs of nodes on consecutive levels. We seek a path through this DAG from a level 0 node to a level n node, where there are $n + 1$ levels in the DAG. We append a special node to each level, a null node, to allow us to skip over a level that for numerical reasons, for example, might yield no correct cross segments. We next define a merit function on the arcs of the DAG. Let $A = A_1 A_2$ be a cross segment from a node at level k and $B_1 B_2$ be a cross segment from level $k + 1$. Then, the merit function for the arc connecting $A_1 A_2$ to $B_1 B_2$ is the weighted sum of three criteria C_i defined as follows.

1. The local normal to A (Equation 3.3) should be nearly vertical, so we set C_1 to be the dot product between the vertical and the normal to A.
2. The slope of the patch containing the two cross segments must be close to vertical. We define that slope to be $A_1B_2 \times A_2B_1$, and we define C_2 to be the dot product between the unit vector in the direction of this cross product and the vertical.
3. The average direction of the two cross segments must be perpendicular to the line joining their midpoints; C_3 measures how well this condition holds.

When the value of any one of these criteria falls below a threshold, the arc is eliminated from the DAG.

The dynamic programming algorithm then simply maintains a set of optimal paths through k levels and extends them to the $(k + 1)$ level by considering the nodes one at a time at the $(k + 1)$ level and retaining only the path from the kth level that is extended with maximal overall merit to the currently considered node on the $(k + 1)$ level. So, at any stage of the optimization, we are only retaining as many paths as there were nodes on the previously considered level (in practice, no more than 4 or 5). The null nodes allow us to skip levels; arcs to null nodes are assigned sufficiently low merit values to deter the optimization from ignoring good cross segments.

3.3.10 Experimental Results

Here, we present the results of road reconstruction using synthetic data only. Additional experimental results on real road images are presented in DeMenthon [8]. Figure 3.6 shows a top and side view of a synthetically generated road. The nominal road width is 4 m, and the road centerline profile (see the side view) is an element from a sinusoid from a crest to a trough. The road slope is modified by varying the sinusoid amplitude. The term *road slope* refers to the slope at the midpoint of the straight segment between the two turns. For the synthetic road, the road slope is $H/38.9$, where H is the difference in meters between the lowest and highest points of the road. In the top view shown in Figure 3.6, the road has a short straight segment, then takes a 45° right turn and then a left turn separated by a short straight segment. The camera's position, orientation and parameters, also shown in the figure, were taken equal to those of the ALV camera at Martin Marietta.

A benchmark was developed for measuring the performance of the proposed matching point algorithm and other road reconstruction algorithms. A reconstructed road is called *navigable* if the tracks of a vehicle of known width (2 m in our examples) following the centerline of the reconstructed road stay within the edges of the actual road over the entire reconstruction and never cross these edges. This requires that no cross segment is shifted sideways by more than one quarter of its width. Notice, that nonnavigable reconstructions are still usable if they are not too far from the actual road. Indeed, a sufficiently fast processor could generate reconstructions in a fixed coordinate

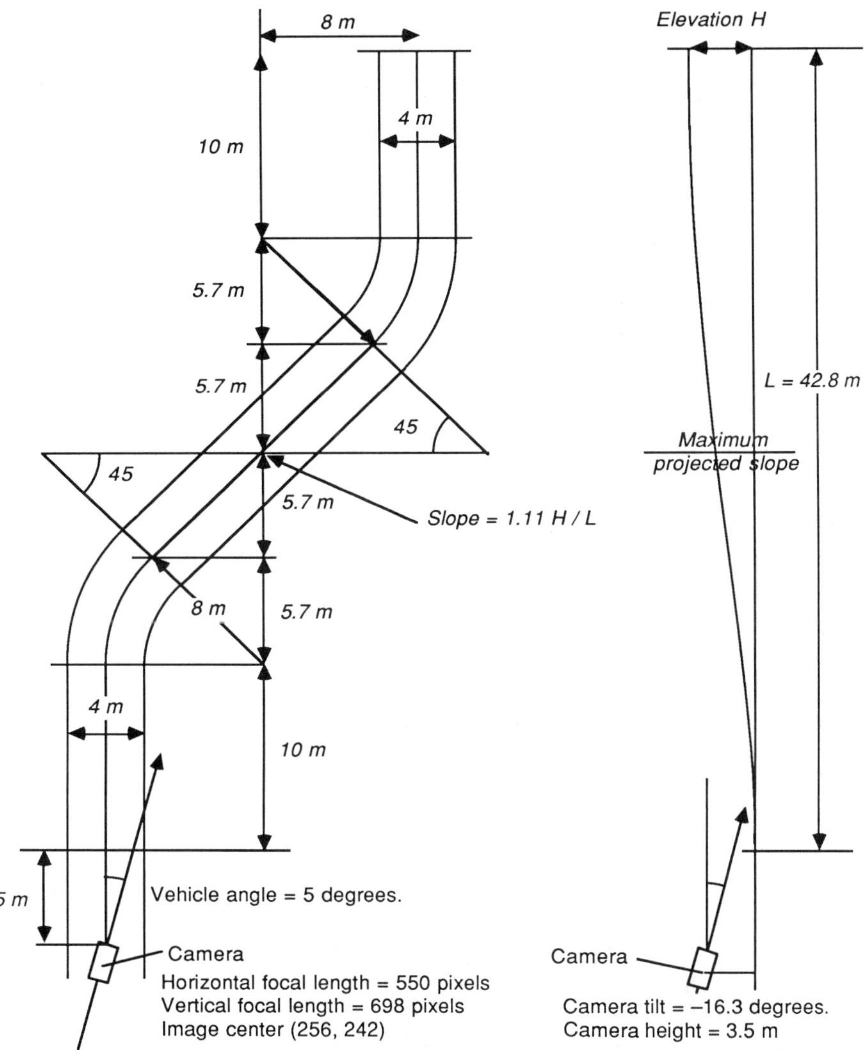

FIGURE 3.6. Synthetic road geometry with two 45° turns and camera position. (a) Top views. (b) Side view showing the $\frac{1}{4}$-period sinusoidal profile of the centerline over the length L. Reprinted with permission from *IEEE Transactions on Robotics and Automation* ("Reconstruction of a Road by Local Image Matches and Global 3D Optimization" by Daniel DeMenthon and Larry Davis). © 1990 IEEE.

system quickly enough such that the composite reconstruction from a sequence of frames is navigable, even though constituent reconstructions may not be. With this in mind, a reconstruction is called *usable* if the centerline of the reconstructed road stays within the edges of the actual road. In other words, a usable reconstruction is a reconstruction in which no cross segment is off the actual road by more than one half of its width. Considering a large number of synthetically generated roads with random variations in their geometry, percentages of navigable and usable reconstructions are computed and characterized by statistics such as the percentage of usable reconstruction.

Specifically, random variations are introduced about the nominal values of road width (4 m) and the road bank (0°). The random width and bank variations are described by Gaussian distributions, and several standard deviations for the Gaussian were employed in the experiments. Five road slopes were chosen: -10%, -5%, 0%, 5%, 10%. Forty roads were produced for 25 combinations of slopes and standard deviations, and the results for these 40 roads were averaged to yield the points plotted in the following graphs.

In Figure 3.7, results are shown for the matching point algorithm. The algorithm produces 100% navigable roads when the road has no irregularities of width and bank, regardless of slope. The rate drops to 5% for the maximal width and bank. Other diagrams display how much navigable length is recovered in reconstructions that are completely navigable and in reconstructions that are only partially navigable. Comparisons have also been conducted between the matching points algorithm, the original iterative zero-bank algorithm, and the new matching points algorithm. Not surprisingly, the Flat Earth model gives 100% reconstruction success if the road *is* flat, but cannot produce navigable or usable road reconstructions once any road slope is allowed. On synthetic data, the principle advantage of the matching points algorithm (its ability to recover from local catastrophic errors caused by, for example, image processing errors in localizing road boundaries) is not much in evidence, although it still led to a higher percentage of usable reconstruction. Details are available in DeMenthon [8], and DeMenthon and Davis [9].

3.4 Detection of Stationary Obstacles on Roads

Here, we provide a description of the obstacle detection algorithm proposed by Veatch and Davis, and we present some experimental results. A more complete description of the implementation (that addresses problems of mixed pixels) can be found in Veatch and Davis [21].

What is an obstacle? Generally, an obstacle is a region that a vehicle cannot or should not traverse. Avoiding regions that a vehicle is physically capable of traversing but for some reason should not go (such as not driving the wrong way down a one-way street) would require a level of artificial intelligence that is beyond the scope of this work.

Excluding places that a vehicle can go but should not, one is left with regions that can be defined by their shape and material properties. Rocks, street signs, and steep slopes are all obstacles whose defining characteristics are their shapes. Swamps and ice patches on the other hand may have sufficiently flat surfaces for navigation but their material properties make them

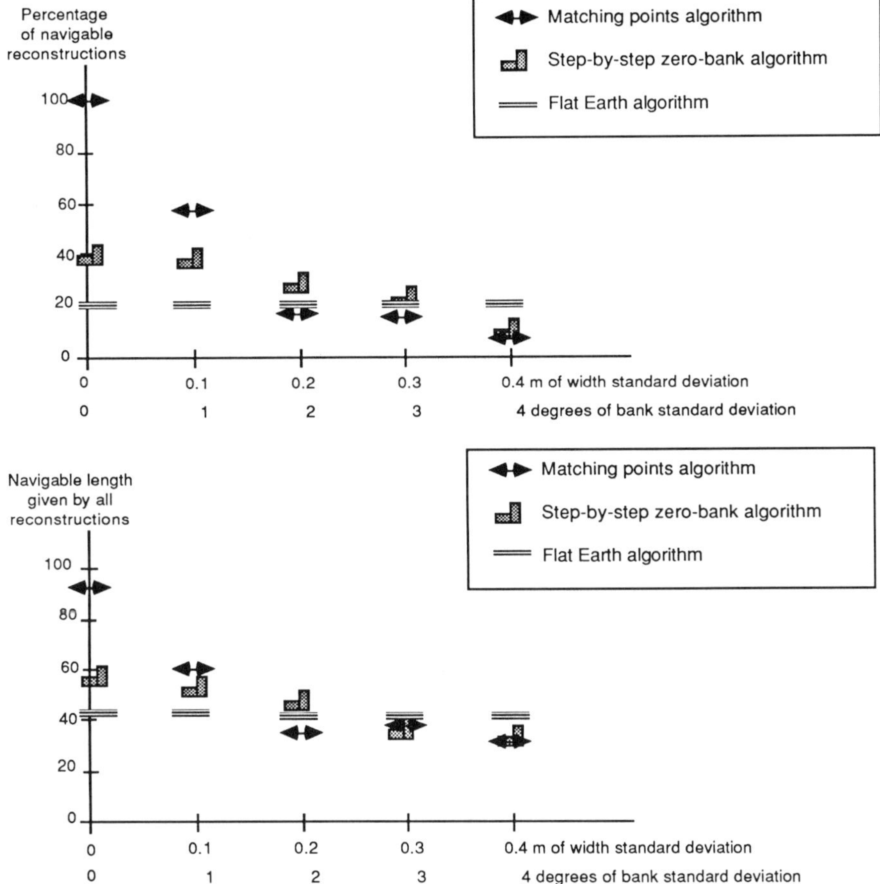

FIGURE 3.7. Comparison of the matching point algorithm with the step-by-step zero-bank algorithm and the flat earth algorithm. Results for different road slopes were averaged. In this figure, a navigable reconstructed road is such that a 2-m-wide vehicle following its centerline will not cross the edges of the actual road. (a) Percentage of navigable reconstructions for various combined width and bank variations. (b) Navigable reconstructed road length.

obstacles for a land vehicle that is not specially equipped. Although material properties are important for determining navigability, they are not readily measured by current remote sensing devices on autonomous vehicles. Here, we make the simplifying assumption that regions can be adequately categorized by their geometry alone.

Given the presumption that obstacles will be defined by their shape, the next issue is how a region's geometry can best be determined by an autonomous vehicle. Perceiving geometry is essentially a depth perception problem. Approaches for creating depth (range) images generally fall into three

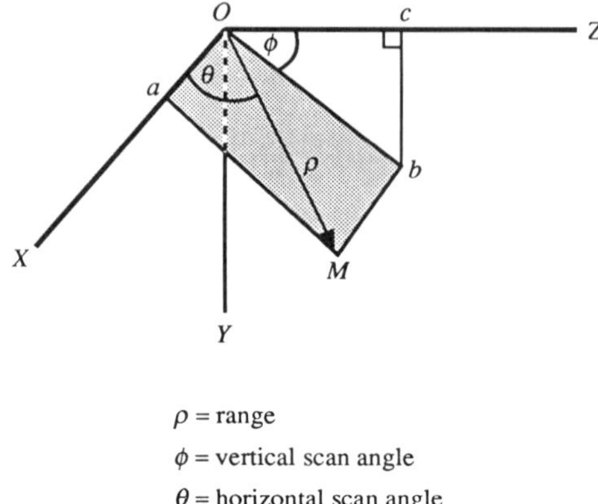

ρ = range

ϕ = vertical scan angle

θ = horizontal scan angle

FIGURE 3.8. Range image coordinate system.

categories: duplicating human visual ranging methods, contriving lighting methods, such as structured light sensors, and applying direct, active ranging technologies.

Direct range sensing methods do not provide insight into human visual understanding, but they are superior to indirect methods for creating fast, accurate range images. The ALV project used the Environmental Research Institute of Michigan (ERIM) [15] range scanner. Figure 3.8 illustrates the spherical coordinate system (θ, ϕ, ρ) that naturally describes range images. The scanner, which was mounted on the ALV, approximately 9 ft above the ground, is at the origin (O) of the system. The positive Y axis points directly down toward the ground. The positive Z axis points out in the direction that the ALV is currently traveling. The length of the line segment OM is the range (ρ) to the point M.

The right triangle Ocb is in the YZ plane. Angle cOb forms the vertical scan angle ϕ. Each row in a range image is taken from a plane that contains the X axis and is ϕ degrees beneath the Z axis. The rectangle $OaMb$ is in this plane. Angle aOM forms the horizontal scan angle θ. Each column in a range image corresponds to a particular θ. This geometry results in the following relationships:

$$x = \rho \cos(\theta), \tag{3.6}$$

$$y = \rho \sin(\theta) \sin(\phi), \tag{3.7}$$

$$z = \rho \sin(\theta) \cos(\phi). \tag{3.8}$$

The 64 rows of the image are at equally spaced values of ϕ, and the 256 columns are at evenly spaced values of θ. An ERIM range image has a 30° vertical field of view in which ϕ goes from approximately 6° to 36°. The 80° horizontal field of view extends from a θ of 130° to a θ of 50°. Although the total magnitudes of the fields of view are fixed, the orientations can be altered either by internal controls in the scanner or by moving the external platform on which the scanner is mounted.

The ERIM scanner has a vertical sampling interval of 0.3125° and a horizontal sampling of 0.46875°. Since the laser beam has an angular divergence of 0.5°, a scene is densely sampled. This removes the need for sophisticated interpolation techniques, such as those proposed by Boult and Kender [3] or Choi and Kender [5] for sparse range data.

Multiple objects at various ranges may occur within the 0.5° solid cone that forms a single pixel's field of view. The signal that returns to the scanner will indicate a range that is a complex average of all the ranges encountered within the cone. For example, if half of a cone intercepts a tree and the other half travels on to the ground, then the returning signal would yield a value that is somewhere between the distance to the tree and the most distant ground that is within the cone. This is called the mixed pixel problem. A strategy for avoiding range errors resulting from mixed pixels is presented in Vealch and Davis [21].

Because of the ambiguity effect, output range values are all between 0 and 64 ft. They are quantized into three-in. units, so that the final output of the ERIM scanner is a 64 × 256 array of 8-bit values ranging from 0 to 255.

The fastest of the ALV obstacle detection algorithms, range differencing, simply subtracts the range image of an actual scene from the expected range image of a flat plane. While rapid, this technique is not very robust. Small errors in the orientation of the scanner or a mild slope in the land will result in false indications of obstacles. We propose using the first derivatives of the range with respect to the vertical and horizontal scan angles as an improved, fast obstacle detector.

3.4.1 The Range Derivative Algorithm for Obstacle Detection

When deciding if a surface is an obstacle or not, the pertinent feature is the change in height across the surface. If the change is too rapid, then the surface is unnavigable. A surface normal contains the necessary information on the change in height, but calculating surface normals is computationally intensive. The surface normal at a point is a function of $\partial y/\partial x$ and $\partial y/\partial z$. Simply calculating the slope, $\partial y/\partial z$, would provide significant information concerning a surface's navigability. However, computing the slope directly from a range image is not much easier than calculating a surface normal. What can be done very quickly, though, is finding $\partial \rho/\partial \theta$ and $\partial \rho/\partial \phi$. The following derivation first shows how $\partial \rho/\partial \phi$ can be closely linked to $\partial y/\partial z$ and then how $\partial \rho/\partial \theta$ can be a measure of $\partial y/\partial x$. Using our knowledge of how range derivatives reflect changes in height across a surface, we can then design a rapid obstacle detection algorithm.

The differential of a function $y(\phi, \rho, \theta)$ can be written as

$$dy = \frac{\partial y}{\partial \phi} d\phi + \frac{\partial y}{\partial \rho} d\rho + \frac{\partial y}{\partial \theta} d\theta. \tag{3.9}$$

If θ is held constant so that the $d\theta$ term is 0, then Equation (3.9) applied to Equation (3.7) gives

$$\Delta y = \rho \sin \theta \cos \phi \, \Delta\phi + \sin \theta \sin \phi \Delta\rho, \tag{3.10}$$

where the infinitesimal terms dy, $d\phi$, and $d\rho$ have been replaced by their finite Δ equivalents. In a similar fashion, Equation (3.8) can be differentiated to yield

$$\Delta z = -\rho \sin \theta \sin \phi \, \Delta\phi + \sin \theta \cos \phi \, \Delta\rho. \tag{3.11}$$

Dividing Δy and Δz yields

$$\frac{\Delta y}{\Delta z} = \frac{\rho \cos \phi \, \Delta\phi + \sin \phi \, \Delta\rho}{-\rho \sin \phi \, \Delta\phi + \cos \phi \, \Delta\rho} = \frac{[(\Delta\rho/\rho)(\tan \phi/\Delta\phi) + 1]}{[(\Delta\rho/\rho)(1/\Delta\phi) - \tan \phi]}. \tag{3.12}$$

If ϕ is held constant, then Equation (3.9) becomes

$$\Delta y = \sin \theta \sin \phi \, \Delta\rho - \rho \sin \phi \cos \theta \, \Delta\theta, \tag{3.13}$$

and Equation (3.6) can be differentiated to obtain

$$\Delta x = \cos \theta \, \Delta\rho - \rho \sin \theta \, \Delta\theta. \tag{3.14}$$

Dividing Equation (3.13) by Equation (3.14) and regrouping yields

$$\frac{\Delta y}{\Delta x} = \frac{[(\Delta\rho/\rho)(\tan \theta/\Delta\theta) \sin \phi - \sin \phi]}{[(\Delta\rho/\rho)(1/\Delta\theta) - \tan \theta]}. \tag{3.15}$$

Excluding the terms in Equations (3.12) and (3.15) that we know a priori, we see that the changes in height in the x and z directions are a function of $\Delta\rho/\rho$. If we used some approximation of ρ, we would have a direct relationship between the easily calculated $\Delta\rho$ for a fixed θ or ϕ at a pixel and the slopes at that pixel. Our experiments with real range data suggest that the following is an adequate approximation:

$$\rho \approx H/\sin \theta \sin \phi, \tag{3.16}$$

where H is the height of the range scanner above the ground. Equation (3.16) comes from substituting H for y in Equation (3.7). In hilly terrain, this approximation is probably not adequate, but it works well for many scenes, and a table will show that the derivative algorithm that uses this approximation is less sensitive to orientation errors than other algorithms of similar simplicity and speed.

Using Equation (3.16), we can calculate what $\Delta\rho$ would be at each pixel if the slopes were zero. The difference between this predicted $\Delta\rho$ and the actual $\Delta\rho$ found in a range image is a measure of the actual slope. Large differences between predicted and actual $\Delta\rho$'s will be formed by edges of objects as well as surfaces with steep slopes. Thresholding the absolute values of these differences yields pixels that are likely to be on obstacles.

One could, of course, simply threshold the actual $\Delta\rho$'s without first subtracting the expected $\Delta\rho$'s and assume that large $\Delta\rho$'s indicate surfaces that

TABLE 3.1. Comparison of obstacle detection algorithms for sensitivity to rotation errors.[a]

Perturbation	Magnitude of errors			
	θ	ϕ	Height	Range
3 degree horizontal	0.8	1.5	5.1	24.7
	0.3	0.4	1.6	6.4
3 degree roll	3.7	13.6	21.2	103.3
	1.1	2.8	6.0	23.1
3 degree vertical	1.1	17.3	25.4	123.7
	0.3	13.8	25.4	98.5
3 degrees in each	9.7	53.5	72.2	351.4
	2.6	21.3	38.9	150.7

[a] © 1990 IEEE.

have steep slopes and hence are not navigable. This approach, however, would severely reduce one's capability to detect obstacles. A perfectly flat surface will yield a $\Delta\rho$ of about 10 if it is 60 ft away, but the same surface at a range of 10 ft only has a $\Delta\rho$ of about 0.3. This wide range in $\Delta\rho$'s leaves any thresholding algorithm in a bind. Small threshold levels would find nearby obstacles, but more distant flat surfaces would be falsely labeled as obstacles. Conversely, larger threshold levels would hide significant obstacles that are near the range scanner. What is needed is a variable threshold setting. This approach points out another way of looking at the range derivative algorithm: we are, in essence, creating a variable threshold that changes across an image based on expected $\Delta\rho$'s. While this simplistic view is a useful description, the derivative algorithm is founded on the mathematical relationships between $\Delta\rho$ and a surface's slopes and is not a randomly chosen heuristic for setting variable threshold levels.

Table 3.1 reveals the advantages of this approach over height or range prediction. The table contains two entries for each combination of algorithm and perturbation of road orientation (horizontal, vertical, and roll). The top entry is the largest absolute value in the entire image and represents a worst case scenario. In many scenes, however, the road will be near the center of the image's horizontal field of view, and large errors at the periphery are not critical. This is captured by the bottom entry, which is the largest absolute error within the central 30° of the image.

Several important trends emerge from Table 3.1. The ϕ derivatives were insensitive to all four rotational perturbations. When the entire image was considered, the maximum ϕ errors for each rotation were always at least 25% less than the maximum height difference errors. Within the central 30° of the horizontal field of view, the maximum ϕ derivative errors were 45–75% less than the maximum height errors. The range difference algorithm was very sensitive to all rotations. In several instances, the range difference errors were a full order of magnitude larger than the derivative errors. These results clearly show that the derivative algorithms are more robust under rotational uncertainties than either the height difference or range difference algorithms.

3.5 Conclusion

This paper provided a summary of the research conducted at the University of Maryland on problems associated with visual navigation of ground vehicles. We focused on algorithms for three modules of the system, the scene model planner, the road reconstruction module from video images, and the obstacle detection module. One of the roles of the scene model planner is to intelligently locate windows of focus of attention in video images. Strategies are represented explicitly as rules, which are easy to modify and extend. For example, some rules would locate the windows farther apart in the image when the road is straight than when it is curved. The road reconstruction module uses points detected in these windows as road edges to build three-dimensional road models. Image points corresponding to endpoints of cross segments of the road are found, and these cross segments are reconstructed. A dynamic programming algorithm weighs the mutual consistency of the cross segments and rejects the cross segments that do not contribute to a consistent road. The obstacle detection module uses images from the vehicle range scanner. It detects local changes of heights in the scene by directly interpreting gradients of range along the range image rows and columns. This method is more direct than approaches based on computing surface normals in the scene and is robust under rotational and vertical perturbations.

Over the past few years, our research has shifted from road navigation to more general navigation problems, with a long-term practical emphasis on cross-country navigation. This has led us to consider many new and interesting problems in the design and organization of autonomous systems. Preliminary results of our research on two navigation systems—RAMBO and Medusa—can be found in Davis et al. [6] and Aloimonos [2].

References

[1] Allen, E. (1983). "YAPS: Yet Another Production System." University of Maryland Computer Science Technical Report 1146, December.

[2] Aloimonos, J. (1990). "Purposive and Qualitative Active Vision." *Proc. Image Understanding Workshop, September 1990.*

[3] Boult, T. B., and Kender, J. R. (1985). "On Surface Reconstruction Using Space Depth Data." *Proceedings Image Understanding Workshop, Miami Beach, Florida, December 1985,* 197–208.

[4] Brooks, R. (1985). "A Robust Control System for a Mobile Robot." *IEEE Transactions on Robotics and Automation* 2, 14–23.

[5] Choi, D. J., and Kender, J. R. (1985). "Solving the Depth Interpolation Problem with the Adaptive Chebyshev Acceleration Method on a Parallel Computer."

Proceedings: Image Understanding Workshop, Miami Beach, Florida, December 1985, 219–223.

[6] Davis, L. S., DeMenthon, D., Bestul, T., Harwood, D., Srinivasan, H. V., and Ziavras, S. (1989). "RAMBO: Vision and Planning on the Connection Machine." *Proc. Image Understanding Workshop, May 1989.*

[7] DeMenthon, D. (1987). A Zero-Bank Algorithm for Inverse Perspective of a Road from a Single Image." *IEEE International Conference on Robotics and Automation*, 1444–1449.

[8] DeMenthon, D. (1988). "Reconstruction of a Road by Matching Edge Points in the Road Image." University of Maryland Center for Automation Research Center Technical Report 368, June.

[9] DeMenthon, D., and Davis, L. S., (1980). "Reconstruction of a Road by Local Image Matches and Global 3D Optimization." *IEEE International Conference on Robotics and Automation*, 1337–1342.

[10] Dickinson, S., and Davis, L. (1990). "A Flexible Tool for Prototyping ALV Road Following Algorithms." *IEEE Transactions on Robotics and Automation* **6** (2), 232–242.

[11] Dickmanns, E. D., and V. Graefe, V. (1988). "Dynamic Monocular Machine Vision" and "Applications of Dynamic Monocular Machine Vision." *Machine Vision and Applications*, International Journal, Springer-Verlag International, Berlin and New York.

[12] Kanatani, K., and DeMenthon, D. (1988). "Reconstruction of 3D Road Shape from Images: A Computational Challenge." Center for Automation Research Technical Report CAR-TR-413, December.

[13] Kanatani, K., Watanabe, K. (1989). "Computing 3D Road Shape from Images for ALV: A Challenge to an Ill-Posed Problem." Computer Science Technical Report CS-89-5, Gunma University, June.

[14] Kanatani, K., and Watanabe, K. (1980). "Reconstruction of 3D Road Geometry from Images for Autonomous Land Vehicles." *IEEE Transactions on Robotics and Automation* **6** (2), 127–132.

[15] Larrowe, V. (1986). "Operating Principles of Laser Ranging, Image Producing (3D) Sensors." Environmental Research Institute of Michigan, Ann Arbor, March.

[16] Liou, S., and Jain, R. (1985). "Detecting Road Edges Using Hypothesized Vanishing Points." Robot Systems Technical Report 18-85, Department of Electrical Engineering and Computer Science, University of Michigan.

[17] Morgenthaler, D., Henessy, S., and DeMenthon, D. (1990). "Range-Video Fusion and Comparison of Inverse Perspective Algorithms." *IEEE Trans. on Systems, Man, and Cybernetics*, Special Issue on Unmanned Systems and Vehicles, November–December.

[18] Ozawa, S., and Rosenfeld, A. (1986). "Synthesis of a Road Image as Seen from a Vehicle." *Pattern Recognition* **19**, 123–145.

[19] Thorpe, C., Hebert, M., Kanade, T., and Shafer, S. (1988). "Vision and Navigation for the Carnegie–Mellon Navlab." *IEEE Transactions on Pattern Analysis and Machine Intelligence* **10**, 362–373.

[20] Turk, M., Morgenthaler, D., Gremban, K., and Marra, M. (1988). "VITS–A Vision System for Autonomous Land Vehicle Navigation." *IEEE Transactions on Pattern Analysis and Machine Intelligence* **10**, 342–361.

[21] Veatch, P., and Davis, L. (1990). "Range Imagery Algorithms for the Detection of Obstacles by Autonomous Vehicles." *Computer Vision, Graphics and Image Processing* **50**, 50–74.
[22] Waxman, A., LeMoigne, L., Davis, L., Srinivasan, B., Kushner, T., Liang, E., and Siddalingaiah, T. (1987). "A Visual Navigation System for Autonomous Land Vehicles." *IEEE Transactions on Robotics and Automation* **2**, 124–142.

4
A Visual Control System Using Image Processing and Fuzzy Theory

HIROSHI KAMADA AND MASUMI YOSHIDA

4.1 Abstract

We developed a visual control system for an unmanned vehicle. The system consists of a dynamic image processor and a fuzzy logic control mechanism. It quickly recognizes markers lined along a road and thereby navigates a driverless vehicle. The markers are detected in real time by pipeline processing in the color identification processor and logical filter; the marker sequence is recognized by an improved Hough transform, then the fuzzy logic control mechanism decides the steering angle. To use the information on the movement of the vehicle, we constructed fuzzy inference rules on how position changes with time. We developed an LSI (large-scale integrated circuit) chip for the logical filter to realize a very compact and practical system (23 × 30 × 9.5 cm). We mounted this system on a vehicle, and it successfully drove around a test track.

4.2 Introduction

Research on unmanned vehicles is being conducted in many countries, in particular, the United States [6] and West Germany [5]. These vehicles include passenger cars and working vehicles, so research on labor-saving intelligent unmanned vehicles is industrially very important.

We developed a practical visual control system using dynamic image processing and fuzzy theory. To develop our system, we simulated the human driving process (Figure 4.1). When a human drives a vehicle, he identifies various markers or reference points along the road and mentally draws a line between them, then steers the vehicle along that line. Our visual control system must have the high-speed functions of

1. checking the surroundings and the status of the moving vehicle and
2. inferring the course of the moving vehicle from the above data and deciding the steering angle.

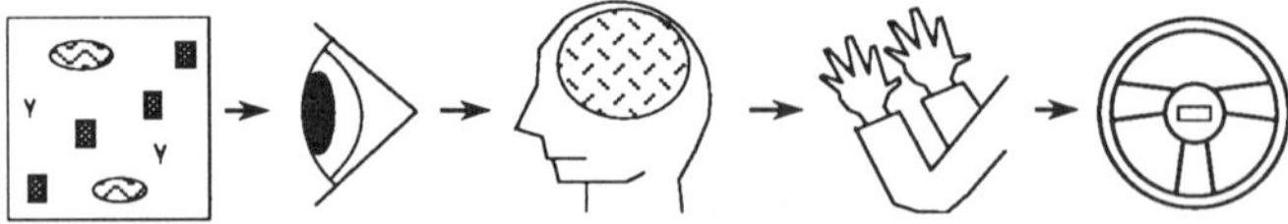

FIGURE 4.1. Human driving process.

Images input from video cameras provide sufficient information for function 1. Since these images consist of a large amount of data, they used to take a long time to process. However, recent developments in processing dynamic images have enabled processing in real time at video speed (30 images per second for a $2:1$ interlace video camera). An example of these innovations is the FIVIS/VIP, a general-purpose, high-speed image processor with a reconfigurable pipeline architecture [3, 4].

For the inference and control function of 2, we studied human driving techniques in developing the calculations. Recently, fuzzy theory has been applied to inference and control using such ambiguous information [2]. Fuzzy theory was first proposed by Zadeh in 1965 [1] as a way to handle ambiguous data mathematically. In inference methods based on fuzzy theory, vague expressions used by humans can be expressed as mathematical rules and used for control. Many industrial applications have already been reported [2].

Since our aim was to develop a practical system, the visual control system had to be small enough to be mounted on a small vehicle, and so we tried to reduce the size as much as possible. Our system quickly recognized a series of markers to navigate the unmanned vehicle. We placed no restrictions on the type of road surface, but to keep things simple, we provided markers that were all the same color and either formed a continuous line or were lined up.

This Chapter reports the problems we faced during development of the system and their solutions. The developed system and its experiment results are also introduced.

4.3 Development of the System

4.3.1 Real-Time Marker Identification

The first problem was to achieve real-time identification of the markers because other objects on the road could easily be mistaken for a marker. In the past, monochrome video cameras have been used for the visual system of unmanned vehicles as the images are easy to process. Unfortunately, they mistake reflected light or other objects for markers, so we decided to use color video cameras.

We then classified these false markers into those that were a different color from the markers and those that were a different size, and then we developed a suitable noise elimination method to eliminate each type.

The elimination method for objects that were a different color from markers was designed to eliminate water pools, mud, and other objects. The system would see the color of the markers in the white light of sunlight reflected by a water pool since white light contains all the colors.

We developed a method of identifying colors using the ratio of the intensity of the specific color to the total intensity of the three primary colors, red, green, and blue (RGB). This means that if the markers were green, for example, and the system detected strong green light, the system would check for strong blue and red light. If the red, blue, and green were all strong, the light was white and was not coming from a marker. The following equation was used:

$$\text{Color ratio} = \frac{\text{Intensity of specified color}}{\text{Sum of intensity of RGB}}. \tag{4.1}$$

The elimination method for objects a different size from markers was designed to eliminate objects of the same color as the markers, but smaller. For example, if the markers were green, grass and weeds were eliminated. This was done by examing the picture elements making up the picture. A picture element was identified as a component of the marker only when adjacent picture elements were the specified color (Figure 4.2).

We developed a color identification processor and logical filter processor and connected them by a pipeline to realize real-time identification (Figure 4.3).

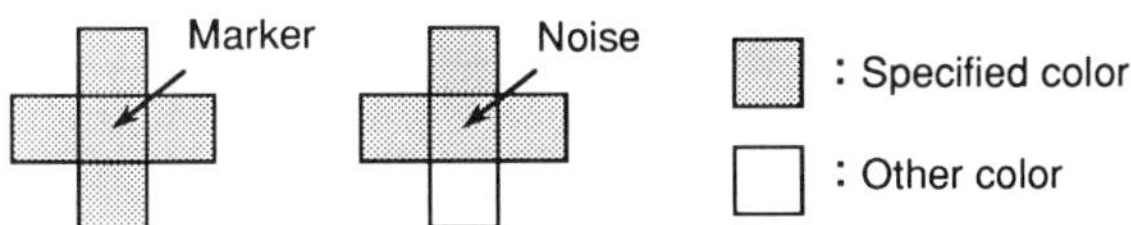

FIGURE 4.2. Size checking.

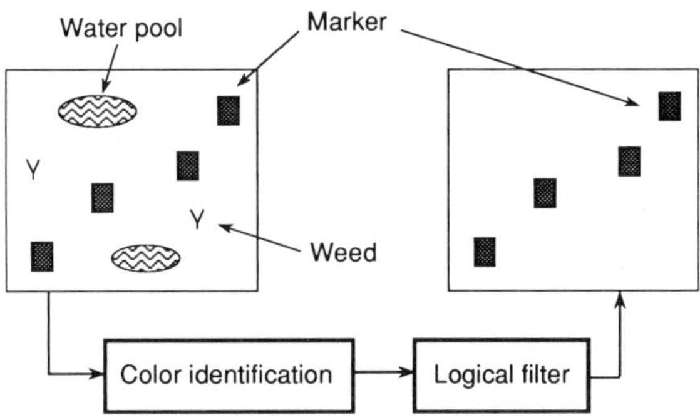

FIGURE 4.3. Real-time marker identification.

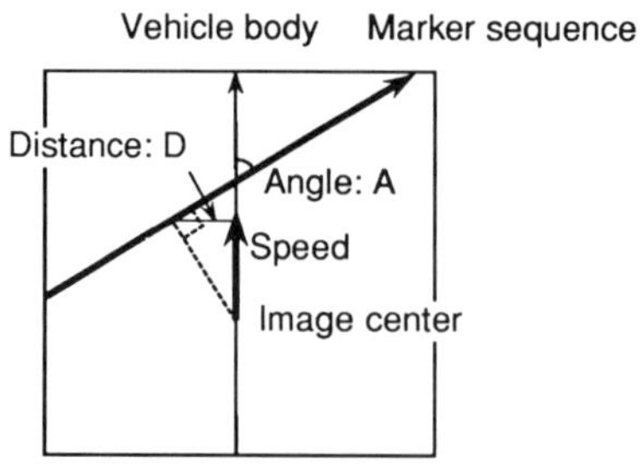

FIGURE 4.4. Case of moving toward marker sequence.

4.3.2 Fast Recognition of Marker Sequence

The second problem was to recognize marker sequence quickly to provide steering information for the unmanned vehicle. To solve this problem, we emulated human driving techniques. When a human drives a vehicle, he identifies various markers or reference points along the road and mentally draws a line between them. He does not see them as individual points but as a continuous line along which to steer, then steers the vehicle along that line.

We used the following two parameters to recognize marker sequence and thus to decide steering (Figure 4.4).

Distance D: This is the horizontal component of a perpendicular line drawn from the image center to the line of markers.

Angle A: This is the angle of the line of markers to the centerline of the image.

Previously, we used the length of the perpendicular as distance D, but this caused a large oscillation in steering. This is because when angle A becomes bigger, the length of the perpendicular does not change, but its horizontal component becomes smaller (Figure 4.4). Even when the image center is distant from the marker sequence, the vehicle will get to the marker sequence so long as it is moving toward it.

In the perpendicular from the image center to the line of markers, the component that is parallel to the direction of travel of the vehicle body is not important, but the one at right angles to the direction of travel of the vehicle (horizontal component in the image) is important.

We therefore adopted the horizontal component of the perpendicular instead, which took into account the inertia of the vehicle. As a result, the vehicle approached the line of markers more directly, with only a small amount of steering.

The system must recognize D and A from the identified marker images. To recognize many points as a line, a conventional Hough transform is generally used. This transform carries out the following processing (Figure 4.5).

1. Calculate the parameters (D, A) of straight lines passing an arbitrary point in a detected marker. This processing matches an arbitrary point to the curve of the parameter space D–A, and it generates the same number of curves as the number of marker points in parameter space D–A.

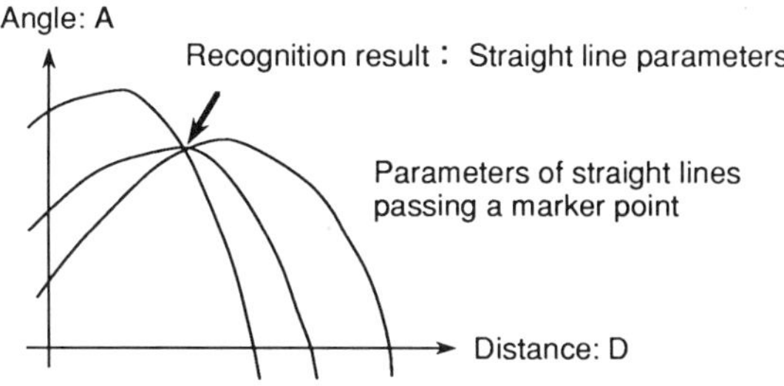

FIGURE 4.5. Conventional Hough transform.

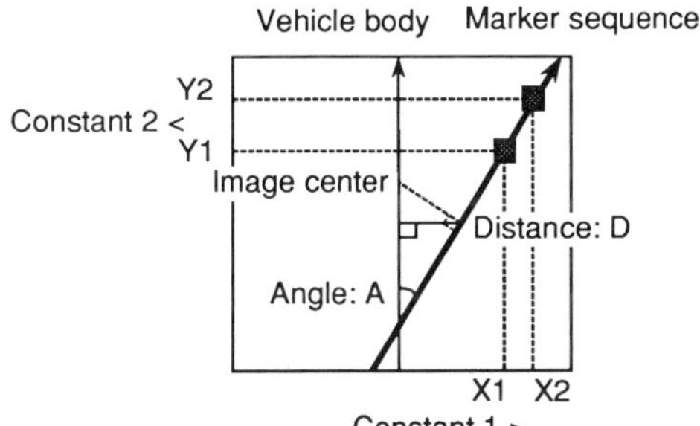

FIGURE 4.6. Hough transform acceleration method.

2. Regard a curve in parameter space D–A as a set of points. Thus, the coordinates of the densest point are recognized as the parameters of the straight lines indicating the marker sequence.

A large amount of calculation is required for 1, making this transform too slow for real-time steering. So, we made the Hough transform faster by replacing 1 with the following (Figure 4.6).

Calculate the parameters D and A of a straight line passing two arbitrary points of an identified marker, but limit the combinations of two points ($X1$, $Y1$) and ($X2$, $Y2$) by the following two expressions to reduce the calculation amount:

$$|X1 - X2| < \text{constant 1}, \tag{4.2}$$

$$|Y1 - Y2| > \text{constant 2}. \tag{4.3}$$

These limitations are based on the fact that the direction of the vehicle is usually close to that of the markers.

Parameters D and A can be calculated from the two points ($X1$, $Y1$) and ($X2$, $Y2$) using the following expressions.

$$A = \text{Tan}^{-1} \frac{|X1 - X2|}{|Y1 - Y2|}, \tag{4.4}$$

$$D = (X1 \cos A + Y1 \sin A) \cos A. \tag{4.5}$$

The modified Hough transform procedure is as follows.

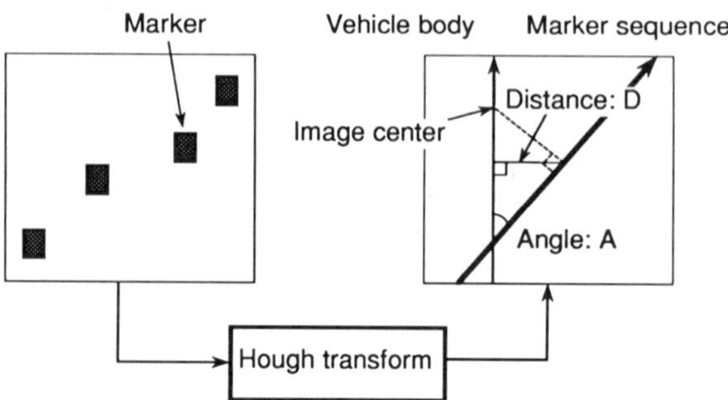

FIGURE 4.7. Recognition of marker sequence.

1. (a) Select the combinations of two arbitrary points of an identified marker
   satisfying Equations (4.2) and (4.3). (b) Calculate the parameters of a
   straight line passing the two points for a single point in parameter space
   $D-A$ using Equations (4.4) and (4.5). (c) Repeat (a) and (b) on all
   combinations of two points of a marker satisfying Equations (4.2) and (4.3).
   This processing generates points in parameter space $D-A$.
2. Regard the coordinates of the densest point as the parameters of the
   straight line.

This improved Hough transform method recognized the marker sequence
from identified marker images (Figure 4.7).

4.3.3 Steering Control System

The third problem was how to control steering. When we steer a car, we are
mostly unaware of making decisions, the action is intuitive. Describing steering to a computer or robot is rather difficult. To solve this problem, we coupled human steering techniques and marker recognition information to control steering by using fuzzy theory.

As the marker recognition information, we added the distance change by
time dD to the distance D and angle A. This is because accurate steering
cannot be controlled without considering vehicle movement. For example, if
the direction of the line of markers changes greatly with time, as shown in
Figure 4.8, steering cannot be decided by D and A at a single time point.

For recognition information D, A, and dD, five types of orientation such as
slightly left or sharp right are expressed by the membership functions shown
in Figure 4.9.

We set fuzzy rules to identify human driving techniques and decide steering
from the recognition information. The rules have the following IF-THEN
format:

IF [(variable is ...) AND ...] -> THEN [steering is ...]

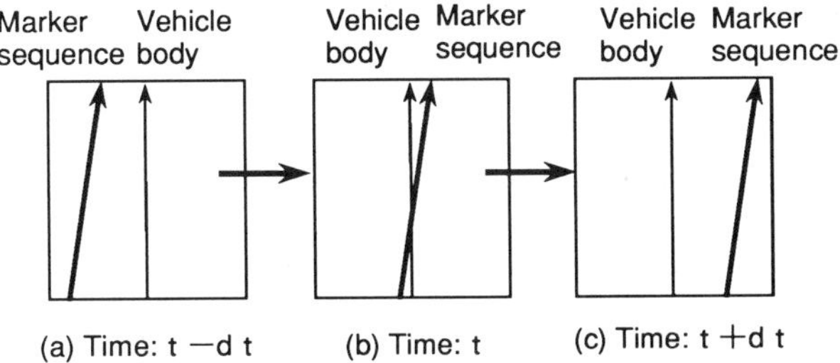

FIGURE 4.8. Change of marker sequence with time.

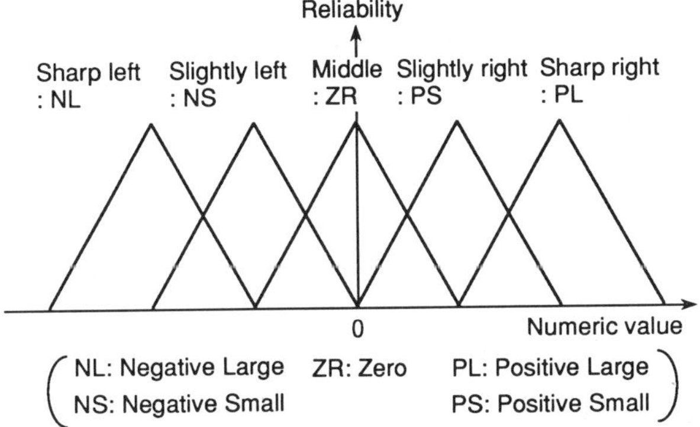

FIGURE 4.9. Membership function types.

TABLE 4.1. Types and numbers of fuzzy rules.

IF clause variable (coupled with AND)		Number
Distance D	Angle A	9
	Distance change by time, dD	9

There is a total of 18 fuzzy rules divided into two groups of 9 rules each, depending on the variables in the IF clause (Table 4.1).

Steering is decided as follows.

1. The vehicle's steering angle is inferred for each fuzzy rule. (a) Enter the marker sequence information in the membership functions of each variable in the IF clause and calculate the reliability. (b) Since the variables in the IF clause are connected by AND, calculate the reliability of the IF clause as the minimum value of the reliability obtained in (a). (c) Restrict the membership functions of the THEN clause into a trapezoidal form using the reliability of the IF clause.

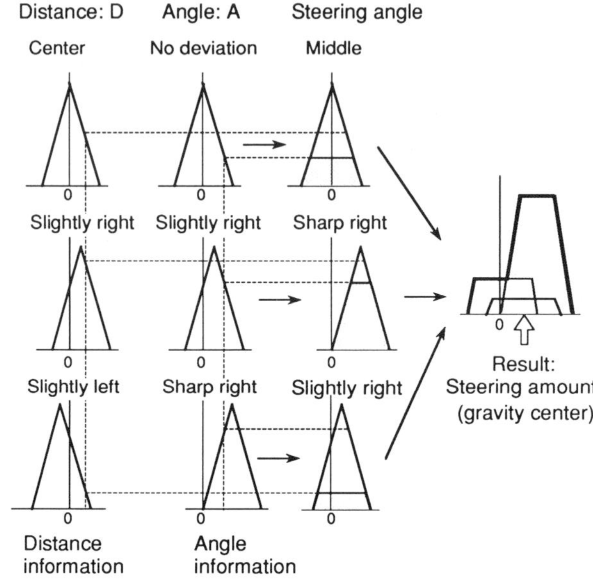

FIGURE 4.10. Steering decision by fuzzy theory.

2. All the inference results of the fuzzy rules are integrated to calculate the steering angle. (a) Overlay the trapezoidal distributions of all rules created by inference to obtain the maximum value. (b) Set the gravity center of the graphic created by overlaying the trapezoidal distributions as the steering angle.

Both 1. and 2. are illustrated by a simple model case in Figure 4.10.

The membership functions shown in Figure 4.9 are actually discrete ones. There are 9 or 11 elements in the table set and the reliability of precision is level 4 or 5. Tables 4.2 through 4.5 list the membership functions.

Tables 4.6 and 4.7 show details of the fuzzy rule bases introduced in Table 4.1. These rules were constructed after consulting a skilled driver. In rule base 1, the two parameters in the IF clause are additive. For example, if both distance D and angle A are slightly right, the steering angle is sharp right.

TABLE 4.2. Membership function of distance D.[a]

Function	−5	−4	−3	−2	−1	0	1	2	3	4	5
Sharp right								3	5	8	10
Slightly right					3	5	8	10	8	5	3
Center			3	5	8	10	8	5	3		
Slightly left	3	5	8	10	8	5	3				
Sharp left	10	8	5	3							

[a] Unit, 1/0.67 pixel.

TABLE 4.3. Membership function of angle A.[a]

Function	−5	−4	−3	−2	−1	0	1	2	3	4	5
Sharp right	10	8	5	3							
Slightly right	3	5	8	10	8	5	3				
No deviation			3		5	8	10	8	5	3		
Slightly left					3		5	8	10	8	5	3
Sharp left									3	5	8	10

[a] Unit, 1/20 rad.

TABLE 4.4. Membership function of distance dD.[a]

Function	−4	−3	−2	−1	0	1	2	3	4
Sharp right							3	7	10
Slightly right						3	7	10	7	3
None				3	7	10	7	3	
Slightly left		3	7	10	7	3			
Sharp left	10	7	3						

[a] Unit, 20/0.67 pixel.

TABLE 4.5. Membership function of steering angle.[a]

Function	−4	−3	−2	−1	0	1	2	3	4
Sharp right	10	7	3						
Slightly right	3	7	10	7	3				
Middle			3	7	10	7	3		
Slightly left				3	7	10	7	3	
Sharp left						3	7	10	

[a] Unit, steering index.

However, the two parameters are not complementary to each other. For example, if angle A is sharp right, even if distance D is slightly left, the steering angle is still sharp right. Angle A is more important to steering angle than distance D.

In rule base 2, the two parameters in the IF clause are not additive. For example, if both distance D and dD are slightly right, the steering angle is also slightly right. The two parameters are, however, complementary to each other. For example, if dD is sharp right and distance D is slightly left, the steering angle is slightly right.

4.3.4 Marker Sequence Selection Method

The fourth problem is to select the correct marker sequence when there are several sequences in the input image. On an actual road, there will often be more than one line of markers, only one of which must be followed.

TABLE 4.6. Fuzzy inference rule base 1.

No.	IF clause variable		THEN clause variable
	Distance D	Angle A	Steering angle
1	Center	No deviation	Middle
2	Slightly right	No deviation	Slightly right
3	Slightly left	No deviation	Slightly left
4	Slightly right	Slightly right	Sharp right
5	Slightly left	Slightly left	Sharp left
6	Slightly right	Sharp left	Sharp left
7	Slightly left	Sharp right	Sharp right
8	Sharp right	No deviation	Sharp right
9	Sharp left	No deviation	Sharp left

Matrix expression

A \\ D	NL	NS	ZR	PS	PL
NL				NL	
NS		NL			
ZR	NL	NS	ZR	PS	PL
PS				PL	
PL		PL			

TABLE 4.7. Fuzzy inference rule base 2.

No.	IF clause variable		THEN clause variable
	Distance D	dD	Steering angle
1	Center	None	Middle
2	Slightly right	Slightly left	Middle
3	Slightly left	Slightly right	Middle
4	Slightly right	Slightly right	Slightly right
5	Slightly left	Slightly left	Slightly left
6	Slightly right	Sharp left	Slightly left
7	Slightly left	Sharp right	Slightly right
8	Sharp right	None	Sharp right
9	Sharp left	None	Sharp left

Matrix expression

dD \\ D	NL	NS	ZR	PS	PL
NL				NS	
NS		NL		ZR	
ZR	NL		ZR		PL
PS		ZR		PS	
PL		PS			

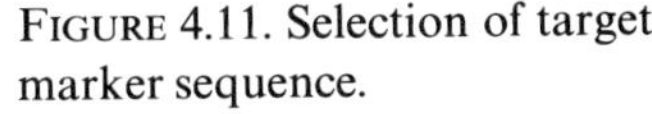

FIGURE 4.11. Selection of target marker sequence.

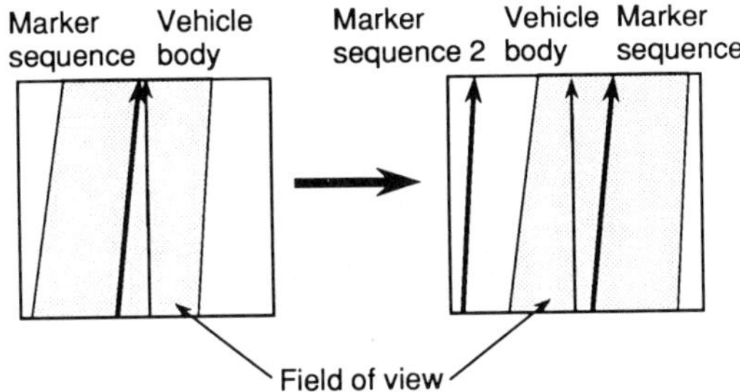

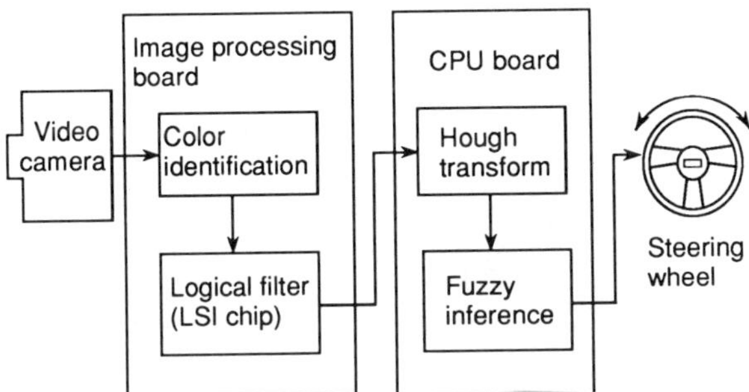

FIGURE 4.12. System configuration.

To solve this problem, we developed a system that masks the image according to the parameters (D, A) of the previous marker sequence so that the target marker sequence can be identified from among several lines of markers (Figure 4.11).

4.4 Developed System

Using the method described previously, we developed a visual control system for an unmanned vehicle. We aimed to create a practical system, and so it had to carry out processing fast enough for real-time steering and yet be compact and light enough for mounting on a vehicle.

To satisfy these conditions, we developed a practical system using two PC (printed circuit) boards (Figure 4.12): an image processing board and a CPU (central processing unit) board. The image processing board executes the color identification and logical filter functions. The logical filter function is provided in an LSI chip. The CPU board executes the improved Hough transform and fuzzy inference as firmware. In this way, we got the size of the system down to about an A4 × 10 cm (Figure 4.13). Details of the sections are given in the following.

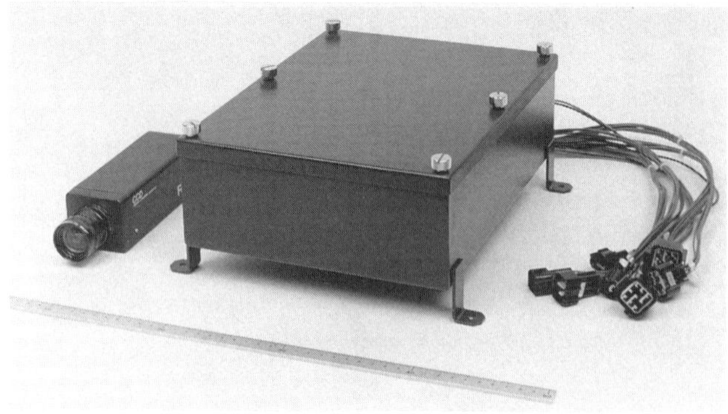

FIGURE 4.13. Visual control system.

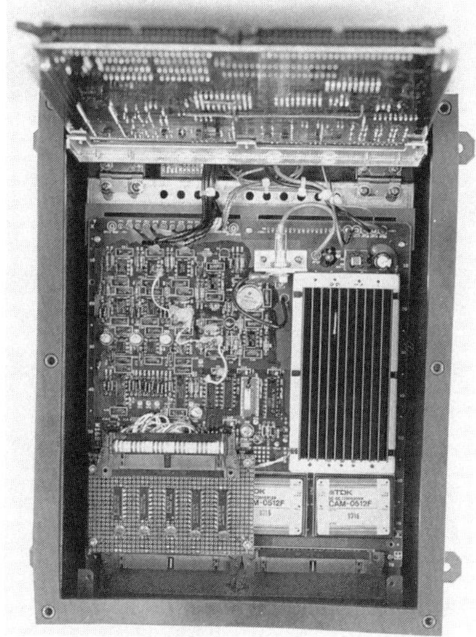

FIGURE 4.14. Image processing board.

4.4.1 Image Processing Board

The A4-sized image processing board includes a color identification section, logical filter section, and power supply section (Figure 4.14).

COLOR IDENTIFICATION SECTION

If a digital circuit were to be used, a large system would be needed to convert all signals of the three primary colors (RGB) from analog to digital. Therefore,

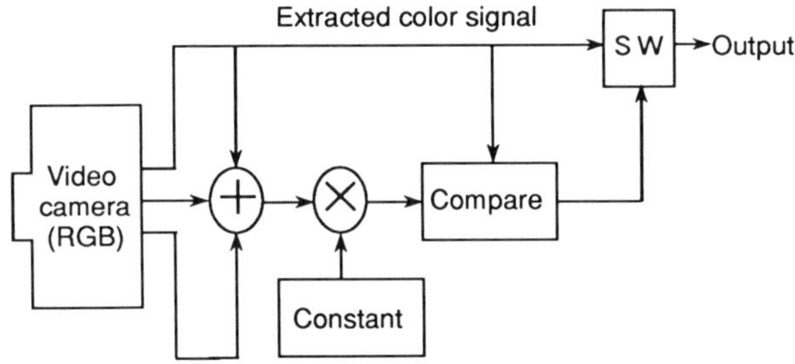

FIGURE 4.15. Configuration of analog color identification section.

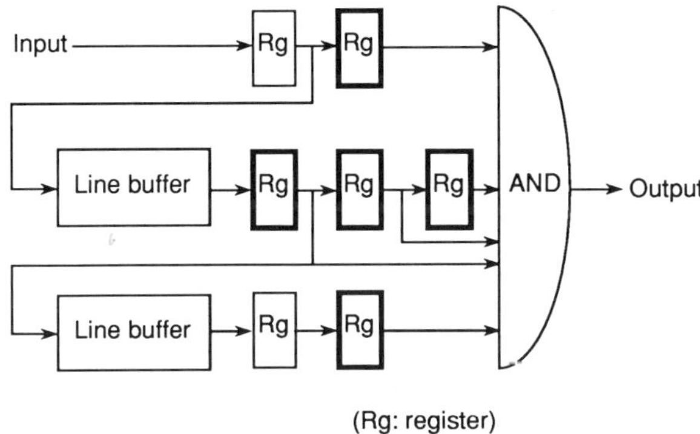

FIGURE 4.16. Configuration of logical filter section.

we created the color identification section using an analog circuit and installed only one A/D conversion section. Since it is hard to identify an arbitrary color with the analog circuit, we chose RGB as the selection colors. The analog color identification section (Figure 4.15) extracts the specified color from the entire image and digitizes the signals.

LOGICAL FILTER SECTION

Objects smaller than the markers are eliminated by the logical filter section. The logical filter section of the local parallel pipeline architecture (Figure 4.16) has the features of

fast processing at video signal speed and
No frame memory.

A video camera inputs images at many frames per second. This architecture does not include memory for storing these frames, thus allowing a reduction in size. If a PC board were used, it would be as large as a sheet of A4-sized paper. So, we created an LSI chip to make this section much smaller (Figure 4.17).

FIGURE 4.17. Logical filter section.

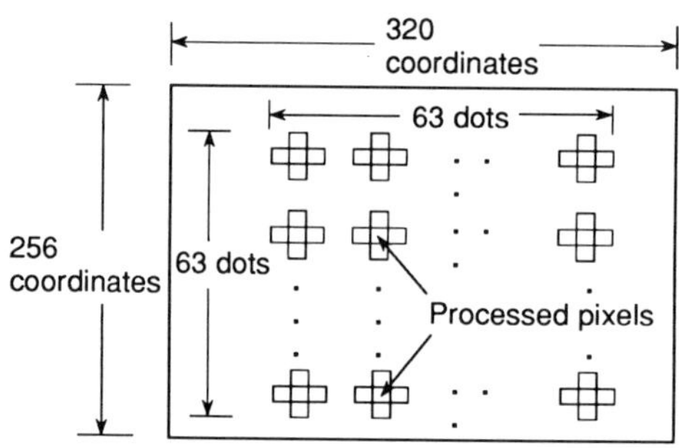

FIGURE 4.18. Logical filter processing position.

We decided to use a standard 2 : 1 interlace video camera, which inputs 30 images (60 frames) per second. However, since detailed information at the pixel level is not necessary for this application, we developed a high-speed processing function that processes 60 images (one field equals one image) a second.

The logical filter section produces a binary image as a result of processing. If each image is processed entirely, however, there will be too much information for real-time processing by the CPU board. So, we decided to compress an image to a binary image of 63 × 63 pixels by processing the horizontal center of the image at every fourth element, using the compression effect of the logical filter (Figure 4.18).

FIGURE 4.19. CPU board configuration.

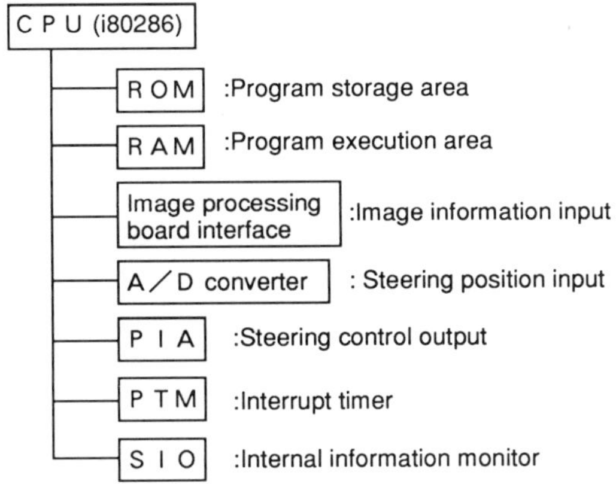

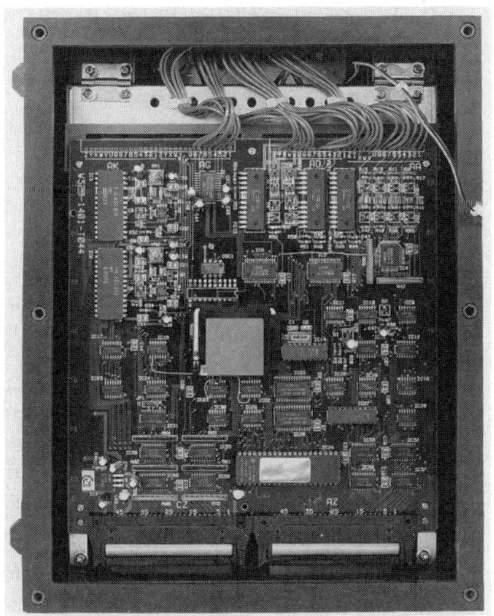

FIGURE 4.20. CPU board.

4.4.2 CPU Board

The CPU board is dedicated to control. The CPU itself is an intel 80286 (8 MHz). Figures 4.19 and 4.20 show the configuration and appearance of the board. The image processing board sends only the coordinate information of identified markers for real-time processing by the CPU board since identified markers account for only a small percentage of an image.

4.5 Experiment

4.5.1 Description

We mounted our system on a vehicle and ran the vehicle unoperated under the working conditions listed in Table 4.8. The vehicle used is a working vehicle that runs in muddy, uneven places. Since only skilled operators can drive this vehicle in a straight line, our experiment is a good test of its capabilities.

We did the experiment outdoors to reflect actual working conditions. We used dry and wet asphalt roads as used by vehicles at a factory site and mud roads as used by construction, agriculture, and forestry vehicles. This system can extract any of the three primary colors, so we used green for the experiment. For markers, we used a centerline and some plants to prove that the system can follow even natural items. The marker sequence was either straight or curved. The vehicle speed was set to 2.5 km/h to mimic the actual work conditions of the above vehicles. Figure 4.21 shows the mounting position and field of vision of the camera.

TABLE 4.8. Specifications of experiment.

Environment	Outdoors, find/cloudy
Road condition	Asphalt, muddy
Marker	Centerline (green); plants (10 cm high, 2 cm in diameter, at 20-cm intervals)
Marker sequence	Straight line; curve (3 m in radius)
Vehicle speed	2.5 km/h

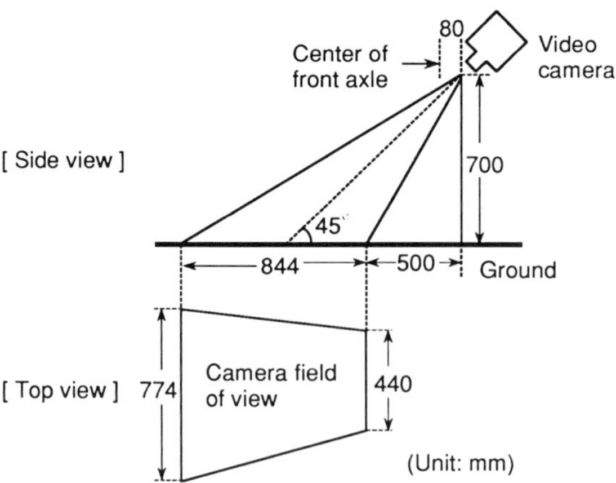

FIGURE 4.21. Camera mounting position and field of vision (unit, mm).

4.5.2 Results

The image processing board identified a marker in 16.7 ms (60 images per second), and the average processing time from image input to steering control was only 100 ms. The vehicle followed the marker sequence within a precision of 5 cm and proved that the new system was very practical.

Although the test was done outdoors on this occasion, we have proved that the system identifies markers even under indoor fluorescent lighting and is suitable for indoor unmanned runs.

4.6 Conclusion

We developed a visual control system for unmanned vehicles by combining dynamic image processing and fuzzy theory. The dedicated image processing LSI chip allowed us to make the system compact. Processing from image input to steering control can be executed very quickly. We set various conditions for the experiment to reflect those that the system may face in actual operation, and the system handled these conditions easily. We believe this system can be used in a wide variety of applications.

4.7 Further Study

In our experiment, we ran the working vehicle at a speed of only 2.5 km/h. With modification, however, our system should be able to used with many other kinds of vehicles operating at very high speeds. To adapt out system for use with such vehicles, we plan to raise the video camera to allow input of images further ahead of the vehicle. Since the marker images will be further away, they will be much smaller, and so the logical filter processing position method (Figure 4.18) would be unsuitable for compressing identified marker image information. Therefore, we also plan to implement more powerful image processing to realize this.

One of the most important features of our system is its compactness, allowing control of various equipment other than vehicles merely by installing suitable firmware. We aim to extend the facilities of our visual control system while retaining its compactness.

Acknowledgment. We would like to express our thanks to Mr. Tanahashi (Manager of Information Processing Division) for his encouragement in research and development of the system.

References

[1] Zadeh, L. A. (1965). "Fuzzy Sets." *Information & Control* **8**, 338–353.
[2] Sugeno, M., ed. (1985). *Industrial Applications of Fuzzy Control*, North-Holland, Publ. Amsterdam.

[3] Sasaki, S., et T. Gotoh, M. Yoshida (1989). "IDATEN: A Reconfigurable Video-Rate Image Processor." In *Advances in Machine Vision* (pp. 356–380). Springer-Verlag, Berlin and New York.

[4] Murano, T., et S. Sasaki, T. Ozaki, Y. Ohta, M. Komeichi (1989). "Time-Varying Color Image Processing System Based on a Reconfigurable Pipeline Architecture." In *Proceedings of Digital Image Processing Applications*, SPIE, 18–25.

[5] Dickmanns, E. D. (1989). "Subject–Object Discrimination in 4D-Dynamic Scene Interpretation for Machine Vision." In *Proceedings of Workshop on Visual Motion*, IEEE, 298–304.

[6] Thorpe, C. E., ed. (1990). *Vision and Navigation—The Carnegie Mellon Navlab*, Kluwer Academic Publishers, Boston/Dordrecht/London.

5
Local Processing as a Cue for Decreasing 3-D Structure Computation

Patrick Stelmaszyk, Hiroshi Ishiguro, Roland Pesty, and Saburo Tsuji

5.1 Abstract

Usually, the extraction of 3-D information from a sequence of images taken by a mobile robot requires a lot of computation because of the complexity of both the 3-D geometry and the robot's environment. If some powerful hardware for computing the 3-D structure in real time exists, its complexity cannot undergo a large-scale duplication. In this chapter, the authors stress that the problem of determining in close real time the 3-D structure of a mobile robot environment can be achieved by local computations. Such an approach implies the computation of local maps surrounding some points of interest selected in the scene, instead of the direct determination of the global one, as far as a local tracking rather than a global matching algorithm. For illustrating our purpose, we show the preliminary results of the 3-D local map reconstruction by an active monocular vision system mounted on a mobile robot. Such a technique computes and updates, while the robot is moving, the local map surrounding some feature points selected in the scene. Although the experimentation has succeeded only on stored data, its real-time implementation can be considered thanks to both the simplicity of equations and the use of a token tracking board that allows the processing of 14 images per second. The algorithmic and hardware aspects of this board are briefly reported, and its robustness is demonstrated by a robotic experiment.

5.2 Introduction

Mobile robot navigation requires an explicit representation of the environment in which it moves, and this research's axis has been abundantly investigated in many laboratories. While many techniques for matching, merging, or computing and updating the 3-D structure exist, it is surprising to realize that many mobile robotic applications are based either on simple vision techniques or special sensors (ultrasonic, laser, sonar, etc.). Although some ambitious programs aim (and are succeeding) to achieve this task in real time, one must admit that the complexity of the specially designed hardware, as far as the unbelievable amount of software, cannot undergo a large-scale duplication.

If we analyze the complexity of these different approaches, it appears that most of them are based on global processing. For example, the important structural change in a stereo pair of images requires us to take into account the topological relations between features in a global matching procedure. The 3-D scene reconstruction is also performed as a global process in the sense that all matched tokens are reprojected in the 3-D scene without any limitation on features for which the uncertainty, constrained by the stereo baseline, is huge. The last consideration concerns the scene updating, which requires, once again, a 3-D global matching between the 3-D observations and the scene model.

The aim of this chapter is to stress that the determination of the mobile robot scene environment can be achieved in close real time by some local processing. This approach is based on the computation of a local map surrounding some feature points selected in the scene. These features are obtained by searching the scene for the points that present some interesting cue for robot navigation or represent a potential obstacle. The robot chooses the closest point and computes, while moving, the local surrounding map. When looking at another point, it reconstructs a new local map. These local maps are inserted into a global one.

Preliminary results that validate the approach on stored data are reported in Section 2, in which we present our robot navigation strategy with regard to the local map computation and some results obtained in a real robotic configuration. Even if this technique has been performed only on stored data, its real-time implementation can be considered by taking the following considerations into account.

The number of tokens in the local map surrounding the point of interest is rather small. This situation implies the camera is equiped with a long focal lens instead of the conventional wide angle.

The computation of the 3-D structure requires, as developed in Section 2.2, a limited number of basic operations.

The matching between two consecutive views can be achieved by tracking features in the temporal sequence of images.

The processing being fast, structural changes between two consecutive views are limited and facilitate the tracking process.

Because of the small number of features for which the 3-D structure has to be computed, as far as the apparent simplicity of equations, this task can be performed in close real time by conventional a digital signal processing (DSP) board. The matching can be achieved by a special token tracker board that allows the processing of at least 17 images per second. The algorithmic aspects and hardware implementation of this board are presented in the Section 3. The robustness of this algorithm has been intensively tested, before its hardware implementation, on a robotic experiment dealing with motion stereo. Tracking edge lines in a sequence of images taken by a camera mounted on a robot arm allows, taking the robot motion parameters into account, a 3-D estimation of edges. The results of this experiment are briefly reported.

5.3 Active Vision

5.3.1 Navigation Strategy

When a human observes a scene from a moving vehicle, his eye automatically follow a feature point moving relative to him, and after a while, they suddenly change attention to find and follow a new feature point. This nonvoluntary eye movement, called optokinetic nystagamus, has been explained as for stabilizing the image at the retina. Our robot perception strategy is based on a similar approach that aims to estimate the local computation of the 3-D structure around feature points tracked by the vision sensor. Such a feature point is called a fixation point.

This concept is closed to a new issue of robotics called active vision. Aloimonos [1] has presented some theoretical works in which a camera gazes a fixation point at its focal center while moving for estimating structure from motion, shading, and contour. Ballard [4, 5] has developed the animate vision system, a stereo pair of cameras in which vergence is controlled in order to assign the projection of the fixation point on the center of both cameras, and he has computed the neighborhood 3-D structure. Sandini [13] has also discussed the constraint imposed by the active motion for computing 3-D information.

The main idea of our technique is that such an active vision system is perfectly suitable for estimating the local 3-D structure in areas that present some interest for the mobile robot navigation. In order to minimize computations and memory space, and as a consequence, to allow easy hardware implementation, it seems important to focus only on one area of attention. The robot navigation can be then considered as a goal-oriented task in which each goal is represented by an area surrounding the fixation points detected during the robot motion.

The robot navigation strategy is based on two basic kinds of motion: a pure translation or a pure rotation around the fixation point. The active visual system, composed of a camera mounted on a rotating table accurately controlled by a computer, gazes a fixation point while the robot moves. During this robot motion, the projection of this fixation point is kept on the camera optical center, and the 3-D structure of tokens surrounded this point is computed. At each acquisition, this local map is updated by combining new data in order to increase accuracy and robustness. As soon this analysis is achieved and no more information is expected, a new fixation point is processed. Figure 5.1 represents a robot navigation scenario in which the robot looks respectively at points A and B before generating a pure rotation around point C and carrying on its trajectory as a set of translations and rotations.

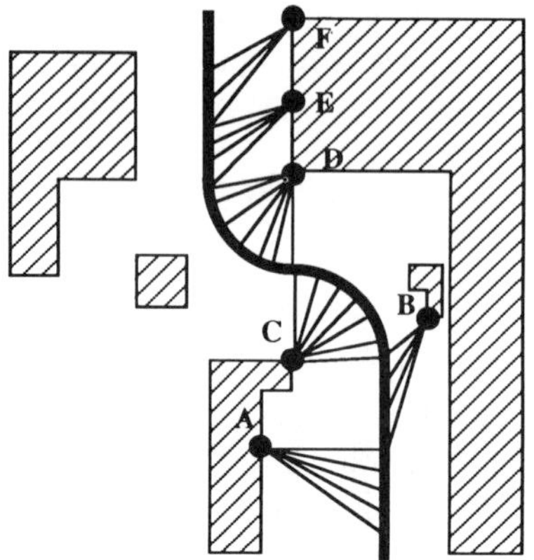

FIGURE 5.1. Top view of robot navigation and perception strategy.

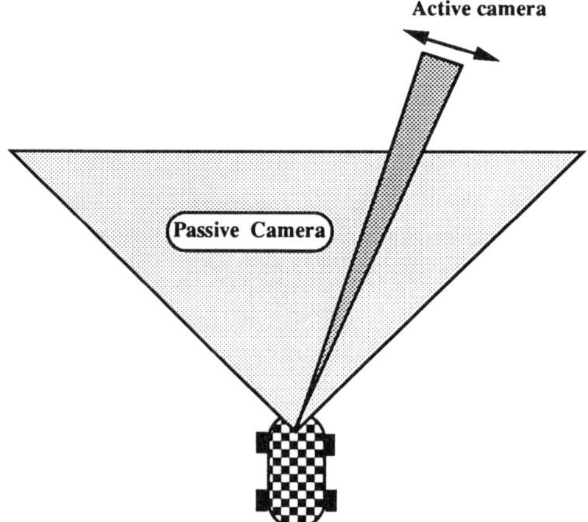

FIGURE 5.2. Collaboration between a passive and an active camera.

In the above-mentioned strategy, the selection of fixation points requires a lot of attention. Such a point must either provide a cue for robot navigation or represent a potential obstacle. With respect to these two constraints, we know that a fixation point has to be selected in the scene with respect to geometrical and photometric properties. Geometrical properties consist in the detection of areas with respect to the robot distance. Such an area is classified for determining the processing order. The closest area will be processed first and, after a while, the second nearest one, and so forth. The photometric properties, based on gradient, texture, and color analysis, consist in pointing out the fixation point inside the area, which allows efficient, robust, and easy tracking under a different angle of view or illumination.

For achieving the fixation point selection with respect to the above considerations, it seems useful to add a static camera to our robot as represented in Figure 5.2. Such a camera, equiped with a wide angle lens, allows the perception of all scene points in the image (the active camera's field of view is constrained to a small scene area by the long focus lens). The 3-D structure of all these scene points (including the scale factor, which corresponds to the distance to the fixation point) can be roughly estimated by a motion stereo technique.

The construction of the global map requires computation from both the active and passive cameras. The active camera provides an accurate 3-D structure estimation in the fixation-point centered coordinates. The global map is represented in a path-centered coordinate system [2]. The main feature of such a representation consists in attaching the origin center directly onto the robot path. Each fixation-point centered coordinate system is then inserted into the path-centered coordinates system. The relations are expressed as the scale factor that represented the distance along an axis perpendicular to the robot path as far as the horizontal distance between two consecutive fixation points.

5.3.2 Local Computation of 3-D Structure

Our method is based on the assumption that the robot moves along linear or circular paths around feature points. The robot is equiped with only one camera mounted on a rotating table, and the 3-D structure of scene points is computed up to a scale factor. The table controller uses visual feedback for keeping the fixation point at the image center while the robot moves. Only the 3-D structure of vertical edge lines is computed because of the lack of a three-axis rotation controller in our experiments. Such a limitation requires us to estimate scene point locations in a 2-D top view representation. Although such information can be sufficient in an indoor scene environment composed of many vertical lines, an extension of equations developed in this chapter in the general case can be performed in a similar way.

Let us assume a coordinate system X, Y, Z attached with respect to the camera, with the Z axis pointing along the optical axis. The translation component of the camera motion is $T = (U, V, W)^T$ and the rotational component $R = (A, B, C)^T$. Let $P = (X, Y, Z)^T$, the instantaneous coordinates of a point in the environment. The perpendicular component u of the instantaneous velocity is expressed by the following well-known equation [6]:

$$u = \frac{fX'}{Z} - \frac{fXZ'}{Z^2} = f\left(-\frac{U}{1} - B + \frac{Cy}{f}\right) - x\left(-\frac{W}{Z} - \frac{Ay}{f} + \frac{Bx}{f}\right), \quad (5.1)$$

in which f represents the camera focus length and (x, y) is the image point position.

Let us assume the robot moves linearly. By controlling the panning of the camera, the fixation point always appears at the optical image center. Locations of scene points relative to the fixation point are estimated while the robot moves. The 3-D structure computation is based on an intermediate cylindrical representation (θ and L), which facilitates the explanations. As

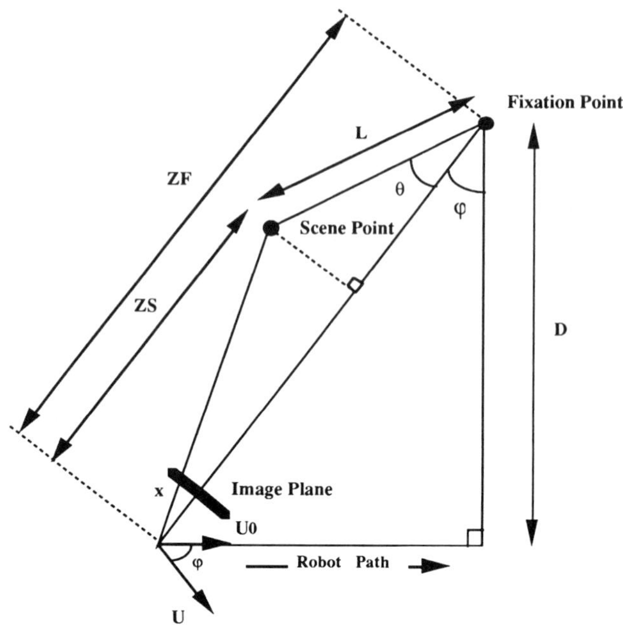

FIGURE 5.3. Geometrical relations among the fixation point, a scene point, and the robot vision coordinate system.

indicated in Figure 5.3, L represents the distance between the fixation point and the object point, and θ is the angle that positions the camera focus, fixation point, and object point. As explained, we consider a top view and do not deal with the height of scene points. Since the robot moves on a flat floor, the camera motion parameters are expressed by the following vectors: $R = (0, \omega, 0)^T$ and $\mathbf{T} = (U_0 \cos \varphi, 0, U_0 \sin \varphi)^T$ in which U_0 represents the linear velocity of the robot, ω is the camera's angular velocity, and φ is the angle between the camera axis and a perpendicular to the robot path. Equation (5.1) can then be written as

$$u = -\frac{U_0(x \sin \varphi - f \cos \varphi)}{Z} - \omega\left(f + \frac{x^2}{f}\right). \tag{5.2}$$

Since the fixation point is fixed at the image center while the robot moves, both the optical flow component and position are equal to zero. From Equation (5.2), we can compute the distance, noted ZF:

$$ZF = -\frac{U_0 \cos \varphi}{\omega}.$$

The distance ZS for all points in the field of view of the camera is also computed from Equation (5.2):

$$ZS = \frac{U_0(x \sin \varphi - f \cos \varphi)}{[u + f\omega + \omega(x^2/f)]},$$

and the computation of the ratio ZS/ZF, noted β, allows the elimination of the robot velocity U_0:

$$\frac{ZS}{ZF} = \beta = \frac{f - x\,tg\varphi}{[(u/\omega) + f + (x^2/f)]}.$$

The scene point cylindrical coordinates are obtained by basic geometric relations (see Figure 5.3.):

$$\theta = tg^{-1}\left(\frac{x}{f}\frac{\beta}{1-\beta}\right), \tag{5.3}$$

$$L = \frac{1-\beta}{D\cos\theta}\cos\varphi. \tag{5.4}$$

As mentioned, the 3-D structure of scene points is estimated relative to the fixation point. If some external sensors can be used for accurately determining the distance between the fixation point and the camera [13], we prefer to consider that all scene point coordinates are expressed up to a scale factor represented by D in Equation (5.4). Although such a scale factor should limit the interest of our technique, it can be estimated with a sufficient accuracy by a motion stereo technique.

In the case of rotational motion, it has been demonstrated [10] that we obtain the same equation for θ, while the angle φ is set to zero when computing L. But, in both the cases, Equation 5.3 cannot be computed when $\beta = 1$. This situation occurs when the scene point is located along a line parallel to the image plane and passing through the fixation point. However, the determination of the distance L can be directly computed by using a basic trigonometric relation ($x/f = L/ZF$), and the sign of the angle θ, which is equal to $\pi/2$, is provided directly by looking at the sign of x.

While moving, the 3-D structures of all visible scene points are expressed in the fixation point centered coordinates system. By looking at Figure 5.4,

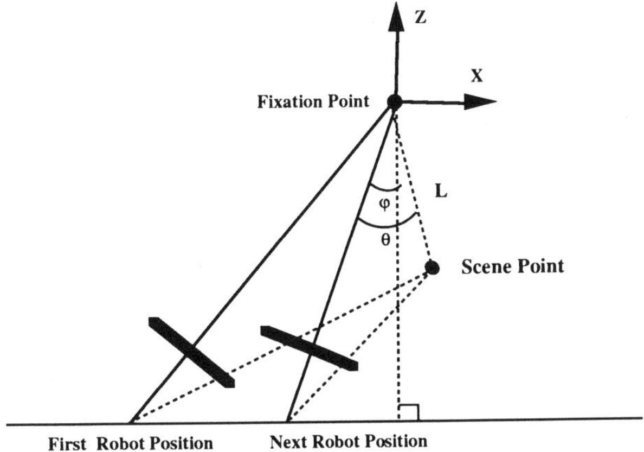

FIGURE 5.4. Geometrical relations between two consecutive positions.

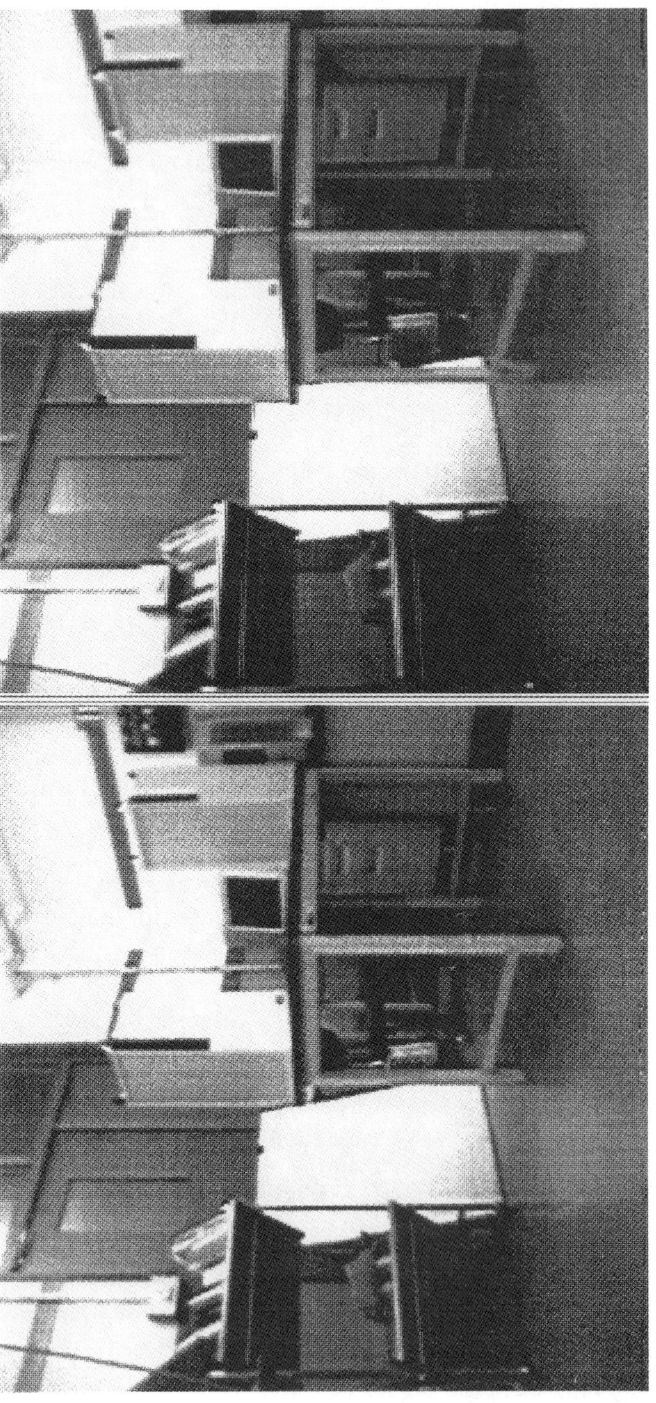
(a)

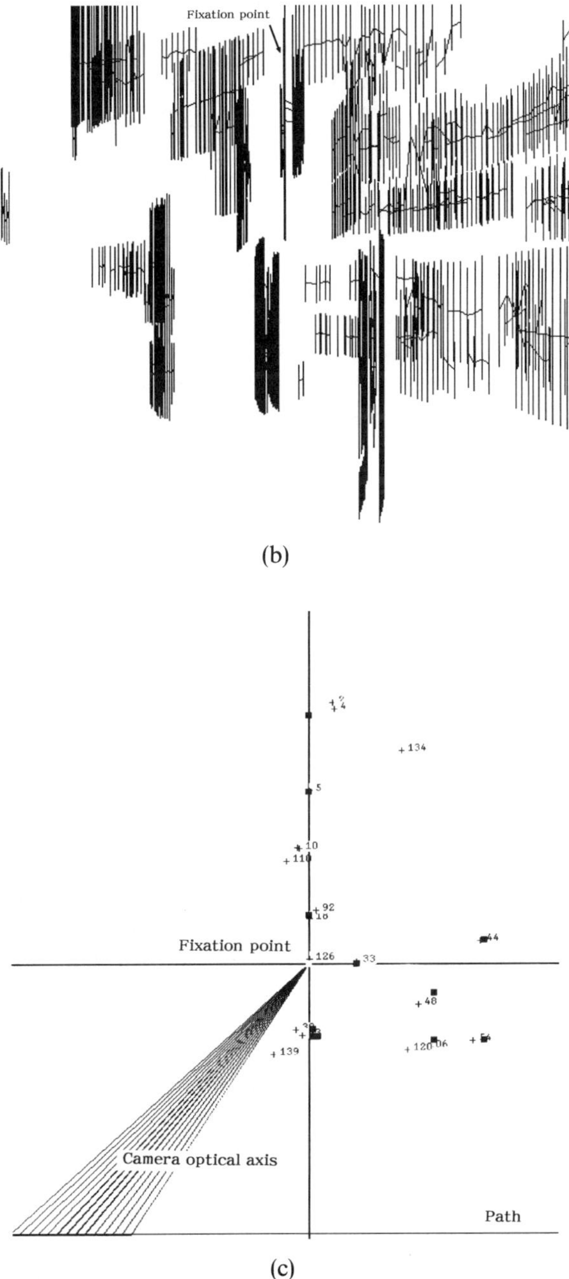

FIGURE 5.5. (a) On page 136, the first and last raw images of a sequence (The fixation center is always located on the center of the image); (b) concatenation of vertical detected in the images; (c) top view representing the robot path, the fixation point, and some scene points.

which represents the robot in two consecutive positions, the Cartesian coordinates can be deduced by basic trigonometric relations:

$$X = L \cos\left(\theta - \varphi - \frac{\pi}{2}\right),$$

$$Z = L \sin\left(\theta - \varphi - \frac{\pi}{2}\right).$$

5.3.3 Preliminary Results

The following experimental assessments have been performed in the case of linear robot motion. A camera is mounted on a rotating platform that swivels with a step of 0.1°. During the robot displacement, the platform is controlled in order for the camera to track the fixation point. The first and last images of a sequence are reprsented on Figure 5.5(a), the robot being successively located at 17 different positions. Between each observation, the robot moves a distance of 5 cm. The concatenation of all edge line segments, obtained by a Sobel operator and a vertical histogram detection, of the sequence (Figure 5.5(b)) points that out.

The fixation point is always located on the image focal center.
Edge line velocity is larger when the scene point is located either far from the fixation point or close to the camera.
Noise and other perturbations considerably modify the edge line representation.

The tracking process is illustrated on figure 5.5(c), which represents all edge lines that have been tracked during the first five images. This figure shows that 25 edge segments have been tracked, the number of false matches being limited to 1 or 2. For each pair of images of the sequence (images 1 and 6, 2 and 7, 3 and 8, …, 12 and 17), we compute the 3-D structure with respect to the equations described in Section 2.2. All these 3-D observations are then merged in order to decrease uncertainty and improve robustness (the false matches will not be integrated in the 3-D local map). Figure 5.5(c) indicates the final top scene view in which both the computed and real scene point positions have been represented. The fixation point is located at the junction of the vertical and horizontal axes, while the bottom horizontal line indicates the robot trajectory. A more qualitative evaluation is reported in Table 5.1, which provides an analytic representation of some scene points and points out that the accuracy is better in the close vicinity of the fixation point as stressed in Ishiguro [10]. In this evaluation, the scale factor has been assigned with the real distance in order to facilitate the comparison. We see that the error does not exceed 1.8 cm when the scene point is located at less than 1 m from the fixation point if we ignore the point labeled 48, which have been merged only 3 times (the column number of observations indicates the number of 3-D observations that have been combined).

TABLE 5.1. Analytic representation of error.

	Linear motion[a]						
Point (label)	X meas (cm)	Z meas (cm)	L meas (cm)	Number of observations	X real (cm)	Z real (cm)	Error (cm)
54	95	−42	141	4	89.5	−42.6	5.5
4	0	137	137	5	14.4	140.5	2
44	95	13	96	3	93.5	12.6	1.5
5	0	95	95	9	0.7	95.7	1
48	68	−16	70	3	59.7	−22.5	10.5
43	−2	−40	40	11	−3.6	−39.7	1.62
18	0	27	27	5	0.8	25.3	1.8
33	26	0	26	5	26.3	1.4	1.4

[a] $D = 104$ cm, $U = 10$ cm/frame, $f = 591$ pixels, and $t = 10$ frames.

5.4 Tracking

5.4.1 Tracking for Matching

As developed in the introduction, the main idea is that local processing can be used for both avoiding the cost of a global stereo matching algorithm and collapsing the cost of matching recovered 3-D segments to update a 3-D model.

In fact, structures from a dense set of images may be matched with a simple linear complexity algorithm. Gennery [9] has demonstrated that the measurement of the motion points in an image sequence can be based on a Kalman filter and Matthies [11] has succeeded to recover the depth from a lateral displacement of points also using a Kalman filter.

The token tracker is based on the assumption that if observations are matched between two consecutive images with sufficiently small time delay the matching can be based only on the prediction of their attributes. Furthermore, if the matching is based only on spatial attributes, then it can be done with a very simple and fast process.

The overview of this process is represented in Figure 5.6. The tokens extracted from each image, referred to as the "observation," are composed of a list of line segments expressed in a suitable parametric representation.

Tokens from each observation are used to update the image flow model composed of active tokens. Each active token is represented by geometric and dynamic attributes (position of the line center, length, orientation, and their corresponding velocity) with regard to its covariance matrixes.

Matching between a new observation and the image flow model requires only a local measurement based on the predicted position, orientation, and length of tokens. For each token, the set of observed tokens is tested for similar orientation, colinearity, and overlap. The similarity is verified if the difference between the observation and the flow model token's attribute is less than a standard deviation of three.

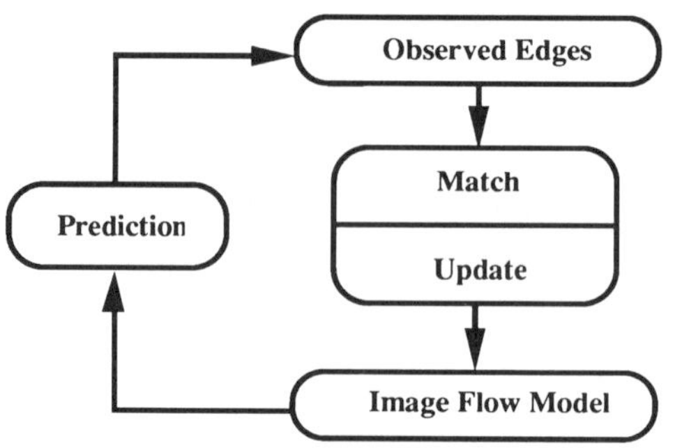

FIGURE 5.6. Overview of the updating process.

The update formula for the attributes and their predictive estimation and uncertainty are based on a simplified form of the Kalman filter. The simplified form of the Kalman filter is made possible by a specific token parametric representation that allows the independency of each geometric attribute.

5.4.2 Hardware Design

The independency of each geometric attribute allows us to consider each token as a vector:

$$\mathbf{T}^t = \{t_1, t_2, t_3, t_4\},$$

in which t_i represents successively the position, orientation, and length. As a consequence, each parameter can be processed independently, and the flow model of a token can be expressed as a set of flow models. The matrix inversions inherent in the classic Kalman filter are then avoided, and computation time is optimized by implementing four Kalman filters processing in parallel.

Such an approach considerably limits the number of basic operations. It has been demonstrated that the implementation of such a basic Kalman filter requires only 3 additions, 1 subtraction and 2 multiplications [14]. Because of the simplicity of these operations, the cost of tracking is very small, and the main bottleneck is the matching. Nevertheless, the matching cost can be also drastically reduced by dividing observed tokens into several subsets. Instead of matching the observed token with the N model tokens, the search is limited only to tokens that belong to the same subset. It has been demonstrated [8] that such a subset can be efficiently defined with respect to edge orientation.

This simple memory organization, as far as the few basic operations in the Kalman filter computation are concerned, allows us to implement this algorithm on a single VME board composed of a digital signal processor. Such a board allows the tracking of 170 tokens between 2 consecutive images in less than 30 ms. By extrapolation, one can consider that 250 tokens can be tracked at a rate of 10 images per second.

5.4.3 Tracking for Motion Stereo

The first application, based on the previously described board, consists of recovering the 3-D structure of a scene from a camera mounted on a gripper of a robot manipulator [7]. In such an application, tracking edge lines viewed by the camera provide some useful capabilities. First of all, the image description of the flow model is less sensitive to image noise and robot vibrations than any individual image. The second property is that the label of edge lines in the flow model provides a correspondence of image tokens for different views of the same scene.

Another advantage of tracking is the capability of avoiding a 3-D matching when updating the 3-D model of the scene. Ayache [3] has proposed such a 3-D matching based on geometrical criteria for combining data provided by a moving robot from two consecutive positions. Ramaparany [12] has developed a retroprojection technique in which data are characterized by a large uncertainty in depth. This uncertainty represents the possible area in which its correspondent in the next images might be found.

For avoiding such a 3-D matching, Suzuki [15] has proposed to solve simultaneously the problem in the 2-D and 3-D space by combining stereo correspondence and robot motion. But, such an approach is still computationally expensive.

In our approach, this 3-D matching is avoided by assigning the same label that characterizes each segment in the token tracking process to the 3-D reconstructed segment. The correspondence between different points of view is immediately available by looking for 3-D segments sharing the same label.

Figure 5.7(a) represents the first and last image of 70 acquisition views taken by the camera mounted on the gripper of a robot arm. The object is located at a distance of about 30 cm from the camera, and the camera displacement between two consecutive acquisitions is less than 1 cm with rotations under 5°. Figure 5.7(b) indicates the image couples 20 and 25 and the matching result. These figures represent tokens, tracked from image 30 up to image 35, which are represented by the same label. The 3-D reconstructed segments for the image couples 30 and 35 and 35 and 40 are displayed in Figure 5.7(c), respectively, on the left and right sides. Each reconstruction corresponds to the same physical scene viewed by a different point of view but represented in the same frame coordinates. One can check that the same physical segment is represented by a similar label in the 3-D reconstructed images.

The superposition of the 14 files obtained by reconstructing each of the 5 images along the sequence of 70 views is represented in Figure 5.7(d). This view points out a slight dispersion of the results but demonstrates the coherence of the obtained data. By merging all 3-D segments that share the same label and without using any matching procedure, we obtain the final model represented in Figure 5.7(e). The validation of the result is demonstrated in

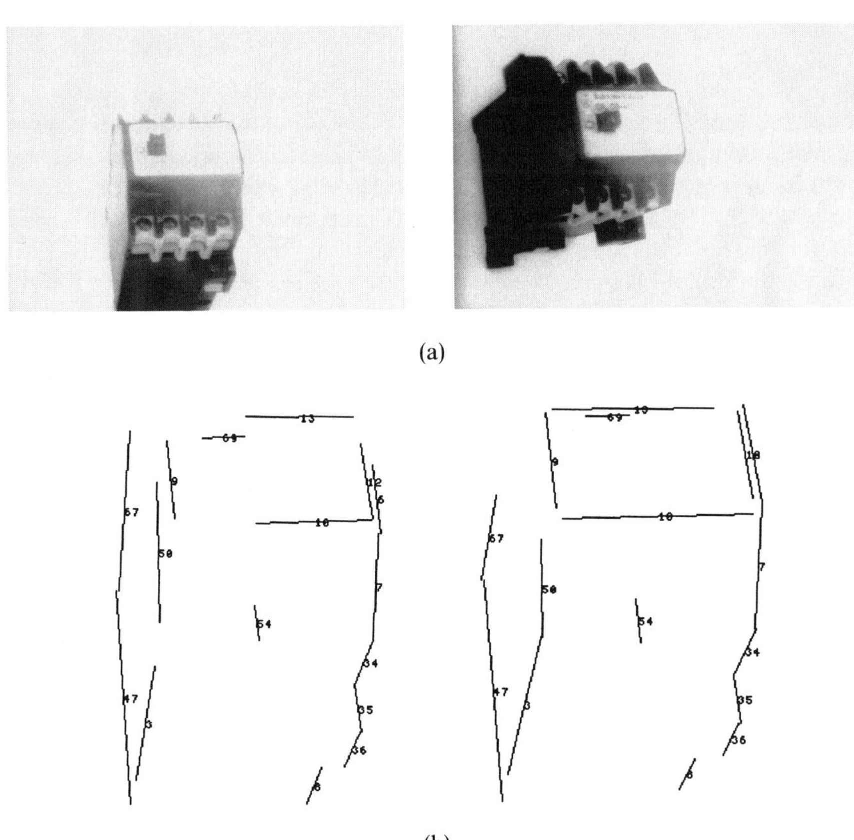

FIGURE 5.7. (a) First and last raw images of the sequence; (b) matching between a couple of images; (c) 30 reconstructed segments; (d) superposition of 14 different 3-D reconstructions in a common coordinate system; (e) merging of the 14 3-D reconstructions; (f) display of the final 30 data on a raw image.

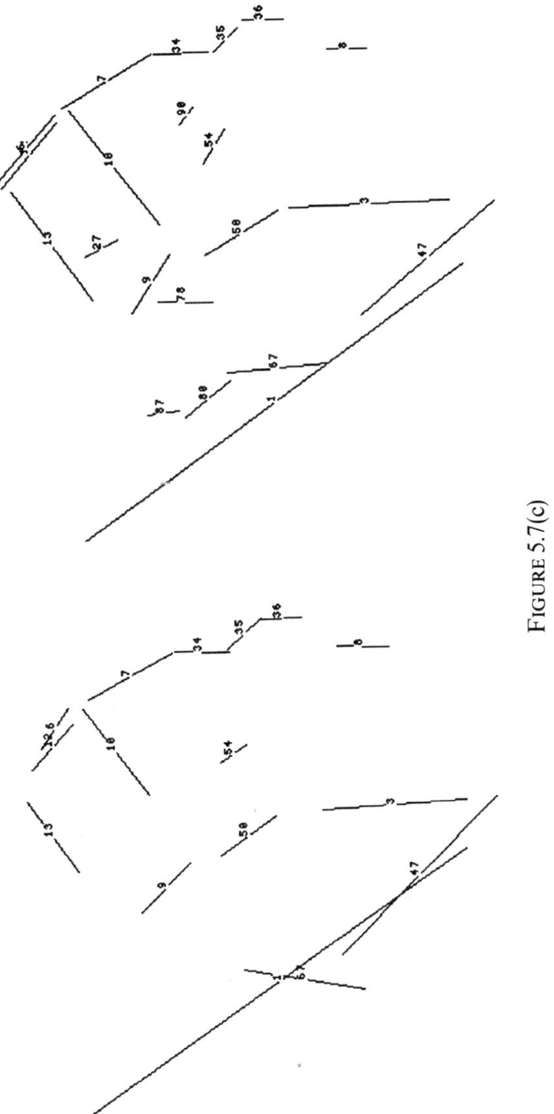

FIGURE 5.7(c)

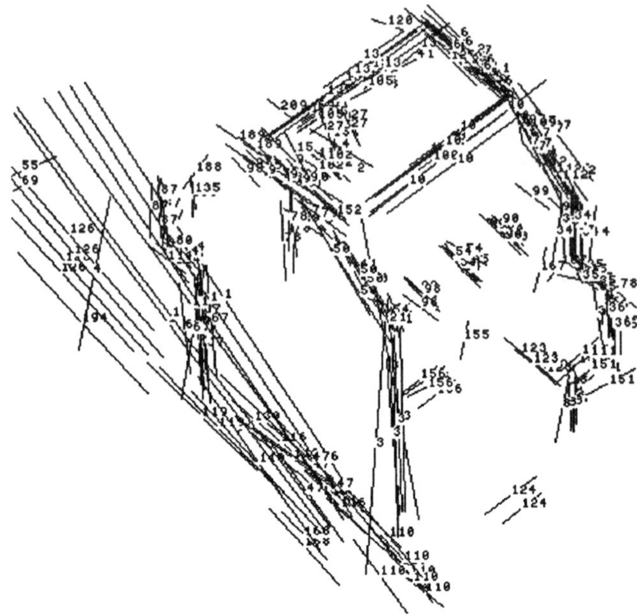

FIGURE 5.7(d)

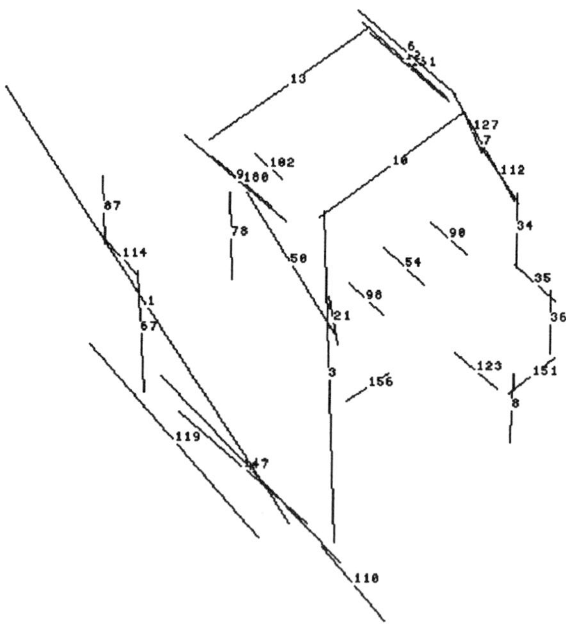

FIGURE 5.7(e)

FIGURE 5.7(f)

Figure 5.7(f), which consists of reprojecting this model on one of the images of
of the sequence.

5.5 Conclusion

In this chapter, we develop the idea that the 3-D structure computation for a
mobile robot can be achieved in close real time by local processing. This
approach is based on computing local maps surrounding some points of interest detected in the scene and by tracking all image features on the basis of
local criteria. Such an approach avoids the expensive computation of stereo
matching correspondences by integrating consecutive images with sufficiently
small time delay. The label attached to each token allows the determination of
correspondences for different views of the same scene. Another property of
such an approach is the capability of avoiding the 3-D matching between the
scene model and the 3-D observation. Indeed, by assigning to the 3-D token
the same label as that in the image, the correspondence is immediately available by looking for 3-D elements that share the same label.

The major contribution of this chapter is related to the computation of
local maps in the vicinity of some feature points instead of performing the
usual overall 3-D reconstruction for all tokens viewed in a stereo pair of
images. Such an approach consists first of all in selecting some points of
interest in the scene by a static and wide angular range camera. The determination of such points, called fixation points, is based on photometric and
geometric properties. The system then accurately computes the 3-D structure
surrounding these points by using an active camera system. This system must
assign the projection of these fixation points on the camera optical center

while the robot is moving. The 3-D structure of scene points surrounding the fixation point is then computed and updated at each acquisition.

By restricting the 3-D computation to a few scene elements, the computational time can be very limited because of the apparent simplicity of equations developed in Section 2.2. Furthermore, by using a special hardware board for tracking edge lines, one can expect to compute the 3-D structure in close real time. The algorithmic and hardware aspects of this board, which tracks 170 tokens in less than 30 ms, is briefly developed. The robustness of the algorithm is demonstrated on a robotic application that aims to determine the 3-D structure in motion stereo.

References

[1] Aloimonos, J. Bandyopadhyay, and Weiss, A. (1987). "Active Vision." *Proc. Image Understanding Workshop*, 552–573.

[2] Asada, M., Fukui, Y., and Tsuji, S. (1988). "Representing Global World of A Mobile Robot with Relational Local Maps." *Proc. IEEE Int'l. Workshop on Intelligent Robots and Systems*, 199–204.

[3] Ayache, A., Faugeras, O. (1987). "Maintaining Representation of the Environment of a Mobile Robot." *Proc. International Symposium on Robotics Research, Santa Cruz, California, August 1987.*

[4] Ballard, D. H., and Ozcandari, A. (1988). "Eye Fixation and Early Vision: Kinetic Depth." *Proc. 2nd IEEE Int'l. Conf. Computer Vision*, 524–531.

[5] Ballard, D. H. (1989). "Reference Frames for Animated Vision.' *Proc. 11th Int. Joint Conf. on Artificial Intelligence*, 1635–1641.

[6] Bruss, A. R., and Horn, B. K. P. (1983). "Passive Navigation." *Computer Vision, Graphics, and Image Processing* **21**, 3–20.

[7] Crowley, J. L., and Stelmaszyk, P. (1990). "Measurement and Integration of 3-D structures by Tracking Edge Lines." *First European Conf. on Computer Vision (ECCV90)*. Antibes, *France, April 23–27, 1990.*

[8] DePaoli, S., Chehikian, A., and Stelmaszyk, P. (1990). "Real Time Token Tracker." *European Signal Processing Conference EUPISCO 90. Sept 18–21, 1990, Barcelona, Spain.*

[9] Gennery, D. B. (1982). "Tracking Known Three-Dimensional Objects." *Proc. of the National Conference on Artificial Intelligence (AAAI-82), Pittsburgh 1982.*

[10] Ishiguro, H., Stelmaszyk, P., and Tsuji, S. (1990). "Acquiring 3-D Structure by Controlling Visual Attention of a Mobile Robot." *IEEE Int. Conference on Robotics and Automation. May 13–18 1990.*

[11] Matthies, L., Szeliski, R., and Kanade, T. (1987). "Kalman Filter-Based Algorithms for Estimating Depth from Image Sequences." *CMU Technical Report CMU-CS-87-185, December.*

[12] Ramparany, F. (1989). "Perception Multi-sensorielle de la Structure Geometrique d'une scene." PhD. Thesis, INPGrenoble.

[13] Sandini, G., and Tistarelli, M. (1990). "Active Tracking Strategy for Monocular Depth Inference over Multiple Frames." *IEEE PAMI* **12**(1).

[14] Stelmaszyk, P., Discours, C., and Chehikian, A. "A Fast and Reliable Token Tracker." *IAPR Workshop on Computer Vision. Tokyo (Japan), October 12–14* 1988.
[15] Suzuki, K., and Yachida, M. (1989). "Establishing Correspondence and Getting 3-D Information in Dynamic Images." *On Electronic Information and Communication Journal,* **J72-D-II**(5), 686–695, May (in Japanese).

6
Object Detection Using Model-based Prediction and Motion Parallax

STEFAN CARLSSON AND JAN-OLOF EKLUNDH

6.1 Motion Parallax and Object Background Separation

When a visual observer moves forward, the projections of the objects in the scene will move over the visual image. If an object extends vertically from the ground, its image will move differently from the immediate background. This difference is called motion parallax [1, 2]. Much work in automatic visual navigation and obstacle detection has been concerned with computing motion fields or more or less complete 3-D information about the scene [3–5]. These approaches, in general, assume a very unconstrained environment and motion. If the environment is constrained, for example, motion occurs on a planar road, then this information can be exploited to give more direct solutions to, for example, obstacle detection [6]. Figure 6.1 shows superposed the images from two successive times for an observer translating relative to a planar road. The arrows show the displacement field, that is, the transformation of the image points between the successive time points.

Figure 6.1 also shows a vertically extended object at time t and t'. Note that the top of the object is displaced quite differently from the immediate road background. This effect is illustrated by using the displacement field of the road to displace the object. A clear difference between the actual image and the predicted image is observable for the object. This fact forms the basis of our approach to object detection (Figure 6.2). For a camera moving relative to a planar surface, the image transformation of the surface is computed and used to predict the whole image. All points in the image that are not on the planar surface will then be erroneously predicted. If there is intensity contrast at those parts, we will get an error in the predicted image intensity. This error then indicates locations of vertically extended objects.

6.2 Image Transformation for Motion Relative to a Planar Surface

With a moving camera, each point in the scene will map to a different point in the image at different times. The transformation of the mapped image point over time is determined by the motion of the camera and the position of the

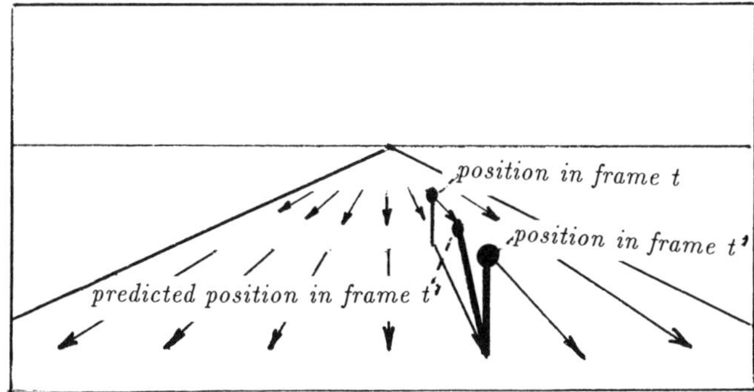

FIGURE 6.1. Displacement field from road with predicted and actual position of vertically extended object.

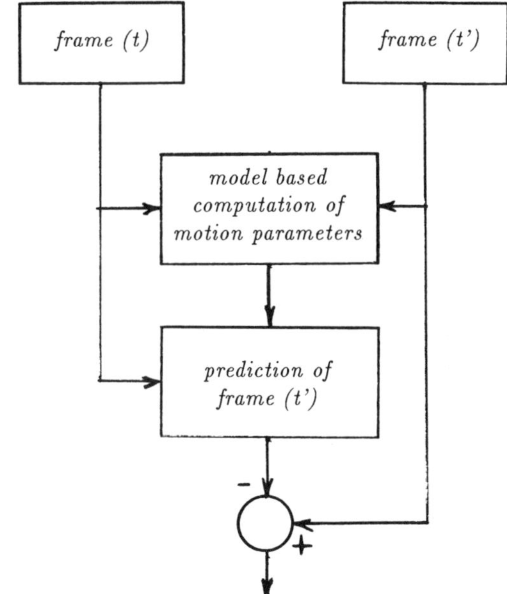

FIGURE 6.2. Block diagram of processing for vertical object detection.

point in the three-dimensional scene. If the point is on a planar surface, the transformation can be computed using the camera motion and position of the surface in space. Figure 6.3 shows the coordinate system of the camera and the image plane. A rigid displacement of the camera can be decomposed into a translation with components D_X, D_Y, D_Z along the coordinates and a rotation around an axis passing through the point of projection, which can be decomposed into rotations around the axis of the coordinate system ϕ_X, ϕ_Y, ϕ_Z. Assuming small rotations, a point in the scene with coordinates X, Y, Z is then transformed to the point X', Y', Z', where

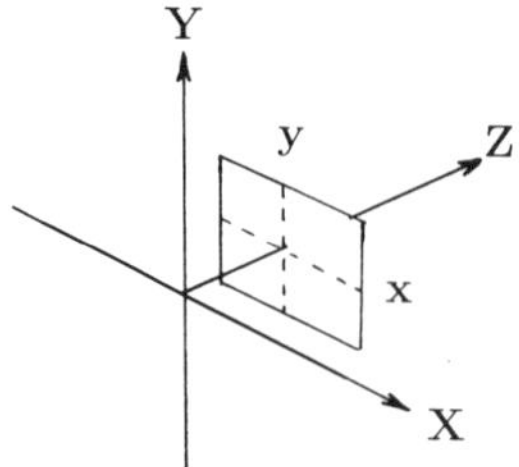

FIGURE 6.3. Coordinate system of camera and image plane.

$$\begin{pmatrix} X' \\ Y' \\ Z' \end{pmatrix} = \begin{pmatrix} 1 & -\phi_Z & \phi_Y \\ \phi_Z & 1 & -\phi_X \\ -\phi_Y & \phi_X & 1 \end{pmatrix} \begin{pmatrix} X \\ Y \\ Z \end{pmatrix} + \begin{pmatrix} D_X \\ D_Y \\ D_Z \end{pmatrix}. \tag{6.1}$$

If the image plane is located at the unit distance from the point of projection, the image coordinates (x, y) of a point X, Y, Z under perspective projection are

$$x = X/Z, \qquad y = Y/Z. \tag{6.2}$$

The transformation of the projected image point of a point in the scene with depth Z is therefore [7]:

$$x' = \frac{x - \phi_Z y + \phi_Y + (D_X/Z)}{1 - \phi_Y x + \phi_X y + (D_Z/Z)}, \tag{6.3a}$$

$$y' = \frac{y + \phi_Z x - \phi_X + (D_Y/Z)}{1 - \phi_Y x + \phi_X y + (D_Z/Z)}. \tag{6.3b}$$

If the point X, Y, Z is located on a planar surface with equation $K_X X + K_Y Y + K_Z Z = 1$, the transformation in the image plane the becomes

$$x' = \frac{(1 + K_X D_X)x + (K_Y D_X - \phi_Z)y + \phi_Y + D_X K_Z}{(1 + K_X D_Z - \phi_Y)x + (K_Y D_Z + \phi_X)y + K_Z D_Z}, \tag{6.4a}$$

$$y' = \frac{(1 + K_Y D_Y)y + (K_X D_Y - \phi_Z)x - \phi_X + D_Y K_Z}{(1 + K_X D_Z - \phi_Y)x + (K_Y D_Z + \phi_X)y + K_Z D_Z}, \tag{6.4b}$$

This is a nonlinear transformation of the image coordinates determined by nine parameters. The actual number of degrees of freedom of the transformation is, however, just eight, since parameters **K** and **D** always occur as products, which means that their absolute values are irrelevant.

6.3 Estimation of Parameters by Minimization of Prediction Error

The transformation of the projected image points due to the motion of the camera will manifest itself as a transformation of the image intensity $I(x, y)$. If t and t' are the time instants before and after the transformation, we shall assume that

$$I(x', y', t') = I(x, y, t) \tag{6.5}$$

where x, y and x', y' are related according to Equation (6.4). That is, we assume that the transformation of the image intensity is completely determined by the geometric transformation of the image points. This is not strictly true in general since we neglect factors as changing illumination, etc.

For points on a planar surface, our assumption implies that the transformation of the intensity is determined by the three vectors $\boldsymbol{\phi}$, $\mathbf{D}$, characterizing the camera motion) and $\mathbf{K}$, surface orientation. The determination of these parameters can therefore be formulated as the problem of minimizing the prediction error:

$$P(\bar{\phi}, \bar{D}, \bar{K}) = \sum_{x,y} \{I[x'(x, y, \boldsymbol{\phi}, \mathbf{D}, \mathbf{K}), y'(x, y, \boldsymbol{\phi}, \mathbf{D}, \mathbf{K}), t'] - I(x, y, t)\}^2 \tag{6.6}$$

where the summation is over image coordinates containing the planar surface.

For the minimization, we use gradient descent, that is, the values of the parameters are adjusted iteratively according to

$$\boldsymbol{\phi}^{(i+1)} = \boldsymbol{\phi}^{(i)} - \mu_1 \frac{\partial P^{(i+1)}}{\bar{\phi}},$$

$$\mathbf{D}^{(i+1)} = \mathbf{D}^{(i)} - \mu_2 \frac{\partial P^{(i+1)}}{\partial \mathbf{D}}, \tag{6.7}$$

$$\mathbf{K}^{(i+1)} = \mathbf{K}^{(i)} - \mu_3 \frac{\partial P^{(i+1)}}{\partial \mathbf{K}},$$

where i denotes the iteration index. The derivatives with respect to the parameter vectors are taken componentwise.

The convergence properties of the gradient descent minimization depend heavily on the structure of the image intensity function $I(x, y)$. Assume that we have computed the approximate values $\hat{\phi}$, $\hat{D}$, $\hat{K}$ for the parameters. These values transform the point (x, y) to the point $(\hat{x}', \hat{y}') = (x' + \delta x', y' + \delta y')$. If the approximate parameter values are close enough to the true values, $\delta x'$ and $\delta y'$ will be small. For the prediction error, we then have

$$e = I(\hat{x}', \hat{y}', t') - I(x, y, t) = I(x' + \delta x', y' + \delta y', t') - I(x, y, t)$$

$$\approx \frac{\partial I}{\partial x'} \delta x' + \frac{\partial I}{\partial y'} \delta y'. \tag{6.8}$$

In order to compute well-defined values of motion and surface orientation parameters, the prediction error e should be sensitive to small changes in these. From Equation (6.8), we see that the size of the image intensity gradient is very important in this respect since it directly amplifies any variations in $\delta x'$, $\delta y'$. Preferably, image points x, y with high-intensity gradients should be used in the computation of the prediction error in Equation (6.6). The selective use of points with high gradients also reduces the volume of the computations involved. For our application, we therefore first applied a version of the Canny–Deriche edge detector to the image [8, 9] and used only the edge points. The choice of edge detector for this problem is probably not critical. What is needed is just a selection of points with high-intensity gradients.

The derivatives of the prediction error P with respect to the parameters were computed by systematically varying the parameters. If the difference between the transformed coordinates x', y' and the original x, y is small, for example, by choosing the time interval $t' - t$ to be small, these derivatives could be computed more efficiently using Equation (6.8). In that case, direct methods of determination of the parameters could be used [10].

6.4 Sequential Estimation Using Recorded Sequence

Since the algorithm assumes motion relative to a planar surface, we first have to select points in the image, projected from the road, to be used for parameter estimation. Under normal driving conditions, the part of the image immediately in front of the car can be assumed to project from the planar road surface. The position of the car relative to the road boundaries can also be considered as relatively stable. The points to be used in the algorithm are therefore selected from a rectangular window in the image chosen so that points from the road immediately in front of the car are contained in the window as shown in Figure 6.4. This window is fixed in time relative to the image. As obstacles are detected, the window could be made adaptive so that these objects are not included in the pixels used for parameter estimation.

The gradient descent algorithm for estimation of parameters can now be applied to successive image frames in the sequence. If the time between the

FIGURE 6.4. Window for selection of points on planar road surface.

successive frames is short enough, we can expect a high correlation in time between computed parameter values. This can be exploited in the gradient descent algorithm by using parameter values computed in the previous frame pair as start values for the next pair. For an ideal planar road surface, there will in fact be a coupling between motion parameters, $\bar{\phi}$, $\bar{D}$ and surface orientation $\bar{K}$, since the change in surface orientation over time is determined by the motion. This can be introduced as an extra constraint in the algorithm in order to build a true spatiotemporal model of the position and orientation of the camera relative to the road. At this stage, however, we did not consider this coupling between parameters.

6.5 Object Detection Using Prediction Error

If the estimated parameters $\hat{\phi}$, $\hat{D}$, and $\hat{K}$ are correct and the road conforms to surface model, the error in the prediction of the according to Equation (6.8) will be 0. Any errors in the estimated parameter values or errors in the model will, however, give rise to a nonzero prediction error. Any vertically extended object in the scene will obviously violate the planar surface model and thereby cause a prediction error. The prediction error is thus an important variable to be used for deciding whether any objects are present in front of the car.

The effect of a vertically extended object on the prediction error is, however, highly dependent on its position in the scene. If we consider the ideal case of a camera aligned with optical axis parallel to the planar surface, translating in the direction of the optical axis only with no rotation, we have for a point in the scene at depth Z the following transformation in the image:

$$x' = \frac{x}{1 + D_Z/Z} \qquad y' = \frac{y}{1 + D_Z/Z} \qquad (6.9)$$

If we choose units so that the height of the camera over the ground, $-1/K_Y = 1$ we have for points on the planar road surface:

$$x^m = \frac{x}{1 - D_Z y} \qquad y^m = \frac{y}{1 - D_Z y}. \qquad (6.10)$$

A point in the image with coordinates x, y at height h above the road will be at depth $Z = -1 + h/y$. For this point, the difference between the actual coordinates and those predicted by the model is

$$x' - x^m = \frac{hD_Z xy}{(1 - D_Z y - h)(1 - D_Z y)}, \qquad (6.11a)$$

$$y' - y^m = \frac{hD_Z y^2}{(1 - D_Z y - h)(1 - D_Z y)}. \qquad (6.11b)$$

Note that this applies only to points x, y in the image that project to the road surface. This means that they are below the horison, that is, $y < 0$ and $h < 1 - D_Z y$.

We see that the error in the prediction of the coordinates grows monotonically with height h above the ground. However, it also grows with distance from the focus of expansion (FOE) $x = 0$, $y = 0$. Objects close to the FOE will therefore give rise to very small errors in the coordinates. This is natural since motion at the FOE is zero, independent of the object's coordinates in space. If we use scene coordinates X and Z instead of image coordinates, we get for the coordinate error

$$x' - x^m = \frac{hD_Z X}{(Z + D_Z)(Z + D_Z - D_Z h)},\tag{6.12a}$$

$$y' - y^m = \frac{h(h - 1)D_Z}{(Z + D_Z)(Z + D_Z - D_Z h)}.\tag{6.12b}$$

From this, we see that the error scales inversely with depth, that is, the distance in front of the camera. It also grows with the distance X to the side of the optical axis and with D_Z, the translatory displacement along the optical axis.

For small errors in the coordinates, we can estimate the error in the predicted image intensity by projecting the coordinate errors on the intensity gradient according to Equation (6.8). This means that errors in the predicted image intensity will only show up at the edges of the vertically extended objects, unless the errors in the predicted coordinates are large enough. Important to note is also the fact that the orientation of the edges of the vertically extended objects influence the size of the error in the predicted image intensity. From Equation 6.11, we see that $\delta x'$, $\delta y'$ will be oriented radially out from the focus of expansion. For maximum prediction error according to Equation (6.8), the gradient should be parallel to this orientation, that is, the edges should be orthogonal to the lines radiating out from the focus of expansion. However, we must emphasize that this only applies when coordinate errors $\delta x'$, $\delta y'$ are small.

Another cause of prediction error is errors in the estimated parameters. For these errors, we also have the dependence on the distance from the FOE. This is important to consider, for example, in the evaluation of false alarm detections.

6.6 Experimental Results and Conclusions

The algorithm for sequential model-based motion estimation and object detection was simulated using a digitized video tape recorded from a camera placed on top of a moving car; 100 frames with a frame rate of 25/s were selected from the video sequence. In this sequence, the car approaches a roadwork where the right lane of the road is blocked by warning signs and fences. In the first frame, the car is about 150 m from the roadwork.

In order to get sufficiently large prediction errors for objects close to the

FOE, the prediction was made three frames ahead. Parameters were computed for every frame, however, that is, the sequence of frame pairs used were 1–4, 2–5, 3–6, etc. A special problem was the initiation of the algorithm. Since only the pixels at the edges from the Canny–Deriche edge detector were used in the prediction error computation, the convergence of the algorithm depended on the initial parameters not being too far away from the correct values. For the first frame pair, therefore, all the pixels in the window were used for computation of the prediction error.

For each frame pair, the gradient descent algorithm was iterated 30 times. No complexity considerations were considered in choosing this number. For more iterations, the improvement of the prediction was found to be negligible.

Figures 6.5(a) and 6.5(b) show the unpredicted difference and the prediction error, respectively, for frames 97–100. We see that a clear reduction of the difference image is obtained in the prediction error image for image points projecting from the ground. For vertically extended objects, the reduction is significantly less, depending on vertical extent and distance from the FOE. The diagrams in Figures 6.5(c) through 6.5(f) illustrate this in more detail. The curves show the image intensity from the difference and prediction error image, respectively. The intensity along two different horizontal lines, indicated in Figure 6.5(a) and 6.5(b) are plotted. The second line containing a white marking from the road is clearly reduced in the prediction error image, while the first line containing the vertically extended objects in the roadwork shows comparatively less reduction from difference to prediction error image.

In Figure 6.6, prediction error images from several different times are shown with white markings thresholded. The images were thresholded at the edge points only and the threshold was increased systematically with distance from the FOE. The threshold was choosen so that at most one marking was obtained from the part of the road containing the parameter estimation window. This means that the threshold was adapted to any errors in the estimated parameters.

From Figure 6.6, we see that as expected the sensitivity of the algorithm increases with distance from the FOE and height over the ground. A very important factor is also the intensity contrast of the objects relative to the background. Some vertically extended objects are not over the threshold in every frame. By combining detections from several frames, the performance should be increased, however. In order to do this, the detections from the different objects have to be grouped and analyzed separately. The main purpose of this prediction-error-based detection can therefore be seen as a preprocessing mechanism that directs resources of the system for further processing.

Acknowledgments. The video recording of the road sequence was made by Lars J. Olsson, Mercel AB, and digitized data were provided by the Computer Vision Laboratory, Linkoping University. The work was performed under

(a)

(b)

FIGURE 6.5. (a) Difference frames 97–100; (b) prediction error frames 97–100; (c) intensity of line 1 in (a); (d) intensity of line 1 in (b); (e) intensity of line 2 in (a); (f) intensity of line 2 in (b).

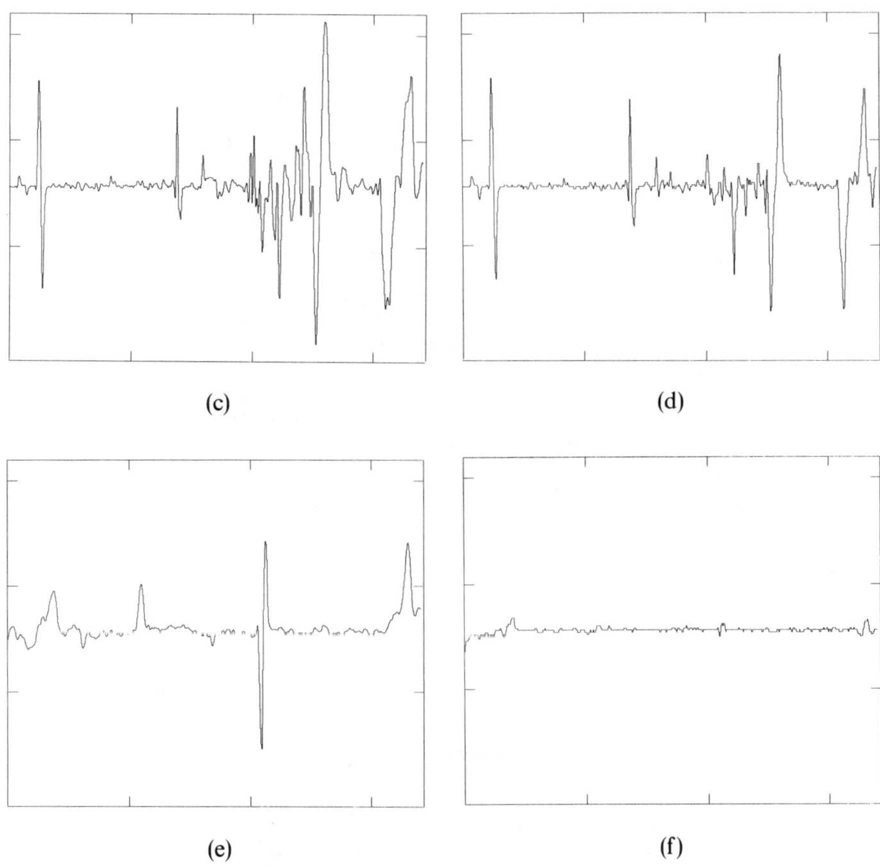

(c)

(d)

(e)

(f)

(a)

(b)

FIGURE 6.6. (a) Thresholded residual frame 10, (b) thresholded residual frame 30, (c) thresholded residual frame 50; (d) thresholded residual frame 70; (e) thresholded residual frame 80; (f) thresholded residual frame 90.

FIGURE 6.6c

FIGURE 6.6d

FIGURE 6.6e

FIGURE 6.6f

contract with the Prometheus-Sweden project. It was also supported by STU, the Swedish national board for technical developement under contract 88-01749.

References

[1] von Helmholtz, H. (1924). *Physiological Optics*, Dover, NY.

[2] Gibson, J. J. (1950). *The Perception of the Visual World*. Houghton, Boston, Massachusetts.

[3] Lounguet-Higgins, H. C., and Prazdny, K. (1980). "The Interpretation of a Moving Retinal Image. *Proc. Roy. Soc. London* **B-208**, 385–397.

[4] Nelson, R. C., and Aloimonos, J. (1988). "Using Flow Field Divergence for Obstacle Avoidance: Toward Qualitative Vision. *Proc. of 2nd Int. Conf. on Computer Vision, Tampa, Florida*, 188–196.

[5] Thompsom, W. B., Much, K. M., and Berzins, V. A. (1985). "Dynamic Occlusion Analysis in Optical Flow Fields." *IEEE Trans. Pattern Anal. Machine Intelligence*, **PAMI-7** (1988). (4), 374–383.

[6] Mallot, H. A., Schulze, E., and Storjohann, K. (1988). "Neural Network Strategies for Robot Navigation." *Proc. of nEuro'88, Paris, June 6–9*, pp. 560–569.

[7] Tsai, R. Y., and Huang, T. S., (1981). "Estimating Three-Dimensional Motion Parameters of a Rigid Planar Patch." *IEEE-ASSP* **ASSP-29** (6), 1147–1152.

[8] Canny, J. (1986). "Computational Approach to Edge Detection." *IEEE Trans. Pattern Anal. Machine Intelligence* **PAMI-8**, 679–698.

[9] Deriche, R. (1987). "Using Canny's Criteria To Derive a Recursively Implemented Optimal Edge Detector." *Int. J. of Comp. Vision* **1**, 167–187.

[10] Horn, B. K. P., and Weldon, E. J. (1988). "Direct Methods for Recovering Motion." *Int. J. of Comp. Vision* **2**, 51–76.

7
Road Sign Recognition: A Study of Vision-based Decision Making for Road Environment Recognition

MARIE DE SAINT BLANCARD

7.1 Abstract

This study is based on an application of vertical road sign recognition by vision. Three types of danger warning signs are recognized. Octagonal stop signs and triangular danger warning signs are distinguished from round forbidding signs on the basis of their outside shape. This recognition is made in "quasi-real-time" in a running vehicle. In addition to the vision algorithm strategy developed, three decision-making softwares are compared: structured programming, expert system approach, and a neural network.

7.2 Introduction

Road environment recognition is more complex than shape recognition in an industrial environment inside buildings. Difficulties in recognition arise for several reasons. First, this environment includes two types of objects that we can call *natural* and *manufactured*. Manufactured objects are buildings on the sides of the roads, cars, and also the road and road markers. The natural objects category includes trees and grass, and a second category of shapes, pedestrians and other forms of natural life.

Second, this environment is *structured*, or rather nonuniformly structured. The most structured road environments are highways, then follow several other types of roads, and in the hierarchy, city streets are last. The city street environment is the most complex because of the superabundance of different manufactured objects along with natural objects of the second category (pedestrians). In fact, the city environment does not have the same nature as do roads and highways and should be considered independently. This study shall consider highways and roads. **Outside conditions** should also be considered. These include the weather (sunny, rainy, foggy, etc.) and local variations, such as the direction of the light and shading.

The choice was made to study the **road environment as it is today** and to define manufactured objects, such as triangular road signs indicating potential dangers on the roads. Highways are not as difficult as other roads because potential dangers, such as crossings, are not usually indicated, but road sign recognition in the different environments will be discussed further.

Finally, the reader must keep in mind that this chapter relates a car manufacturer's study, the goal of which is more general than road danger sign recognition. This application is based on actual vision techniques in a running vehicle, such as in a common car. The second goal is the use and comparison of decision-making techniques: structured programming, an expert system (in LeLisp language), and a neural network.

7.3 Technology and Car Equipment

7.3.1 Road Signs

As mentioned in the introduction, road environment recognition is based on actual road signs [1]. The most interesting is the STOP sign. It indicates an absolute interdiction and requires a full stop of the vehicle. Furthermore, it is the only octagonal road sign [see Figure 7.1(a)]. The sign is red, as are all signs indicating a danger, with a white border and white letters. The white border is used very efficiently as a highlight against the background, and since it works as well for the camera as it does for the human eye, it is used in our detection process.

The second most interesting sign is the triangular danger sign [see Figure 7.1(c)]. It is posted to warn drivers of potential dangers ahead of the vehicle, such as crossings, loss of priority, and stop signs (at 150 m on roads and 50 m in towns). One type of triangular red warning sign points upward rather than downward [see Figure 7.1(c)]. The other red signs are round and mostly

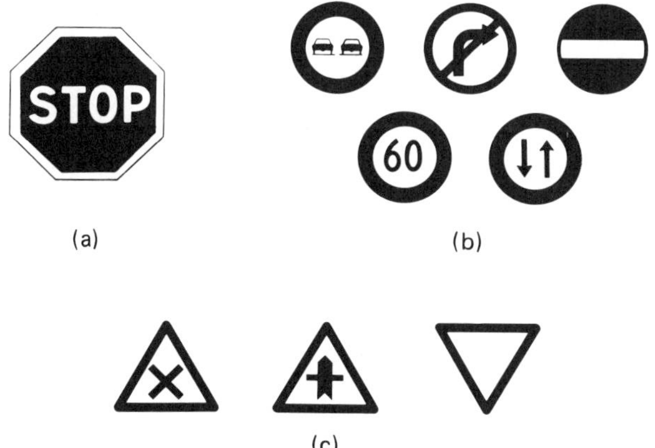

FIGURE 7.1. Red road signs: (a) stop, (b) interdictions, and (c) danger warnings.

indicate interdictions, such as speed limits or one way streets [Figure 7.1(b)]. We can also find red and blue round signs for parking interdictions, but usually only in towns.

Stop signs and triangular downward danger signs have to be distinguished from the others. Road sign size is not standard. Usually, signs are much bigger on highways, and the smallest ones are found in towns. This is why no indication of the distance of the sign is deducted from a one-camera analysis.

7.3.2 Hardware

Hardware constraints are to be exposed very early because of the applicative aspect of this study, the first goal being to have a recognition application in a running vehicle.

The choice was made to use PC-based hardware. The PC (personal computer) is a small and powerful microcomputer with sufficient scientific capacities and other developmental tools. **The PC is used as a host.** Specific add-on boards are hosted by the PC, which only performs the user interfaces.

7.3.3 Vision

This approach is based on a **black and white** vision sensor, such as a CCD (charge coupled device) camera. The camera is placed next to the driver mirror looking through the windshield; **256×256 pixel images** are sufficient for this application.

After a first phase of development on a general-purpose software run on a PC microcomputer, the vision algorithm was transferred onto an add-on PC-based vision board. This vision board, the GIPS25, manufactured by the French company EIA, is based on a digital signal processor by Texas Instruments, Inc., the TMS320C25 [2].

7.4 Vision Algorithm

The goal of this algorithm is to extract from the pixel image the least amount of information, yet the most relevant, in the shortest possible time to help the decision-making system. The actual process is described in Figure 7.2.

7.4.1 Description

The vision algorithm is based on one black and white image analysis at a time. There is for now no image sequence analysis or time coherence checking [3].

The first operation to be performed is a gradient. Several gradient algorithms were compared for this application. Three were performed with success.

Using *mathematical morphology* operations such as dilation and substraction on the original image [4]. This gradient avoids connections of the contour

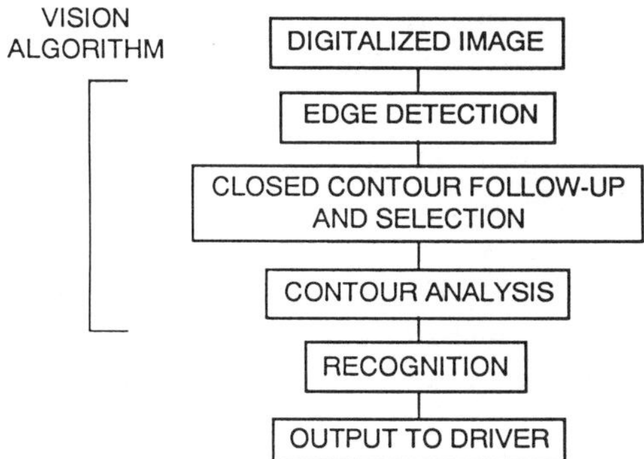

FIGURE 7.2. Detection process.

with background artifacts [see Figure 7.3(b); Figure 7.3 is found in the color plate section that follows this chapter]. It does not give the best results for round signs where inside black letters (generally numbers) connect to the contour.

The *sobel gradient* performs very well, but it is a little time consuming.

In the *differential Nagao* gradient, the first operation consists of a type of Nagao filtering that smoothes the image while enhancing its contrast. A differential operator provides the gradient contour [see Figure 7.4(b); Figure 7.4 is found in the color plate section that follows this chapter].

These gradients enhance the edges of manufactured objects. Natural objects appear with unconnected, smaller, and weaker edges. These results are used in the second step to eliminate the natural objects.

The second operation consists of following the edges of the gradient using Freeman coding. The image is totally scanned and the follow-up follows these rules:

A follow-up starts any time the pixel value is higher than a given threshold.

The contour follows the highest edge computed from the pixel values ahead. This computation involves several pixel values and a direction of follow-up close to the present one. The length of search may vary. This computational criterion allows jumping above pixel gaps, if any.

The follow up stops (a) when the computed criterion gets too small below a given threshold (the last point of this open contour is marked); *or* (b) if the contour joins an already detected contour.

See Figures 7.3(c) and 7.4(c).

The last step of the vision algorithm consists of eliminating

all contours that are too small or too big, that is, below or above a given length or perimeter, and

every contour that is not closed, easily spotted by their end of follow-up marking.

See Figures 7.3(d) and 7.4(d).

As described, the detection is based on a closed contour search in the image. In the case of connected closed contours, two approaches were implemented.

The first was the selection of all possible combinations of closed contours. This approach is very complete, but it is time consuming in the case of complex contours, such as grid-like forms.

The second was the selection of the outside contour. Because of the choice of gradient, cases in which the road sign contour is connected to the background, or included in another, bigger contour, are very rare. They are still possible, but they were never encountered during the latest experiments.

7.4.2 Performances

At this point, it should be noted that color information is used for the vision algorithm performances. A cutband filter was placed in front of the camera lens. This filter cuts off the spectral bandpass corresponding to the color red, thus making the red areas darker in the black and white digitalized image. It increases the contrast of the sign on the background.

The parameters mentioned in the description of the algorithm are to be fixed or adapted for all test conditions when possible.

The minimum and maximum perimeter values of a closed contour are fixed along with the camera lens and the desired detection distance. Details on this choice are not described here because they are common knowledge.

The depth of research was experimentally fixed.

The threshold values of start and stop follow-up searches are fixed for constant outdoor conditions.

Drawbacks of thresholding are well known. Adaptation to outside weather conditions of start and stop follow-up search thresholds is possible. But, difficulties come from the near background. The *local contrast* cannot be estimated, and it is the main cause of failure when combined with sign that is too small in size within the image.

Having the constraint of a short processing time, a certain amount of non-detection can be allowed, rather than impairing it with more sophisticated algorithms. To reduce the processing time, a window of interest is defined by experiment, depending on the focus of the camera lens.

The best compromise to detect octagonal, triangular, and round signs was accomplished by using the differential Nagao gradient, which runs in 350 ms in the window of interest mentionned previously (about 200×150 pixels). The edge closure and selection run in 300 ms, with almost no variation of edge number. Figures 7.5(a)–(c) shows typical contours that were obtained.

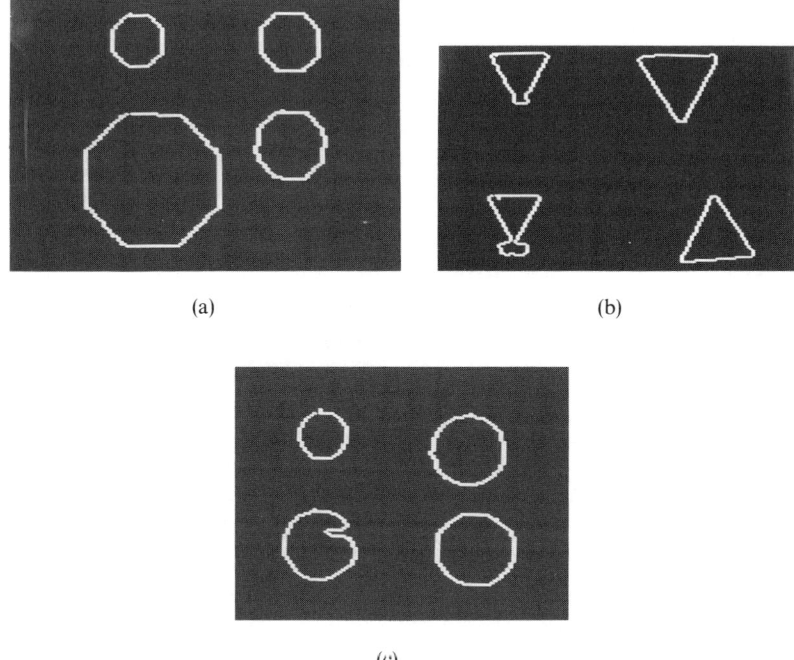

FIGURE 7.5. Typical sign contours: (a) stop signs, (b) danger warning, and (c) round signs.

7.5 Decision Making

7.5.1 Criteria

Once the vision algorithm gives the closed contours, a decision-making system must correctly identify them. For that purpose, the closed contours must be analyzed to feed the decision-making system with relevant data. [5]

The following list contains all the criteria that will be used by either decision-making system:

1. perimeter, length in pixels;
2. outside surrounding box, coordinates of the four corners;
3. surfaces, inside contour and outside within the surrounding box;
4. center of gravity, x and y coordinates;
5. compactness, length over width of the surrounding box;

6. polygonal approximation, number of sides and coordinates of each corner, with minimal distance error (3.5 pixels);
7. Freeman code, complete contour coding;
8. histogram of the Freeman code, addition of the number of occurrences of the eight directions, histogram of the changes in directions, or histogram of the changes in chosen directions; and
9. average gray level, computed within the surrounding box.

7.5.2 Structured Programming

This first method comes directly from the preceding steps and was written in C language. It examines different criteria from the list and determines if their values are valid within given intervals. These intervals were strictly chosen from the test experiments.

This works well under the usual circumstances, but shows its weaknesses for limit cases or special conditions. It is not flexible handling exceptions and requires special programming. Execution time is multiplied by a minimum factor of 2 when handling these special conditions, and it does not ensure a regular detection time. Furthermore, although this kind of programming seems easy at the beginning of a study, it requires special skills and can even become a drawback itself.

7.5.3 Expert System Classification

An *expert system approach* was also undertaken. [6] [7]. At that time, no software generator choice satisfactorily fulfilled certain requirements, such as PC execution, performance, and interfacing with the vision software, nor did it fulfill technical requirements, such as full object description and possible heritage. Vision data might not always be complete or totally reliable. Thus, the final decision is associated with a confidence coefficient.

These last requirements led to the choice of specific LeLisp programming software. Several software tools were created, but they are not detailed here.

Figure 7.6. illustrates the structure of the expert system. Each class contains

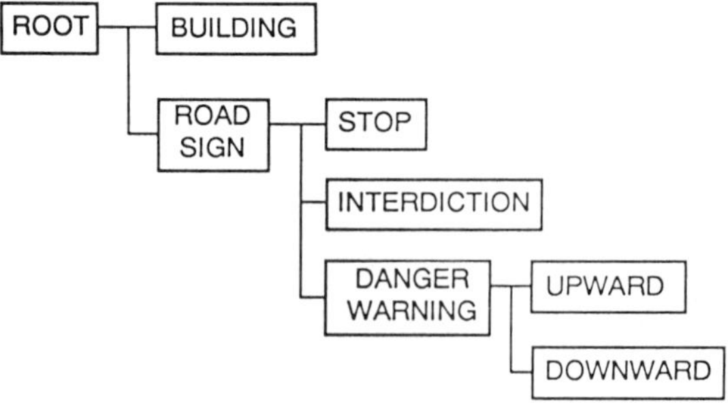

FIGURE 7.6. Class hierarchy example.

a variable number of description fields. Each daughter class inherits the fields and can have additional fields.

A *model* is described with possible values corresponding to the fields. An object passes from one class to another if it satisfies the given model. Models are described by *attribute* values that are associated to an *error function*. This function is a probability described by a percentage of success or a repartition function. Input data from the object coming from the vision are also associated with an error function. [8]

An expert system to classify road signs was tested with the simple class structure described in Figure 7.6. This approach is potentially the most complete and allows future development, such as a higher level of description for environmental parameters (type of road, weather, driver, etc.), thus creating a blackboard structure with data coming from other sensors or to be updated at different frequencies. This approach was not pursued here for short-term goals. It requires a lot of specific programming with real-time capacities that will be provided by the coming generation of hardware and firmware.

7.5.4 A Neural Network

Comparison of different types of networks is not the subject here. The network consists only of software and was arbitrarily chosen. The development was realized with NDS (Nestor Development System), a product by Nestor, Inc. This development system includes NLS (Nestor Learning System), based on **restricted Coulomb energy, RCE**. [9] [10]

Five types of contours have to be classified:

circles,
octagons,
triangles pointing upward,
triangles pointing downward, and
counterexamples

For the learning process, 1100 forms were used, and for the testing phase, 614 forms with the following repartitions were used.

288/198 circular signs,
271/178 octagonal signs,
68/19 upward triangular signs,
239/183 downward triangular signs, and
225/36 other objects.

Different choices were tested for encoding the contour. They are not discussed here, but some are detailed in Tables 7.1–7.3. General rules can be inferred.

Geometrical criteria that are too precise are detrimental to good recognition because the contours that are obtained by vision do not correspond to an exact shape. The general idea obtained with Freeman codes gives better results.

TABLE 7.1. NDS results using the histogram of Freeman's codes.

| | Identified (%) | | | |
Contour	Correct	Incorrect	Uncertain (%)	Unidentified (%)
Circles	60	12	12	16
Octagons	69.7	7.8	10.7	11.7
Triangles up	100			
Triangles down	100			
Other	55.5	10.1	3.8	30.5
Total	68	11.5	9.9	10.4

TABLE 7.2. NDS results using the histogram of directions in Freeman's codes.

| | Identified (%) | | | |
Contour	Correct	Incorrect	Uncertain (%)	Unidentified (%)
Circles	65.6	11.6	12.5	10.1
Octagons	74.1	7.3	8.9	9.5
Triangles up	100			
Triangles down	99.4		0.5	
Other	80.5	8.3	8.3	2.7
Total	80.1	6.3	7.2	6.1

TABLE 7.3. NDS results using the normalized histogram and average gray level within the surrounding box.

| | Identified (%) | | | |
Contour	Correct	Incorrect	Uncertain (%)	Unidentified (%)
Circles	92.4	0.5	5.0	2
Octagons	98.8		1	
Triangles up	89.4		10.5	
Triangles down	99.4		0.5	
Other	64.4	11.1	5.5	13.8
Total	94.9	0.8	2.6	1.4

Triangular signs are easy to distinguish. The difficulty is in sorting octagons from circles or vice versa.

Because of the quality of the contours, prototypes for recognition are numerous, over 100. These prototypes are very close. There exists some overlapping, but prototypes are clearly differentiated. The best result obtained is summarized in Table 7.3, with 94.9% correct and 0.8% incorrect answers. These false answers correspond to mismatches between circles and octagons.

A human operator cannot see the difference in these cases. The reader can check this difficulty by looking at Figure 7.5.

To conclude the discussion on this neural network approach, the industrial user should known about the ease in programming and of analyzing the results given by NDS. These results were obtained with a single network, the NDS 500, without specific hardware or software. The use of a multiunit system was studied, but it would not have gained significantly more than the 94.9% of the success already obtained. This solution was chosen for its reliability under test conditions, its *constant run time*, about 0.1 s, and its capability to enrich the experience.

7.6 Performances and Conclusions

This application of road sign detection is actually running in a car. It uses a black and white CCD camera, a specific vision board, and a PC. The vision performs contour detection in approximately 0.6 s, and a neural network classifies the results. When a sign is recognized, it is displayed to the driver. This process takes 0.7 s. Experiments were done at medium speed, 40 to 60 km/h, and signs were recognized 2 to 4 times before they were reached.

Because cars move fast on highways, road signs are bigger to make them visible earlier. This is why the author thinks that the experiment described in this chapter could still work. But, road signs on highways do not have varied shapes: they are all round. There are no STOP signs or triangular warning signs on highways; and where there are no cross roads, there are less traveler deaths.

This application educated us richly. It now requires improvement of the vision detection part. Several areas are under study, such as color segmentation (danger signs are always red), use of other sensors, such as telemeters, and earlier use of a neural network.

Acknowledgments. I would like to acknowledge the helpful contributions of three students, D. Germa, J. Granger (Ecole Nationale Supérieure des Télécommunications, Paris), and F. Baucher (Université de Technologie de Compiègne).

References

[1] Akatsuka, H., and Imai, S. (1988). "Road Signposts Recognition System." 0096-736 X 88/19601-0936 $02.30, copyright 1988, Society of Automotive Engineers, Inc.

[2] EIA (1990). "GIPS Vision." Manufacturer's manual, Bièvres, France.

[3] Germa, D. (1989). "Détection de panneaux de STOP." *PSA-ER Internal Report*, January.

[4] Serra, J. (1982). *Image Analysis and Mathematical Morphology*. Academic Press, San Diego, California.

[5] Granger, J. (1990). "Paramètres de formes de contours pour vision par ordinateur.' *PSA-ER Internal Report*, January.

[6] Baucher F. (1989). "Réalisation d'un Système de Classification Orienté Objet Temps Réel." *PSA-ER Internal Report*, June.

[7] Granger, C. (1985). "Reconnaissance d'objets par mise en correspondance en vision par ordinateur." PhD. Thesis, Université de Nice.

[8] Zadeh, L. A. (1978). "Fuzzy Sets as a Basis for a Theory of Possibility." *Fuzzy Sets and Systems* **1**, 3–28.

[9] Reilly, D. L., Cooper, L. N., and Elbaum, C. (1982). "A Neural Model for Category Learning." *Biological Cybernetics*, *45*, 35–41.

[10] Nestor, Inc. (1990). "Introduction to Nestor Development System NDS." Manufacturer manual, Providence, Rhode Island.

Color Plates for Chapter 7

FIGURE 7.3. (a) Image; (b) gradient with mathematical morphology; (c) contours after follow-up; (d) selected candidate for sign recognition.

(a)

(b)

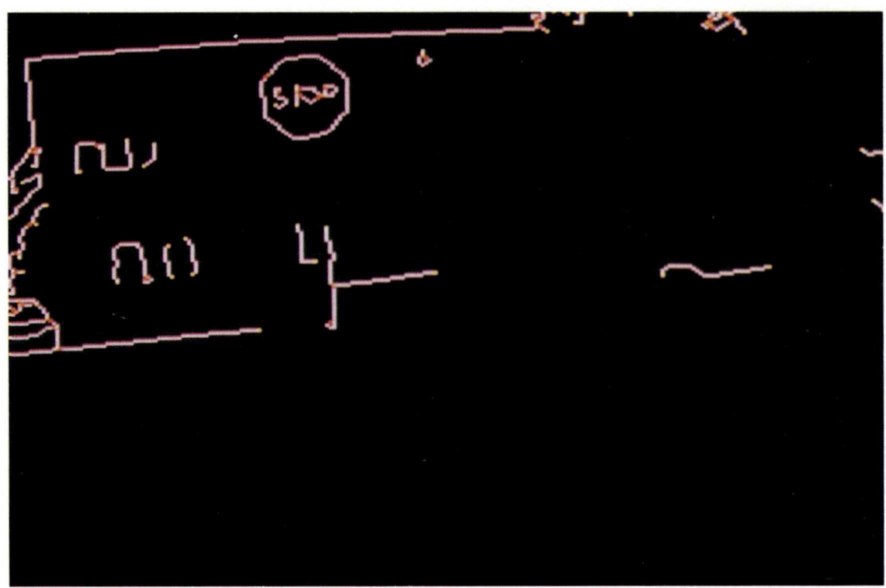

(c)

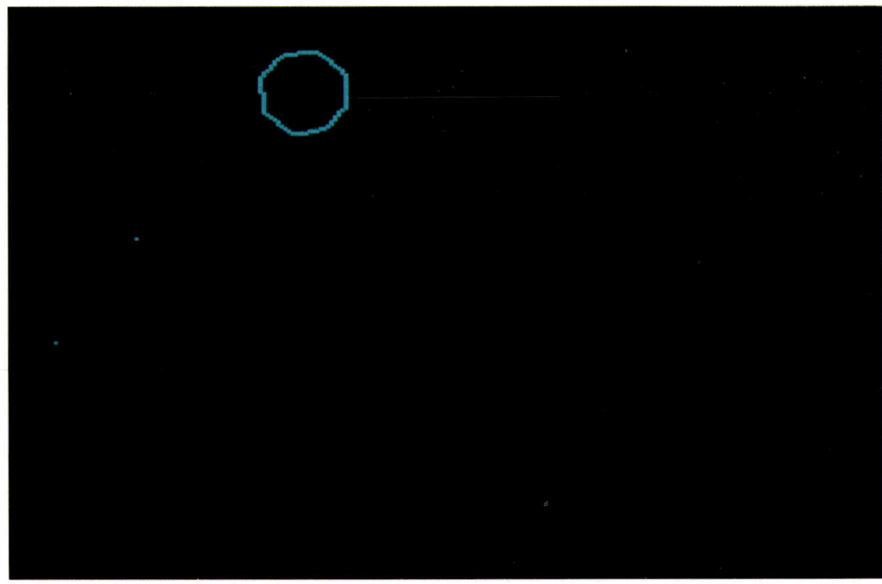

(d)

FIGURE 7.4. Detection and recognition: (a) image; (b) gradient image with differential Nagao; (c) contours after follow-up; (d) candidates for recognition; (e) danger warning sign recognition and display.

(a)

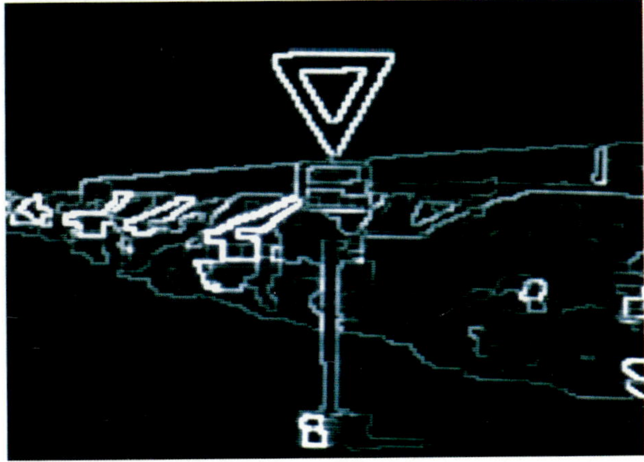

(b)

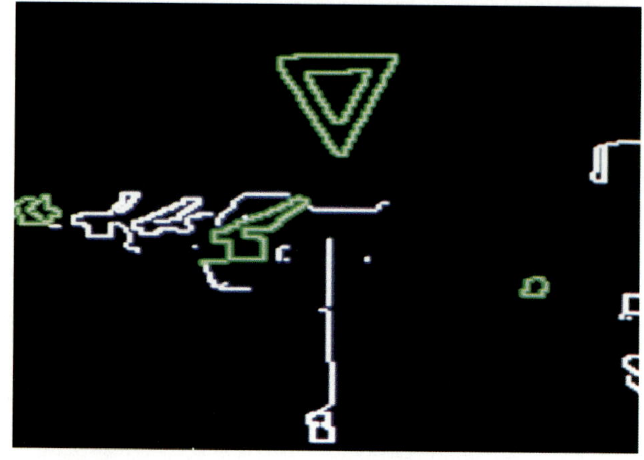

(c)

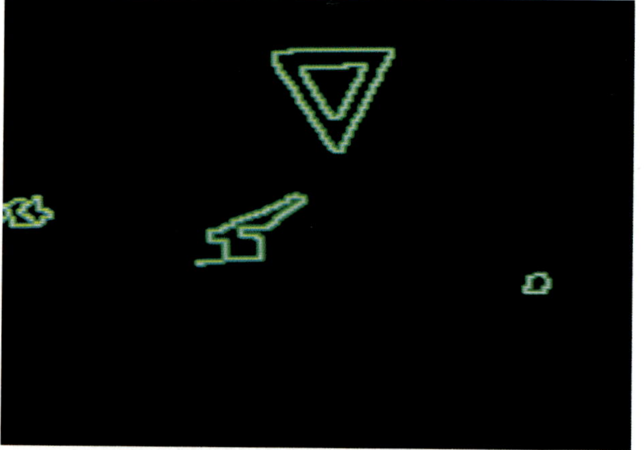

(d)

(e)

Figure 7.4. (*continued*)

8
From Self-Navigation to Driver's Associate: An Application of Mobile Robot Vision to a Vehicle Information System

KOHJI KAMEJIMA, TOMOYUKI HAMADA,
MASAHIRO TSUCHIYA, AND YURIKO C. WATANABE

8.1 Abstract

The concept of a vehicle information system, called the driver's associate, is presented for identifying a driving environment in cooperation with human drivers. Sharing the image feature extracted and segmented in observed imagery, a set of frustration resolution schemes for top-down processing and 2-D syntax analysis for bottom-up processing are integrated into a dynamic perception mechanism that is implemented by a set of LSIs (Large Scale Integration chips). The schematics of the software are verified through simulation studies to demonstrate the basic operation scenario of the driver's associate.

8.2 Introductory Remarks

Despite the difficulty in control and adaptation to a city plan, transportation by means of privately owned vehicles will continue to increase because the information-intensive society requires highly flexible mobility. Since each privately owned vehicle is used strictly for the convenience of the owner, the driver is responsible for optimizing his own driving plan as an element of the transportation system. However, the dynamic behavior of a transportation system is governed by the decision making by individuals who control privately owned vehicles that cannot be directly controlled by other drivers. Thus, the transportation process must be optimized by integrating the traffic management process and the vehicle control process. Although human drivers believe that they spontaneously control their own vehicles, each individual driving process is essentially governed by a global transportation process. For example, consider the situation illustrated in Figure 8.1, in which another vehicle is going to enter the crossing from the road behind the building, and an obstacle, perhaps the other vehicle, is in the way. At this crossing, the driver is forced to select a route, whether straight ahead or to the left or right, to arrive at the planned destination in the shortest time. In this situation the local roadway scene, consisting of landmarks and obstacles, must be con-

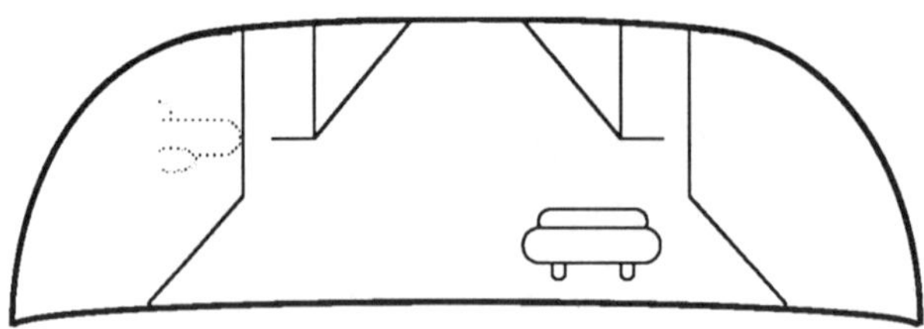

FIGURE 8.1. Scene to be observed.

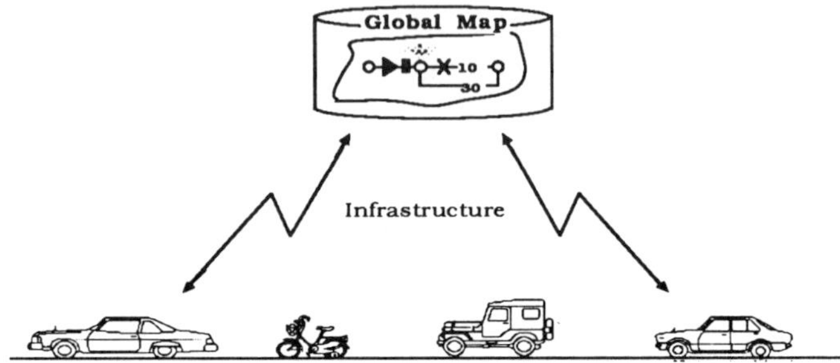

FIGURE 8.2. Infrastructure for the driver's associate.

tinuously recognized for smooth vehicle control. Simultaneously, global traffic circumstances, including traffic distribution and roadway capacity, must be evaluated for route decision and trip scheduling when the crossing is recognized.

The global transportation process can be represented in terms of the distribution of traffic data, such as density and average velocity of vehicles, on a driving map on which self-motion is also identified. Hence, dynamic gathering and broadcasting of traffic data, as illustrated in Figure 8.2, is a promising source for drivers to use to select the correct route. For this purpose, and also for improving mobility in the roadway network as a result of individual optimization, various road–automobile communication systems have been proposed and tested to develop the infrastructure for information intensive transportation [30, 37]. For example, the IVHS (intelligent Vehicle/Highway System) in the United States [3], AUTOGUIDE in the United Kingdom [10], and PROMETHEUS (Program for a European Traffic with Highest Efficiency and Unprecedented Safety), and DRIVE (Dedicated Road Infrastructure for Vehicle Safety in Europe) in the European Community (EC) [2] are investigated. In Japan, following the pioneer project for vehicle information systems, called the Comprehensive Automobile Control System (CACS) [21], two types of infrastructure systems are planned: the Advanced Mobile Traffic Information and Communication System (AMTICS) [27] for traffic control and the Road/Automobile Communication System (RACS) [29] for dynamic driving information service.

As far as dynamic optimization of the transportation process and minimization of a driver's memory load, the guidance message should be specified based on the history of the transportation process up to the decision making process and clearly indicated just in time when the message is required to attract the driver's attention. However, the driver's concentration is hampered by having to directly observe the message indicated on the scene if the message is not matched with the scene and the driving context. Thus, the dynamic driving information service yields an environment identification problem, that is, how to understand the driving context based on the guidance message and how to make local observations for adjusting the message transmission channel to match a dynamically varying scene. Since the early stage of robotics research, the autonomous maneuver has been studied as one of the most fundamental problems. Many machine vision systems have been developed for both indoor applications [35, 26] and outdoor applications [9, 22]. As a result of these mobile vision studies, the dynamism in visual perception has been revealed, and the role of a priori information on the environment has been clarified [36]. This implies that the information processing mechanism can be applied to a computer implementation of various aspects of a driving information service.

In this chapter, an environment identification system, called the driver's associate, is presented for enhancing the driver's visual perception capability. First, a mobile robot vision system is reconstructed into an environment identification system to complement the driver's subliminal perception in complicated and dynamic driving situations. Next, the environment identification process is designed in terms of dynamic integration of top-down image analysis for resolving symbolic and geometric frustration and bottom-up image analysis for triggering the top-down process. These image analysis mechanisms are implemented by a set of TV rate VLSIs and examined in experiments. Finally, the operation scenario of the driver's associate is verified through simulation.

8.3 The Concept of Driver's Associate

Consider the situation illustrated in Figure 8.1 again. In maneuvering the vehicle, the driver must detect many objects, including the hidden vehicle, obstacles, and the crossing to determine his relative location to these objects. Since these objects move independently and the situation is dynamically altered, the visual perception channel of the driver is activated incessantly and extended to cover the scene. Although the human eye cannot focus throughout the scene, the concentration area of human visual perception is dynamically controlled by a marginal visual stimulus accepted in the view with the range of a fish eye lens. Hence, human visual perception is of sufficient capability for driving vehicles in a normal situation. However, the visual perception channel of a human driver is seriously restricted because of distractions such as conversation, nervousness, and tediousness. In addition, the guidance message provided by the infrastructure may cause selective attention to concentrate the visual perception capability in order to detect the specific moment at which the route must be altered. The result of this restriction and concentration is a significant decrease in driving safety, and in recent years, this has inspired the research and development of various automated driving systems [4, 25, 32, 33]. Although these automated driving systems have been designed to control vehicle mechanisms without interaction with human drivers, the machine vision, particularly environment identification systems developed for autonomous maneuvering of mobile robots [12, 19], exhibits essentially the same capabilities of human visual perception before selective attention. In fact, these environment identification systems are equipped with layered image analysis schemes sharing an environment model as the basis of collaboration. The model-based architecture seems to be quite similar to the process of human perception based on a cognitive map [23].

The structural consistency between an environment identification system and human perception, particularly in the earlier stages, implies that a machine vision system can be designed to cooperate with a driver's visual perception channel. In what follows, a mobile robot vision system is reconstructed into a driving information system, called the driver's associate. The purpose of the driver's associate is to complement the capability of the driver's limited subliminal perception by analyzing the scene and indicating the results to the driver. The design policy of the driver's associate is illustrated in Figure 8.3. The driver's associate dynamically matches global knowledge of the transportation process provided by the infrastructure and local information acquired through direct observation. As the result of this matching procedure, the driver's associate paraphrases the abstract global map, provided from the infrastructure, into the environment model. On this environment model, the driver's associate also generates a dynamic representation of the transportation process with a structure consistent to the system of knowledge evoked in the human driver, called the mental image. As illustrated in Figure 8.3 the mental image evocation mechanism is modeled as a scene analysis process in which a global map, consisting of the symbolic representation of locations of auxiliary vehicles and the distribution of traffic parameters, and observation of the scene are combined through a prediction–detection link. Through this link, the prediction of the roadway scene, as well as blind spots and shortcuts,

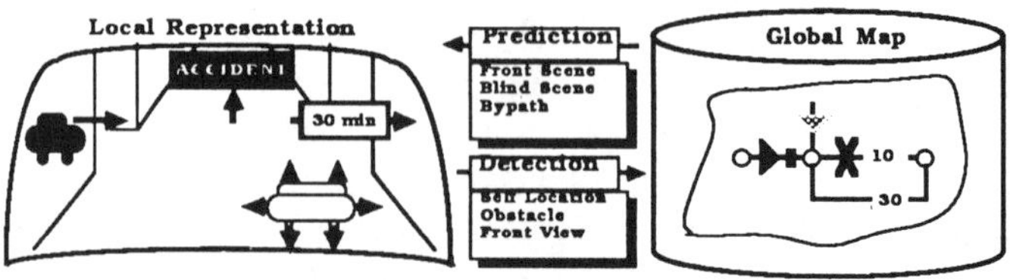

FIGURE 8.3. Mental image of transportation process.

TABLE 8.1. Comparison of the driver's associate with self-navigation.

	Self-navigation	Driver's associate
Infrastructure	A computer system	Multimedia network
Knowledge	Hierarchical	Distributed
Base line	Road following	Database access
Decision maker	An operator (a central system)	Many drivers (peripheral systems)
Effector	Vehicle mechanism	Display device
Mission	Substitution	Extension

is provided by the global map. The results of this scene analysis are edited into a set of data. The data, consisting of self-location, obstacles and the frontal view, are indicated to the driver through the head-up display in this example and marked on the global map. In Table 8.1 the concept of the driver's associate is compared with the features of the self-navigation system for autonomous mobile robots. Thus, the driver's associate captures essentially the same information from the scene as a vision system designed for autonomous maneuvering. The only difference is that the driver's associate utilizes captured information to enhance the capability of the human driver by providing nonstop subliminal visual perception capability although the self-navigation system decides the route and controls the vehicle mechanism.

8.4 An Environment Identification Problem

By introducing the dynamic link between the infrastructure and the vehicles, as illustrated in Figure 8.3, the dynamic behavior of traffic is comprehended on the global map. This implies that the transportation plan of each vehicle is decided individually by solving route optimization problems for each vehicle. Let this decentralized route optimization problem be solved on a shared global map and let the resulting maneuver plan be available in an on-board computer system with a copy of the global map. Then, the mission of the driver's associate is to solve an environment identification problem consisting of

- identifying objects to be detected in the scene to trace a route safely,
- analyzing the observed imagery to detect and classify objects, and
- displaying the detection results to complement the driver's subliminal perception.

The environment identification problem is of essentially the same structure as the visual guidance system for an autonomous mobile robot [14], in which an environment model, consisting of a node map and a multiviewpoint geometric map, was introduced as the basis of object detection and path planning. In this environment model, the node map is a graph representation of the environment, and the maneuvering cost is represented as the attribute of links of the graph. By using the node map, the optimal route is computed as

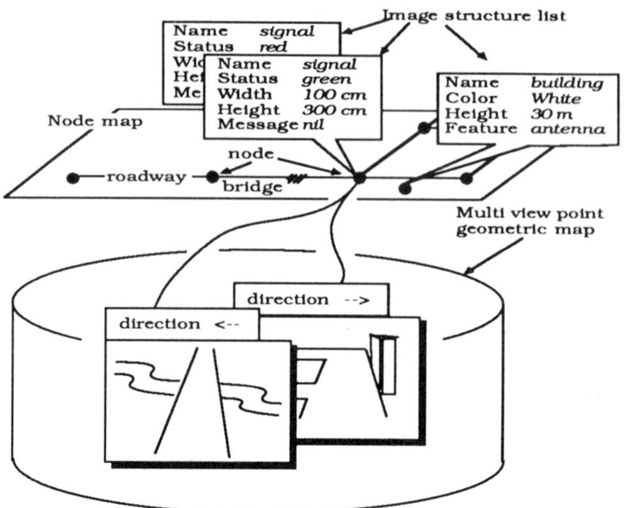

FIGURE 8.4. Transportation model on digital road map.

the chain of links connect the origin and destination nodes with a minimal maneuvering cost. To each node, a set of image structure lists is attached for specifying the scene. In this multiviewpoint geometric map, a set of visible 3-D vectors specifying landmark objects located near the node is classified according to the maneuvering direction and stored in association with nodes. By this association mechanism, the multiviewpoint geometric map is utilized to generate the reference image to be matched with the observed image for evaluating guidance error.

Consider the environment model for the driver's associate. Since various environmental features, including locations of landmark buildings and distances between node points, are stored in a currently distributed digital road map, it is not difficult to generate the node map based on a digital road map. This implies that the dynamic behavior of the transportation process can be described on the digital road map as illustrated in Figure 8.4. In this figure, the transportation process is described by node maps and a multiviewpoint geometric map, as well as in the visual guidance system for mobile robots. In fact, the locations and features of landmark objects are stored in the node map as in the mobile robot visual guidance system. However, the existence of many obstacles exhibiting complicated behavior on the roadway implies that the digital road map can no longer be exploited as the basis of a scene analysis without continual updates of the digital road map. Thus, in order to exploit as the information basis of the driver's associate, the image feature list attached to the node map must be frequently updated using the scene analysis result itself. In addition, since the roadway scene is far more complicated then an interior building scene in which mobile robots maneuver, it is not practical to a priori specify visible contours of landmark objects, as in the mobile robot visual guidance system. Hence, the 3-D vector description in the multiview-point geometric model for the driver's associate should be generated as a result of scene analyses.

From the incompleteness of the a priori node map and the emptiness of the a priori multiviewpoint geometric map, the environment identification problem in the driver's associate is formulated as follows:

- self-positioning on node map,
- description of scene, and
- interpretation of frontal view.

In what follows, a set of image analysis schemes is integrated to design the driver's associate by simultaneously solving the above-mentioned problems.

8.5 Image Feature Extraction and Segmentation

The driver's associate extends the human visual perception channel by combining two image analysis mechanisms: multiagent surveillance on a 2-D segment array and successive resolution of frustration on an image feature. The driver's associate is required to recognize the 3-D location of geometrically specified landmark objects, for example, sign posts, signals, and message boards, in a continually varying and complicated scene for the identification of self-location on the node map. However, landmark objects can no longer be uniquely determined only by 3-D contour matching in a practical driving situation. In addition, the driver's associate must continually detect cueing objects not yet geometrically specified, such as obstacle vehicles, obstructive buildings, and bifurcations, in successively observed scenes for dynamic completion of the multiviewpoint geometric model in the environment model. Thus, the environment identification in the driver's associate should be initiated by the specification of the matching area, in which landmark objects and "something" are expected to be observed before activating the geometric matching process. For this purpose, a parallel syntax analysis scheme is introduced for specifying the region that contains the segments matched with a part of an object's contour. On this attracted region, the driver's associate evaluates the disparity between the object model and the observed edge pattern. For this purpose, a set of successive resolution mechanisms is designed for reducing the disparity between the edge pattern in noisy observed imagery and the line drawing generated through the perspective projection of the 3-D contour model.

Consider the image feature extraction in image analysis triggered by a non-geometric specification of an attractant region associated with landmark and cueing objects. Let the shape of the landmark and cueing objects be approximated by a circumscribing box. Then, the attractant region can be described in terms of the combination of rectangles that may be deformed. Introducing

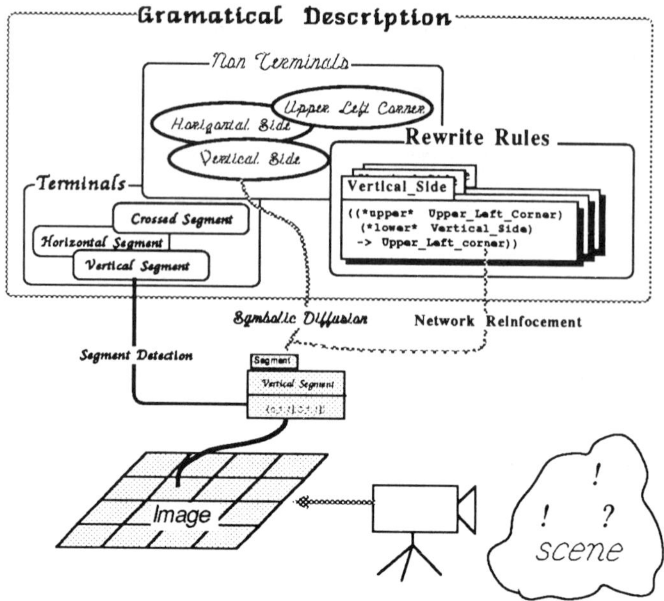

FIGURE 8.5. Detection of grammatically defined pattern.

a grammatical description of a rectangle [13], the image feature extraction mechanism is organized as illustrated in Figure 8.5. In this mechanism, the grammatical model is matched with a 2-D array parameter list defined by

$$\Xi[i,j] = [\xi_1[i,j]\xi_2[i,j]\xi_3[i,j]\cdots], \tag{8.1a}$$

$$\xi_{(\cdot)}[i,j] = [\theta_{(\cdot)}[i,j], \rho_{(\cdot)}[i,j]], \tag{8.1b}$$

where $\theta[i,j]$ and $\rho[i,j]$ are direction and distance parameters of the contour image detected in the ijth partition, designated by $\Sigma[i,j]$ of the image field Σ.

The edge pattern is said to be the image feature if a parameter $\xi[i,j]$ is uniquely determined in at least one partition $\Sigma[i,j]$. The image feature is said to be segmented under the grammatical model if the set of partitions $\Sigma[i,j]$ containing a hierarchy of subdescriptions of the model is specified, as illustrated in Figure 8.6. Let T be the totality of a priori specified parameter lists associated with terminal symbols for a grammatical mode and assume that the list Ξ is matched within an inclusion-based list matching concept [26], as illustrated in Figure 8.7, that is Ξ is said to be identified with $\tau \in T$ if $\Xi \supset \tau$, that is, τ is a sublist of Ξ. In this segmentation procedure, terminal symbols defined in terms of the distribution of linear segments in a small image field should be associated with a set of nonterminal symbols and, finally, with the start symbol "RECTANGLE" through successive and parallel applications of a set of rewrite rules [12]. Despite the flexibility in model description and the robustness in parsing, it is very difficult to apply the grammatical pattern

FIGURE 8.6. Schematics of image feature segmentation.

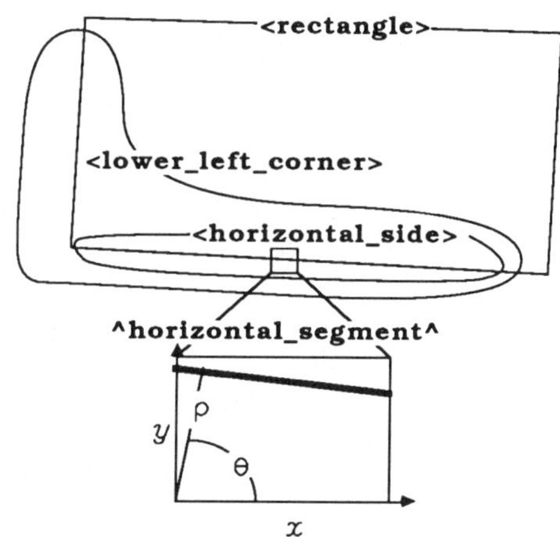

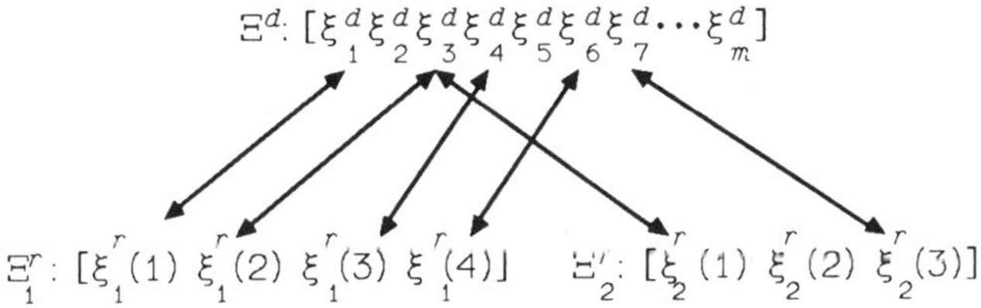

FIGURE 8.7. A list matching scheme.

recognition technique to a noisy edge pattern because the ambiguity in terminal symbol association, due to the noisy pattern, yields an intolerable number of backtrackings in accordance with the progress of parsing. In order to reduce the noise pattern in observed imagery, which is provided by a single-eye TV camera attached to a randomly vibrating vehicle mechanism, a linearized version of a Mach effect mechanism is applied to extract temporally smoothed and spatially enhanced edge patterns as the image features [19].

$$d\varphi/dt + \alpha\Delta\varphi = \beta[f - \varphi], \qquad \alpha, \beta > 0, \tag{8.2a}$$

$$|\varphi| \leqq M, \tag{8.2b}$$

$$\eta = \{s \in \Sigma \,||\Delta\varphi(s)| > \lambda\}, \tag{8.2c}$$

where Δ is the Laplacian operator, f is the gray level distribution of the observed image, φ is the gray level distribution model, η is the extracted image feature, and λ is the threshold.

The left-hand side of Equation (8.2a), combined with the limitation described by Equation (8.2b), means that the steadystate distribution of model φ exhibits the spatial "overshoot" limited by the boundary M and intensified by the coefficient α. On the other hand, the right-hand side of Equation (8.2a)

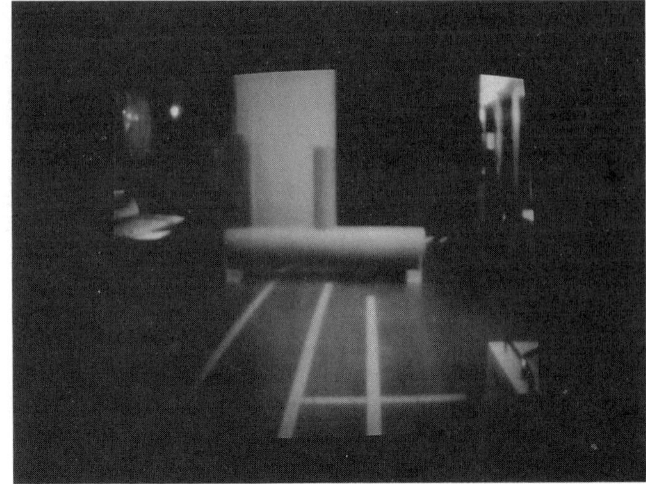

(a)

(b)

FIGURE 8.8. Experimental results of image feature extraction: (a) gray level distribution; (b) extracted image feature.

implies that the time evolution of the distribution of the model is smoothed at each point in Σ through negative feedback to the gray level distribution f. The noise reduction by the extraction algorithm [Equation (8.2)], called a linearized Mach effect (LME) algorithm, is examined through experimental studies. The results are shown in Figure 8.8, where the application of the LME algorithm is verified to yield a clear edge pattern of ambiguously contoured objects under poor lighting conditions. In addition, the smoothing procedure is also verified to effectively reduce the flicker noise in edge patterns that causes parameter estimation bias in the voting process for segmentation.

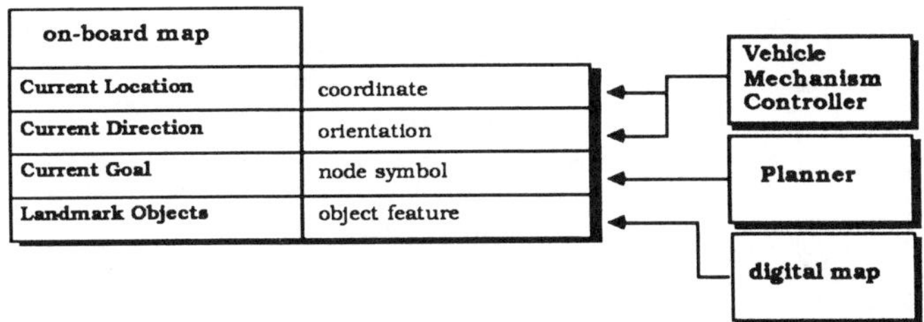

FIGURE 8.9. On-board road map generation scheme.

8.6 Frustration Resolution Schemes for Top-Down Processing

The edge pattern is transferred to the geometric matching process in which two kinds of top-down image analysis mechanisms are activated for resolving two kinds of frustration: symbolic and geometric. Symbolic frustration is evoked by the generation of a copy of an environment model as the on-board map in the driver's associate that is not yet completed. The generation procedure is illustrated in Figure 8.9, where the on-board map is shown to be completed by binding together information concerning positioning, that is, the current location and direction; the route to be traced, that is, the current goal; and the environment to be recognized, that is, the landmark objects. As far as the information on current positioning and the current goal, the completion procedure of the on-board map is easily implemented by a simple data transfer from the vehicle mechanism controller to the route planner. However, since the landmark objects are associated with typical but not specific, contour models in the digital road map, the specific model of landmark objects in the scene, called the object feature, should be determined through an implicit database access. An implicit database access mechanism has been implemented in the interactive work space model generation system for robot operation [16, 18]. In this access mechanism, symbolic frustration is induced on the frame-based world model, consisting of knowledge of the environment and objects. The scenario for frustration resolution is illustrated in Figure 8.10. In this figure, symbolic frustration is shown to be evoked by generating an a priori scene model, designated by "Roadway_scene," with the completion method "Identify_All_Visible_Rectangles." The frustration is resolved by activating the contour matching procedure "Analyze_Edge_Structure" according to the scenario described in the "Identify_All_Visible_Rectangles" procedure. An example of the frustration evocation and resolution scheme is illustrated as follows:

$$identify\_a\_scene(\_scene,\_data\_r,\_data\_o,\_status\_r,\_status\_o) \quad :-$$
$$identify\_a\_roadway(\_scene,\_data\_r,\_status\_r),$$
$$identify\_objects(\_scene,\_data\_o,\_status\_o). \qquad (8.3a)$$

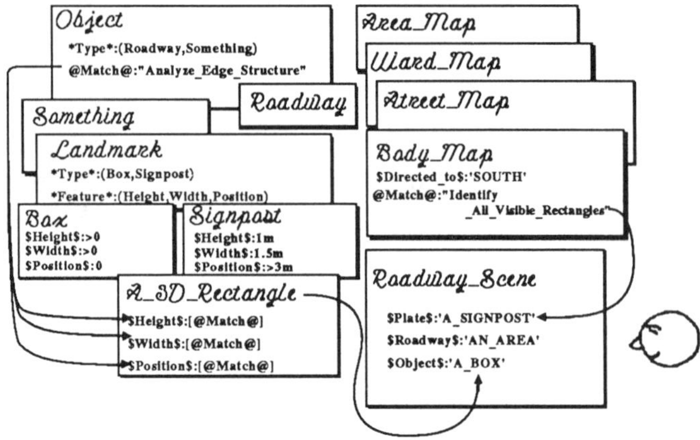

FIGURE 8.10. A scenario for roadway model generation.

identify_a_roadway(_scene,_data,_status) :–
fremove_all(_scene, $roadway$,'value),
inquire_vision(_scene, $roadway$,'default,_data,_status),
get_process(_scene, $roadway$,'if_need,_process),
process(_process,_scene, $roadway$,_status). (8.3b)

identify_objects(_scene,_data,_status) :–
fremove_all(_scene, $objects$,'value),
inquire_vision(_scene, $objects$,'default,_data,_status),
get_process(_scene, $objects$,'if_needed,_process),
process(_process,_scene, $objects$,_status). (8.3c)

This evocation-resolution scheme [(8.3)] is described using the prologue syntax as well as the motion schema utilized for autonomous robot operation [6]. In the interpretation of the motion schema, backtracking is not permitted because the nonuniqueness of the mechanism motion in a complicated environment causes a combinatorial explosion. However, since the convergent algorithm can be designed for the evocation-resolution process, Equation (8.3) is implemented by a standard prolog system. An example of the results of the database access experiment is illustrated in Figure 8.11. In this experiment, a human operator, instead of a planning system, activates the evocation process by pointing the target object, a pipe, in the view of the scene, as shown in Figure 8.11(a). The evocation process generates an empty map and activates the resolution process to complete the empty map. The completed map is indicated to the human operator, as illustrated in Figure 8.11(b). In this figure, the result of the process is displayed by a set of line drawings associated with the matched model on the scene image.

Geometric frustration is caused by 2D disparity between the extracted image feature and the reference generated by the perspective projection of the

(a)

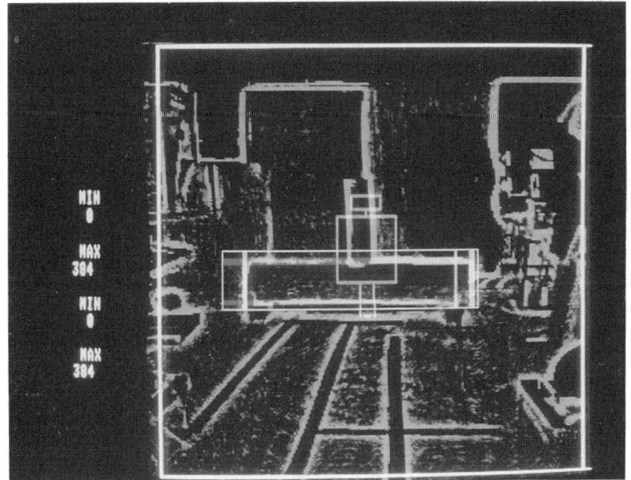

(b)

FIGURE 8.11. Experimental results of database access: (a) activation by pointing; (b) indication of access result.

3D contour model, which is specified as the result of symbolic frustration resolution. Let π and η be the reference and extracted image features, respectively. Assume that the reference is generated as a function of parameter θ. Additionally, let the geometric frustration be evaluated in terms of an interaction between π and η satisfying the following properties:

$$\Phi[\eta, \pi(\theta)] \geqq 0 \quad \text{for any} \quad \eta, \pi \in \Sigma, \tag{8.4a}$$

$$\Phi[\eta, \pi(\theta)] = \Phi[\eta, \pi(\theta)], \tag{8.4b}$$

$$\Phi[\eta, \pi(\theta)] \leq \Phi[\eta, \pi'(\theta)] \quad \text{if} \quad \rho[\eta, \pi'(\theta)] \leq \rho[\eta, \pi(\theta)], \qquad (8.4c)$$

where $p[\cdot, \cdot]$ denotes the distance in Σ.

By introducing the Newton type gravity for evaluation of the interaction between patterns η and π, the interaction $\Phi[\eta, \pi(\theta)]$ can be successively approximated by the solution to the following 2D diffusion system:

$$\partial\phi/\partial t = \Delta\phi + \chi[\pi(\theta)], \qquad (8.5a)$$

$$\phi(s) = 0, \qquad s \notin \Sigma, \qquad (8.5b)$$

$$\psi(s) = \chi[\eta]\nabla\phi(s), \qquad (8.5c)$$

$$\Phi[\eta, \pi(\theta)] = \int_\Sigma \psi[\eta, \pi(\theta)](s)\, ds. \qquad (8.5d)$$

where ∇ is the gradient operator, $\chi[(\cdot)]$ is the characteristic function of a set $(\cdot)$, and ψ is the gradient vector on π of the field ϕ generated by η. Assume that $\pi(\theta)$ is generated near η and let the optical flow between η and $\pi(\theta)$ be approximated by the vector field $\psi[\eta, \pi(\theta)]$. Then, the geometric frustration $\Phi[\eta, \pi(\theta)]$ is reduced by updating the parameter θ as follows:

$$d\theta/dt = \int_\Sigma \Omega(s)\psi[\eta, \pi(\theta)](s)\, ds, \qquad (8.6)$$

where, Ω is a 2×3 distributed gain specifying the gain for the direction, distance, and rotation parameter update. The update algorithm, Equation (8.6), has been proved to be asymptotically stable if the gain matrix $\Omega(s)$ satisfies the following condition, named the structural consistency condition:

$$\alpha(s)[\partial u(s)/\partial\theta] = \Omega(s), \qquad 0 < \kappa < \alpha(s), \qquad (8.7)$$

where u is the optical flow associated with the reference and image feature patterns. An example of the gain matrix $\Omega(s)$ is illustrated in Figure 8.12, where $\Omega(s)$ is described in terms of the computation mechanism for the vertical and horizontal forces, F_x and F_y, respectively, and the moment N under the contex "BOX is on the ROADWAY" [7].

8.7 2-D Syntax Analysis for Bottom-Up Processing

Consider the bottom-up surveillance process to simultaneously recognize grammatical models of generalized objects, that is, "RECTANGLE" and "ROADWAY" on the segment array. Let the totality of nonterminal symbols associated with partial description of start symbols be designated by N. The start symbols, "RECTANGLE" and "ROADWAY," are elements of N and a set of rewrite rules on N is denoted by Π. A recursive syntax analysis mechanism is illustrated in Figure 8.13. In this mechanism, the recognition result is represented as a network on the terminal symbol array, designated by $\mathcal{T}$, and

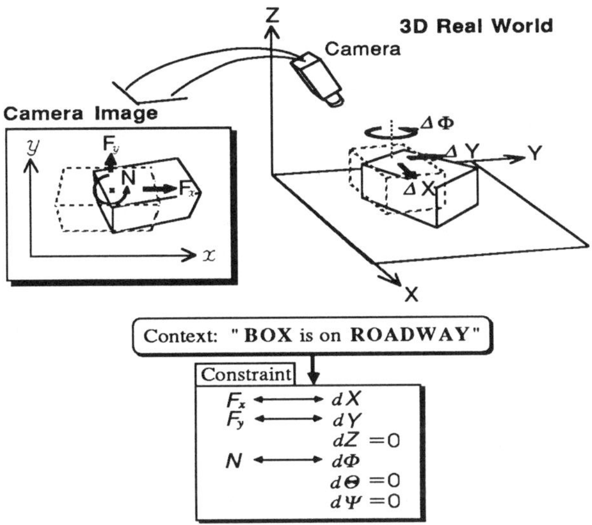

FIGURE 8.12. A recursive 3-D parameter estimation scheme.

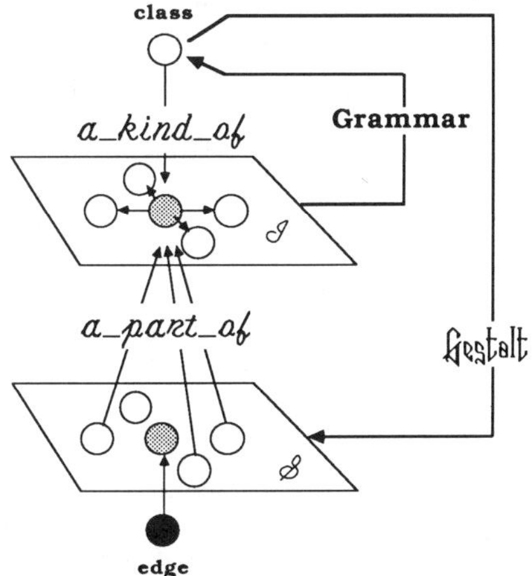

FIGURE 8.13. Locally parallel parsing algorithm.

defined on the segment array, denoted by $\mathscr{S}$. The grammatical description is applied to the local connection in $\mathscr{T}$ for hypothesizing a nonterminal symbol as a class representation of the possible recognition result. The hypothesized nonterminal symbol is verified through a consistency test with respect to the connectability between segments, called a Gestalt constraint, to recognize for the local connection to be a part of the instance of the recognition result associated with the hypothesized nonterminal symbol. Let the mutual location in a 2-D array be specified in terms of four local connections, labeled by upper, left, lower, and right. By introducing these local connections into the segment array, the parsing mechanism can be designed, without explicit *ij* addressing in the rule description, in terms of successive applications of localized rewrite rules described as follows:

$$p: v \to \sigma v', \quad v, v' \in N, \tag{8.8a}$$

$$\sigma = [(recognition\ result\ of\ upper\ segment)$$
$$(recognition\ result\ of\ left\ segment)$$
$$(recognition\ result\ of\ lower\ segment)$$
$$(recognition\ result\ of\ right\ segment)]. \tag{8.8b}$$

For example, a rewrite rule to recognize a corner of a rectangle is shown as follows:

$$p: \langle upper\_left\_corner \rangle \to \sigma \langle vertex \rangle, \tag{8.9a}$$

$$\sigma = ((upper\ segment\ is\ dont\_care)$$
$$(left\ segment\ is\ dont\_care)$$
$$(lower\ segment\ is\ \langle vertical\_side \rangle)$$
$$(right\ segment\ is\ \langle horizontal\_side \rangle)), \tag{8.9b}$$

where dont_care means an arbitrary nonterminal symbol is allowed. The context of grammatical analysis is locally determined by the path connecting nonterminal symbols associated with each terminal symbol. For integrating the local context into a scenario for global pattern analysis, let an ordering rule be introduced in the set of nonterminal symbols as follows:

$$if\ there\ exists\ a\ rule\ p^* \in \Pi[v_1]: v_1 \to \sigma^* v_2,$$
$$then\ v_1 > v_2, \tag{8.10a}$$

$$if\ there\ exists\ a\ rule\ p^* \in \Pi[v_1]: v_1 \to \sigma^* n_2,\ for\ v_2 > v_3,$$
$$then\ v_1 > v_3. \tag{8.10b}$$

For example, an intermediate recognition set of rectangles is ordered as follows:

$$\langle rectangle \rangle$$

$$> \{\langle upper\_left\_threequarters \rangle,$$
$$\langle upper\_right\_threequarters \rangle,$$
$$\langle lower\_right\_threequarters \rangle,$$
$$\langle lower\_left\_threequarters \rangle\}$$

$$> \{\langle left\_half \rangle, \langle upper\_half \rangle, \langle right\_half \rangle,$$
$$\langle lower\_half \rangle\}$$

$$> \{\langle upper\_left\_corner \rangle, \langle upper\_right\_corner \rangle,$$
$$\langle lower\_right\_corner \rangle, \langle lower\_left\_corner \rangle\}$$

$$> \{\langle vertical\_side \rangle, \langle horizontal\ side \rangle, \langle vertex \rangle\}. \tag{8.11}$$

Noting that the totality of rewrite rules, Equation (8.8), can be partitioned into the sets of rules $\{\Pi|v[1], \Pi|v[2], \dots\}$, where $\Pi|v[i]$ is the totality of rules generating $v[i]$, the 2-D attributed rewrite rules, Equation (8.8), and ordered nonterminal symbols are combined in a recursive procedure for parallel grammatical analysis described in terms of the following abstract dynamical system:

$$p \in \Pi[v]: \mathit{if}\,(\sigma v)\,\mathit{then}\,(v'),\, v \in N[\sigma], \tag{8.12a}$$

$$\sigma = (\mu[i, j - 1]\mu[i - 1, j]\mu[i, j + 1]\mu[i + 1, j]), \tag{8.12b}$$

$$\mu[i, j] = ((\mathit{nonterminal\ symbols\ associated\ with}\ \Sigma[i, j]) \\ (\mathit{terminal\ symbols\ labeled\ to}\ \Sigma[i, j])), \tag{8.12c}$$

where $N[\sigma]$ is the totality of nonterminal symbols associated with the segment. In this procedure, the recognition steps are initiated by the excitation provided through the variation of terminal symbols directly linked to the image feature. At the same time, the recognition steps autonomously progress through successive renewal of nonterminal symbols by activating rewrite rules in accordance with the scenario described in terms of the recognition result itself. An abstract dynamical system, Equation (8.12), specifies a successive generation mechanism of a spatially sensitive network in essentially abstract nonterminal symbols without any global positioning mechanism. In a dynamical system, Equation (8.12), the nonterminal renewal algorithm, Equation 8.12(a), is of the same form as that of an ordinary production system [24], so computer implementation is achieved by attaching the following rule interpreter to each segment:

$$p \in \Pi[v]: \mathit{if}\,(\sigma v)\,\mathit{then}\,(v'),\, \mathit{for\ highest\ order}\ v \in N[\sigma]. \tag{8.13}$$

In Equation (8.13), the highest ordered nonterminal is copied into the working memory at each recognition step to avoid the reactivation of rules already used. Although a rule interpreter, Equation (8.13), is easily implemented in conventional architecture computer systems, the selection of the highest nonterminal requires an additional procedure because of the nonuniqueness of the highest ordered nonterminal set, as shown in Equation (8.13). In addition, since the highest ordered nonterminal is not necessarily on the path to the nonterminal associated with the final recognition result, it is not easy to design a generalized conflict resolution algorithm in image structure analysis. The difficulty in conflict resolution can be overcome by activating all nonterminals associated with each segment. This procedure is as follows:

$$p \in \Pi[v]: \mathit{if}\,(\sigma v)\,\mathit{then}\,(v'),\, \mathit{for\ arbitrary}\ v \in N[\sigma]. \tag{8.14}$$

In the chaining mechanism, Equation (8.14), the linkage of associated nonterminals is reinforced at each recognition step. Additionally, the associated region in the nonterminals, $N[\sigma]$, is successively expanded through the feedback of newly associated nonterminals as follows:

$$N[\sigma] \rightarrow N[\sigma] + v'. \tag{8.15}$$

Let the associated region in the nonterminals, $N[\sigma]$, be successively

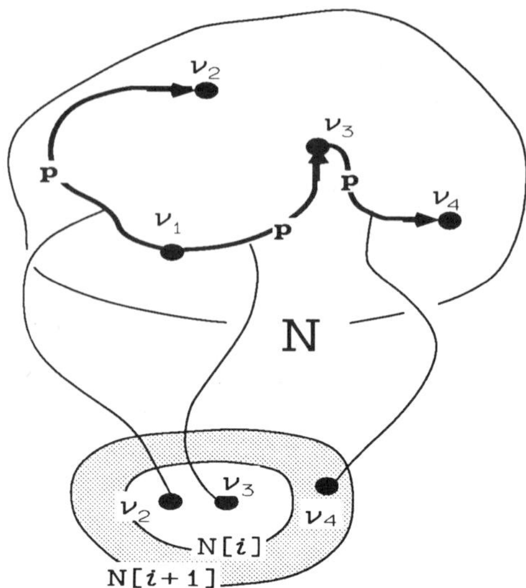

FIGURE 8.14. Stability of network reinforcement mechanism.

expanded through the feedback of newly associated nonterminals [5], as illustrated in Figure 8.14. Then, the network reinforcement procedure is as described in the following recursive N-update algorithm [17, 20]:

$$p \in \Pi[n]: \textit{if } (\sigma v) \textit{ then } (v'), \qquad v \in N[\sigma], \tag{8.16a}$$

$$N[\sigma] \rightarrow N[\sigma] + v'. \tag{8.16b}$$

Since algorithm (8.16b) implies that the associated nonterminal set, $N[\sigma]$, monotonically increases as the progress of recognition steps, the N-update algorithm converges only if the totality of nonterminal symbols is a finite set. However, the convergence is weak in the sense that the maximum networked region in the nonterminals attains a steady state, and the steady state of $N[\sigma]$ is not ensured to be minimal. A stronger version of the convergence concept is introduced based on a subset of $N[\sigma]$ consisting of the most effective nonterminals in $N[\sigma]$: A nonterminal symbol $v^* \in N[\sigma]$ is said to be most effective if v^* is of highest order in $N[\sigma]$ and there exists a rule p^* that rewrites v^* according to Equation (8.8). Let the most effective nonterminals be denoted by $M/N[\sigma]$. Then, the network reinforcement algorithm with an M/N update process is formulated as follows:

$$p \in \Pi[v]: \textit{if } (\sigma v) \textit{ then } (v'), \qquad v \in M/N[\sigma], \tag{8.17a}$$

$$N[\sigma] \rightarrow N[\sigma] + v', \tag{8.17b}$$

$$M/N[\sigma] = \{v^* | v^* > v \in \{N[\sigma] - M/N[\sigma]\} \quad \textit{and} \quad \Pi[v^*] \neq \phi\}. \tag{8.17c}$$

A block diagram of the M/N update algorithm is shown in Figure 8.15, where

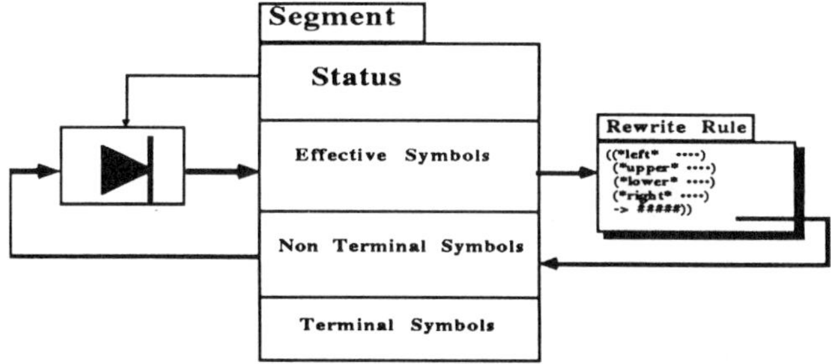

FIGURE 8.15. A network reinforcement mechanism.

the generation procedure of the $M/N[\sigma]$ set is symbolized by a semantic diode. In this M/N update algorithm, since the associated nonterminal set, $N[\sigma]$, monotonically increases, as in the N update algorithm, Equation (8.16), the recognition process is weakly convergent, as in the N update algorithm. In addition, because of the restriction of rule activation into a subset $M/N[\sigma]$ in $N[\sigma]$, the steady state of $M/N[\sigma]$ indicates the final recognition result. However, the M/N update algorithm yields backtracking caused by a static constraint described by Equation (8.17a). The difficulty in the implementation of the backtracking mechanism can be overcome by replacing the algorithm with static constraint, Equation (8.17c), with the following equivalent full recursive algorithm, called the dynamic effective firing algorithm:

$$p \in \Pi[v]: \textit{if } (\sigma v) \textit{ then } (v'), \qquad v \in M/N[\sigma], \tag{8.18a}$$

$$N[\sigma] \rightarrow N[\sigma] + v', \tag{8.18b}$$

$$\Lambda/N[\sigma] = \{v^*|v^* > v \in N[\sigma]\}, \tag{8.18c}$$

$$N[\sigma] = \{N[\sigma] - \Lambda/N[\sigma]\}, \tag{8.18c}$$

$$M/N[\sigma] = \{v^*|v^* \in \Lambda/N[\sigma] \qquad \textit{and} \quad \Pi[v^*] \neq \phi\}. \tag{8.18d}$$

8.8 Hardware Structure

These top-down and bottom-up algorithms have been evaluated through a series of experimental and simulation studies. Based on these evaluations, a set of VLSIs have been designed as key devices for the hardware implementation of these algorithms.

To solve the 2-D diffusion equation, which is commonly utilized in both the linearized Mach effect algorithm, Equation (8.2), and geometric frustration resolution, Equation (8.5), a local pattern processor (LPP) has been developed. The LPP is a LSI with 136-K transistors on an 8.5 × 9.1 chip, as illustrated in Figure 8.16. The LPP is essentially a TV-rate (6 MHz) locally parallel processor that transforms a digital image field into another image field by

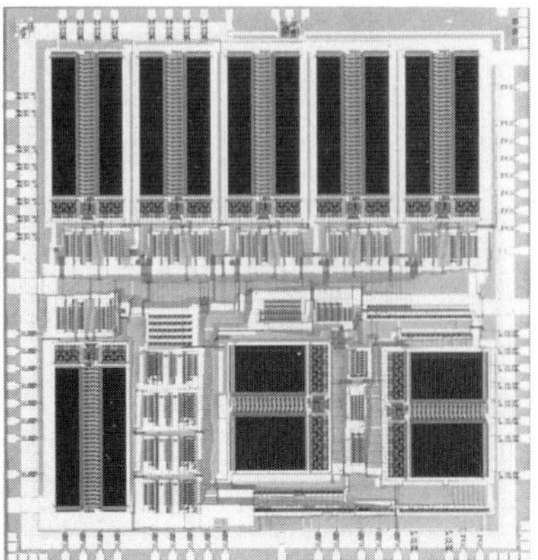

FIGURE 8.16. VLSI implementation (1) pattern processor.

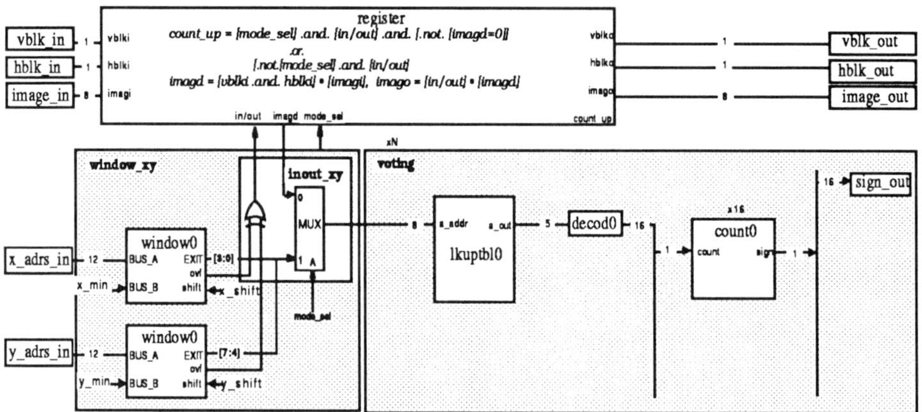

FIGURE 8.17. VLSI implementation (2) volting processor.

aggregating the brightness intensity distribution in four neighboring pixels. In the LPP, neighboring pixels are linked through a lookup table so as to implement arbitrary linear and nonlinear 2-D differential operators. Adding to this generality, the LPP also provides the timing signal, that is, the horizontal and vertical clocks, synchronized with transformed pixel data for the convenience of direct LPP–LPP linkage needed in algorithms (8.2) and (8.5).

For the specification of terminal symbols in 2-D window arrays, a Hough parameter estimator (HPE) has been designed, as illustrated in Figure 8.17. The HPE is also a TV-rate voting processor equipped with a window operation. The window operation is crucial for both detectability and model simplification [34]. By restricting the voting area into a small window, the sensitivity of the image feature in the (ρ, θ) parameter space is improved. In addition, the hardware implementation of the window control mechanism implies

FIGURE 8.18. An Image processor implemented by four LPPs.

that the communication between segment detection and 2-D parsing, which may cause a bottleneck in a parallel distributed version of a bottom-up process, is no longer the computation load in a hypothesis-testing intensive algorithm, Equation (8.18). The HPE is designed to decode an input timing signal into the (x, y) address needed in the voting process so as to operate using only the set of scanned image data and timing signals provided by LPP.

These VLSIs can be applied to a 256×256 pixel digital image with an 8-bit gray level for image feature extraction and segment detection by voting within 16 ms. Additionally, these devices are equipped with interface circuits for mutual linkage. Direct linkage capability of image processing devices is crucial for compact system design. For example, a TV-rate recursive image processor for solving 2-D diffusion equations is configured as illustrated in Figure 8.18, where four LPPs are compactly allocated in a standard IEEE-796 bus board.

8.9 Dynamic Image Analysis Mechanism

Using hardware described in the previous section, top-down and bottom-up algorithms are integrated into a unified dynamic system, that is, the driver's associate. The conceptual design of the image analysis mechanism for the driver's associate is illustrated in Figure 8.19. In this figure, the driver's associate system is shown to be of a version of the perception-control architecture that plays a crucial role in continuous navigation of autonomous mobile robots [11]. In the driver's associate system, the image feature extracted by the linearized Mach effect (LME) processor is transferred to the display controller through the location parameter estimator, which is implemented by

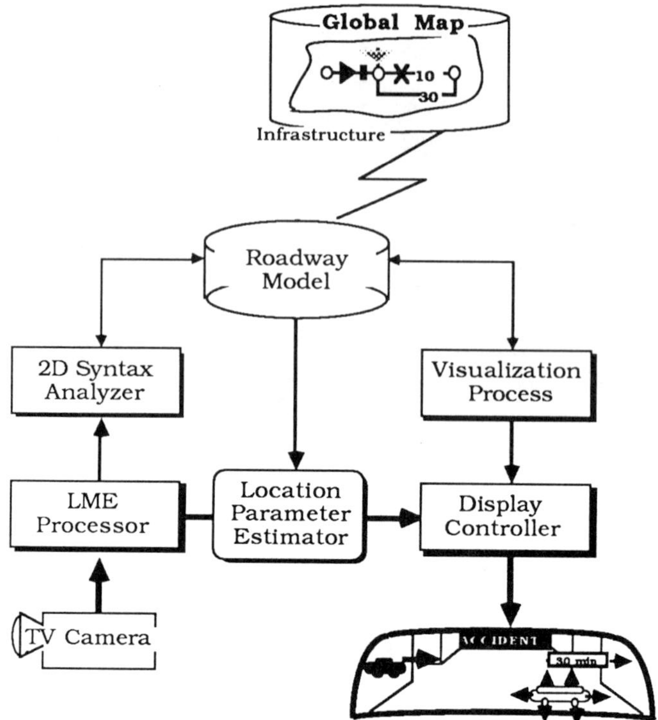

FIGURE 8.19. Perception-control architecture.

the recursive image processor illustrated in Figure 8.18. In this location pattern estimator, 3-D geometric data for landmark and obstacle objects are matched with the image feature to update simultaneously the self and object locations by solving a 2-D diffusion equation excited by the image feature pattern. At the same time, the TV-rate voting processor, implemented using a set of HPEs, is applied to the image feature pattern, generated by the LME processor, to specify a 2-D segment array for the 2-D parsing processor. In the 2-D syntax analyzer, the structure of the image feature pattern is extracted through the 2-D recursive parsing process, Equation (8.18), and the obstacle object model in the roadway map is identified. The object model is identified through a nondeterministic algorithm, as stated in Section 7. The 2-D syntax analyzer successively updates the roadway model.

The roadway map is edited into a mental image of the transportation process in its visualization process. The visualization process activates a set of rules describing the association process between observed scenes and the memorized symbolic roadway map to generate a design scenario for the view, as in the complicated manufacturing process [8]. The design scenario is transmitted into the display controller and combined with location information provided by the location parameter estimator to indicate a set of iconic patterns on a display device. An example of the view is illustrated in Figure 8.19, in which the objects and messages are displayed on the front glass. The physical design of this overlay indication can be implemented easily using a head-up display device. However, of the display effect, the efficiency in message transmission and any distractions caused to the driver by the display should be carefully examined through psychological studies.

As illustrated in Figure 8.19, the driver's associate, as well as the mobile robot navigator [15], is designed on a simplified TV-rate data line. From this high-speed data path, symbolic information is also extracted through an inclusion-based association process, and into this high-speed data path, a control parameter is provided through two channels, the 3-D object model into the location parameter estimator and the view design into the display controller. This asynchronous connection of a high-speed deterministic data processing mechanism with a nondeterministic low-speed reasoning mechanism contributes to the implementation of the driver's associate by small-sized and flexibly structured hardware for an on-board system. Adding to the asynchronism, the driver's associate is implemented, essentially, on the basis of recursive algorithms that also exhibit great advantages in direct coupling with other control devices. Therefore, even though the driver's associate is not intended for automatic vehicle control, it is not difficult to apply the information captured by the driver's associate to various kinds of vehicle mechanism control systems. In extending the scope of the driver's associate, the symbolic and explicit representation of the transportation process, implemented in the roadway map, the 2-D syntax analyzer, and the visualization process, can also be a useful guideline in the design of a control policy, as well as in reliability improvement of the control signal.

8.10 Basic Operation Scenario

As illustrated in Figure 8.3, the transportation process is essentially represented by the interactive generation and dynamic verification of a symbolically described plan, as is the manufacturing processes [6]. This implies that the software system of the driver's associate is expected to be designed in a way similar to that of an interactive robot operation for intelligent manufacturing [8]. Thus, although the control program has not yet been completely written, the basic operation scenario for the driver's associate is planned as follows.

Initially, the grammatical model of "something" and roadway are loaded into a parsing processor for a two-sided examination of the situation, that is, the vehicle is correctly positioned on the roadway, and at the same time, there is nothing that requires attention. In this normal situation, the 3-D contour model for a null object is matched with the image feature pattern, and a null frame is generated for the scene model. Thus, neither geometric nor symbolic frustration is caused, and no message is indicated to the driver. An example of

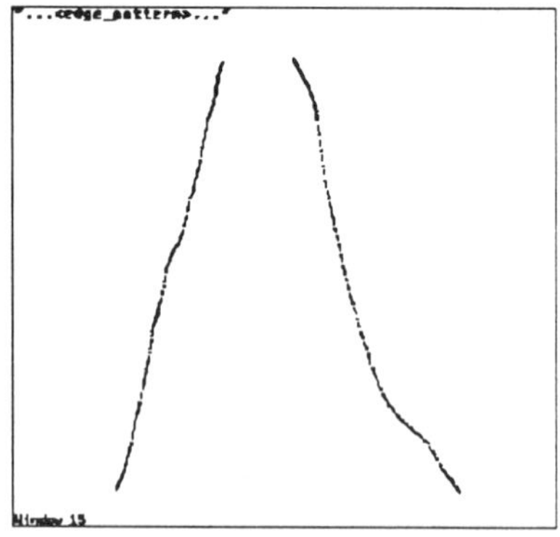

(a) Image Feature Pattern

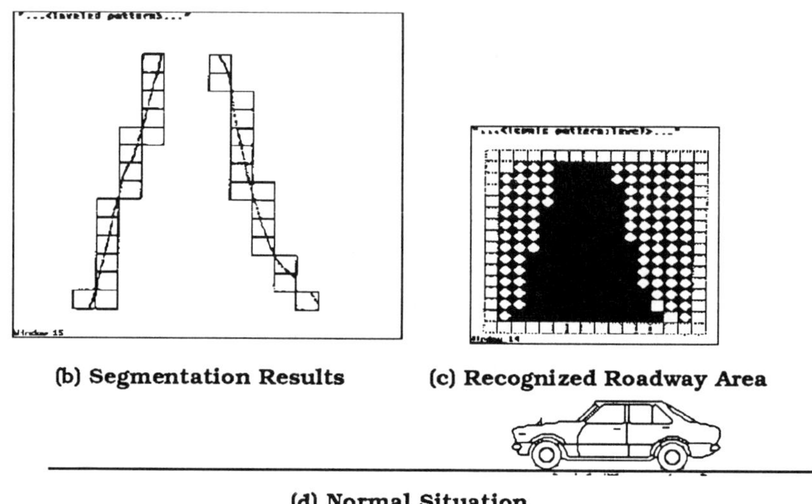

(b) Segmentation Results   **(c) Recognized Roadway Area**

(d) Normal Situation

FIGURE 8.20. A scene analysis scenario, normal status: (a) image feature pattern; (b) segmentation results; (c) recognized roadway area; (d) normal situation.

the simulated operation of the driver's associate in the normal situation is illustrated in Figure 8.20, where a roadway that is not geometrically specified is detected in the scene. In this simulation, the rule for the roadway detection is designed within the following recursive definition

> *The bottom-center window is always a part of the roadway.* (8.20a)

> *A 2-D window connected with the window already recognized*
> *as a part of the roadway is also a part of the roadway if*
> *no edge pattern is detected in the window.* (8.20b)

The definition is hand-compiled into a set of 2-D rewrite rules and applied to each window concurrently. The results of roadway detection are also illustrated in Figure 8.20.

When the bottom-up process detects objects on the segment array, the object shapes are approximated by a circumscribing box with the image feature described by a rectangle. A specification process is activated for the 2-D location and 2-D size of the rectangle. An example of simulated behavior of the driver's associate in detecting "somethings" in the image feature is illustrated in Figure 8.21. The 2-D recognition result, that is, the size and

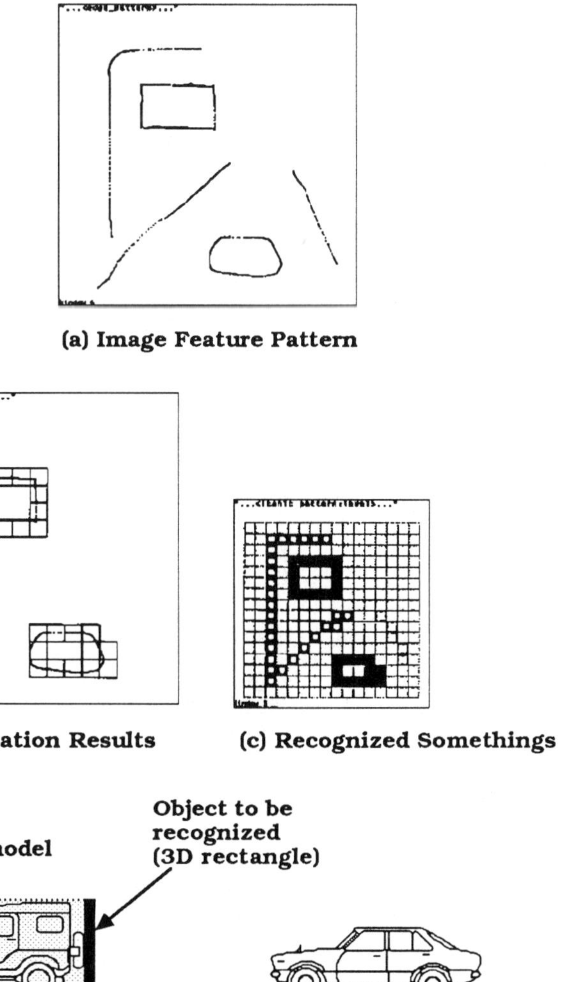

(a) Image Feature Pattern

(b) Segmentation Results   **(c) Recognized Somethings**

(d) Complicated Situation

FIGURE 8.21. A scene analysis scenario, frustration resolution process: (a) image feature pattern; (b) segmentation results; (c) recognized somethings; (d) complicated situation.

location of rectangles in the image feature, is transferred to the geometric frustration resolution mechanism, and the pattern is locked on the view of the scene. At the same time, the relative location of the rectangle on the simultaneously recognized roadway area is examined on the 2-D segment array to classify whether the rectangle is associated with an object located on the roadway. For the case of an object on the roadway, the 2-D form of the rectangle is interpreted as a 3-D view of a box based on the location of the bottom line. The reference of the geometric frustration resolution process is substituted by the perspective projection of a 3-D box. Following this substitution, the geometric frustration resolution process calculates precisely the relative 3-D velocity between the vehicle and the 3-D box.

8.11 Discussion

Although the driver's associate does not directly control the vehicle mechanism, it captures essentially the same information from the scene as a vision system for autonomous maneuvering. This information equivalence implies that the driver's associate can be utilized to recognize the insufficiency in understanding the environment and inquire the transportation management system to determine scenes not yet visible for well-informed decision making. The reduction in distractions due to this autonomous information retrieval is expected to increase driving safety significantly because it was pointed out that collision accidents are reduced to $\frac{1}{10}$ normal levels only by notifying the driver of the situation just 1 in advance [3]. The driving efficiency of each privately owned vehicle is seriously dependent on the behavior of the global transportation process. On the other hand, the traffic load is considerably reduced if vehicles are guided to optimal routes. Thus, the benefits to each vehicle and transportation management will be simultaneously maximized through dynamic updating of the global road map by aggregating local environment models generated in each driver's associate. According to experiments in the United Kingdom and Japan, a reduction of approximately 10% of the transportation cost is expected by dynamic route guidance [10].

Although remarkable benefits are expected, reciprocal linkage between the driver's associate and the traffic management systems requires innovations in the design and maintenance of an essentially cooperative system consisting of a very large number of independent agents, that is, privately owned vehicles, and a hierarchy of dynamic information storage systems. In this cooperative system, a dynamic, comprehensive, and, at the same time, clearly understandable transportation process model should be introduced as the basis of the system design. By describing the transportation process as a dynamical system with a structured state variable, that is, the traffic parameter, on a graph, that is, the node map, and by sharing this process model in both the driver's associate and traffic management systems, the dynamic binding mechanism is implemented in terms of the identification scheme for the environment model, as in the case of the self-navigation system for mobile robots. However, the superiority of peripheral vehicles in a cooperative system requires an on-board vision system of a much more extended capability than in an indoor mobile robot in aspects of

- geometric map generation through local scene analysis under an a priori symbolic-roadway-model provided transportation management system, and
- decision making on the objects to be recognized through reasoning on the global symbolic map controlled by a driving context model.

To solve these problems, the introduction of a parallel distributed processing paradigm [28] is expected for the development of an intensive and flexible reasoning mechanism.

The comprehensive information aggregation on the network generated by the cooperation between the traffic management system and the driver's associate also contributes to the driving process itself by providing reliable prediction and blind information to the vehicle mechanism controller. Adding to this enhancement of physical capabilities, the ability to comprehend the scene is also expected to extend the adaptability of the driving process to the driver's behavior. This is essentially a result of the driver's understanding of the global transportation process. Currently, traffic management and the intelligent vehicle are recognized to be closely correlated problems that are finally solved through dynamic optimization of the transportation process [1]. However, since the transportation process is one of the largest systems and includes vehicle mechanism control, large-scale database management, and, particularly, human–computer interaction, the complete specification of the design target is not easy. In fact, the transportation process should be clarified only based on the experience using an infrastructure of a practical size. This chicken-or-egg paradox will be solved by a step-by-step buildup of the transportation process consisting of a partially developed infrastructure and partial implementation of an on-board computer system. For this step-by-step implementation, it is crucial that the system architecture is explicitly described and the interface is clearly specified between subsystems. In Table 8.2, the interface between the infrastructure and the driver's associate is indicated for cooperative identification of the transportation process. Although

TABLE 8.2. Cooperation of on-board system with infrastructure.

Aspect or facet	On-board system	Infrastructure
Scope	Visible scene	Global network
Roadway description	3-D plane	Link
Recognition level	Geometric	Symbolic
Service	Real view	Traffic parameter
Operation	Dynamic, continuous	Static, periodic

the driver's associate has a unique internal architecture, as shown in Figure 8.19, the interaction with the infrastructure is clearly designed based on local representation of a roadway map, which can be designed based on various ALV research [31] and on-board navigation systems. In fact, a first version of the driver's associate system can be designed exploiting a conventional navigation system as an on-board copy of a node map and low-speed telecommunication system for adding traffic parameters to this node map. However, in the earlier stage of system development, the introduction of a full-size on-board recognition system may be required, at least in a portion of official vehicles, to maintain the global road map system.

8.12 Concluding Remarks

The concept of the driver's associate was presented for environment identification in cooperation with human drivers. The self-navigation system for autonomous mobile robots is reconstructed into a driving information system to extend the subliminal perception of human drivers. Sharing the image feature extracted and segmented into observed imagery, top-down and bottom-up processes were integrated into a unified dynamic perception mechanism. The image analysis mechanism was designed within the framework of the perception-control architecture and implemented by a set of TV-rate VLSIs. The interface with the infrastructure was designed as a local map that can easily be shared by an on-board system and an infrastructure system not yet completely specified. The schematics of the software was verified through simulations to demonstrate that the basic operation scenario of the driver's associate can be implemented within the framework of proposed image processing methods.

Acknowledgments. The authors would like to express their sincere gratitude to Professor Yoshifumi Sunahara of the Kyoto Institute of Technology for his helpful advice in the application of the distributed parameter system theory to image analysis. Thanks are extended to Professor Hironao Kawashima of Keio University for valuable discussions on the introduction of machine vision in a vehicle information system in cooperation with an infrastructure.

References

[1] Aono, S. (1989). "Technology for the Intelligent Car of the Future." *Proc. JSK International Symposium—Technological Innovations for Tomorrow's Automobile Traffic and Driving Information Systems*, 109–119.

[2] Behrent, J. (1989). "Development of Automobile Traffic Information System in the FR Germany and Its Importance." *Proc. JSK International Symposium—*

Technological Innovations for Tomorrow's Automobile Traffic and Driving Information Systems, 41–52.

[3] Betsold, R. J. (1989). "Intelligent Vehicle/Highway Systems for the United States —An Emerging National Program." *Proc. JSK International Symposium— Technological Innovations for Tomorrow's Automobile Traffic and Driving Information Systems*, 53–59.

[4] Davis, L. S., and Kushner, T. (1985). "Road Boundary Detection for Autonomous Vehicle Navigation." Technical Report CAR-TR-140/CS-TR-1538, University Maryland, July.

[5] Fauconnier, G. (1985). *Mental Spaces*. MIT Press, Cambridge, Massachusetts.

[6] Hamada, T., Kamejima, K., and Takeuchi, I. (1988). "Knowledge Representation for Image Based Robot Operation." *Proc. IEEE Int. Workshop on Artificial Intelligence for Industrial Applications*, 417–422.

[7] Hamada, T., Kamejima, K., and Takeuchi, I. (1989). "Dynamic Work Space Model Matching for Interactive Robot Operation." *Proc. IEEE Int. Workshop on Industrial Application of Machine Intelligence and Vision*, 82–87.

[8] Hamada, T., Kamejima, K., and Takeuchi, I. (1989). "Image Based Operation: A Human-Robot Interaction Architecture for Intelligent Manufacturing." *Proc. IEEE IECON'89, 15th Annual Conference of IEEE Industrial Electronics Society, Philadelphia, November 6–10*, 556–561.

[9] Herbert, M., and Kanade, T. (1985). "3-D Vision for an Autonomous Vehicle." *Proc. IEEE International Workshop on Machine Vision and Machine Intelligence*, 375–380.

[10] Jeffery, D. J. (1989). "In-Vehicle Route Guidance. The Future for Automobile Transport and Traffic Operation in Europe." *Proc. JSK International— Technological Innovations for Tomorrow's Automobile Traffic and Driving Information Systems*, 61–79.

[11] Kamejima, K., Ogawa, Y. C., and Nakano, Y. (1984). "Perception-Control Architecture in Image Processing for Mobile Robot Navigation System," *Proc. SICE-IEEE IECON'84*, 52–57.

[12] Kamejima, K., Ogawa, Y. C., and Nakano, Y. (1986). "Image Structure Detection with Application to Mobile Robot Navigation." *Proc. IEEE IECON'86*, 713–718.

[13] Kamejima, K., Ogawa, Y. C., and Nakano, Y. (1986). "Image Structure Identification Using Parallel Production Systems." *Proc. 18th JAACE Symp. on Stochastic Systems Theory and Its Applications*, 245–248.

[14] Kamejima, K., Funabashi, M., and Ichikawa, Y. (1986). "Knowledge Processing for Mobile Robots." *Hitachi Review* **35** 9–12.

[15] Kamejima, K., Ogawa, Y. C., and Nakano, Y. (1987). "A Fast Algorithm for Approximating 2D Diffusion Equation with Application to Pattern Detection in Random Image Fields." Futagami, T., Tzafestas, S. G., and Sunahara, Y. eds., *Distributed Parameter Systems: Modeling and Simulation, Proc IMACS/IFAC Int. Sym on Modeling and Stimulation of Distributed Parameter Systems*, 557–564.

[16] Kamejima, K., Takeuchi, I., Ogawa, Y. C., and Hamada, T. (1988). "Interface for Interactive Robot Operation." *Proc. IAEA International Conference on Man-Machine Interface in the Nuclear Industry*, 559–564.

[17] Kamejima, K. (1988). "A Recursive Network Reinforement Algorithm for Stochastic Reasoning in Segment Detection." *Proc. 20th ISCIE Symp. on Stochastic Systems Theory and Its Applications*, 223–226.

[18] Kamejima, K., Takeuchi, I., Hamada, T., and Watanabe, Y. C. (1988). "A Direct Mental Image Manipulation Approach to Interactive Robot Operation." *Proc. IEEE Int. Workshop on Intelligent Robots and Systems*, 673–678.

[19] Kamejima, K., and Watanabe, Y. C. (1988). "Environment Identification for Autonomous Mobile Robot Operation." *Proc. IEEE Int. Workshop on Intelligent Robots and Systems*, 645–650.

[20] Kamejima, K. (1989). "Generic Model: Nonlinear Interaction of Potential Fields for Micro-Pattern Understanding." *Proc. 21th ISCIE Symp. on Stochastic Systems Theory and Its Applications*, 161–164.

[21] Koshi, M. (1989). "Development of the Advanced Vehicle–Road Information System in Japan—the· CACS Project and After." *Proc. JSK International Symposium—Technological Innovations for Tomorrow's Automobile Traffic and Driving Information Systems*, 9–19.

[22] Moigne, J. Le, Waxman, A. M., Srinvasan, B., and Pietikainen, M. (1985). "Image Processing for Visual Navigation of Roadways." Technical Report, CAR-TR-138/CS-TR-1536, University Maryland, July.

[23] Neisser, U. (1976). *Cognition and Reality—Principle and Implications of Cognitive Psychology*. Freeman, San Francisco, California.

[24] Newell, A. (1973). "Production Systems—Models of Control Structures." Chase, W. G., ed., *Visual Information Processing*, Academic Press, San Diego, California, 463–526.

[25] Nitao, J. J., and Paradi, A. M. (1986). "A Real-Time Reflexive Pilot for an Autonomous Land Vehicle." *IEEE Control System Magazine*, February, 13–23.

[26] Ogawa, Y. C., Kamejima, K., and Nakano, Y. (1987). "Syntactic Image Analysis for Environment Understanding." *Proc. IEEE Int. Workshop on Industrial Application of Machine Vision and Machine Intelligence*, 266–271.

[27] Okamoto, H. (1989). "Traffic Control in Japan and Development of the Advanced Mobile Traffic Information Communication System—AMTICS." *Proc. JSK International Symposium—Technological Innovations for Tomorrow's Automobile Traffic and Driving Information Systems*, 21–27.

[28] Rumelhart, D. E., McClelland, J. L., and the PDP Research Group (1986). *Parallel Distributed Processing-Exploration in the Microstructure of Cognition. I-Foundations, II-Psychological and Biological Models*. MIT Press, Cambridge, Massachusetts.

[29] Shibata, M. (1989). "Road Traffic Management in Japan and Development of the Road/Automobile Communication System—RACS." *Proc. JSK International Symposium—Technological Innovations for Tomorrow's Automobile Traffic and Driving Information Systems*, 29–37.

[30] Spreizer, W. M. (1989). "Technology, Vehicle, Highway and Future Transportation." *Proc. JSK International Symposium—Technological Innovations for Tomorrow's Automobile Traffic and Driving Information Systems*, 93–105.

[31] Stefik, M. (1985). "Strategic Computing at DARPA: Overview and Assessment." *Communication of the ACM* **28** 690–704.

[32] Tsugawa, S., Hirose, T., and Yatabe, T. (1984). "An Intelligent Vehicle with Obstacle Detection and Navigation Function." *IEEE-SICE IECON'84*, 303–308.

[33] Tsugawa, S., and Tabei, S. (1984). "Computer Aided Instruction System of Automobile Driving." *IEEE-SICE IECON'84*, 318–323.

[34] Tsuchiya, M., Hamada, T., and Kamejima, K. (1990). "A Real Time Generalized

Hough Transformation Processor." *40th Information Processing Society of Japan*, 1255–1256.

[35] Tsuji, S. (1984). "Monitoring of a Building Environment by a Mobile Robot." *Proc. 2nd International Symposium of Robotic Research*, 349–356.

[36] Tsuji, S.: Continuous Image Interpretation by a Moving Viewer, Proc. Int. Conf. on Pattern Recognition, Rome, (1988), pp. 514–519.

[37] Weidemann, W. (1989). "Future Automotive Technologies for Traffic Safety and Efficiency." *Proc. JSK International Symposium—Technological Innovations for Tomorrow's Automobile Traffic and Driving Information Systems*, 81–89.

9
Recent Progress in Mobile Robot Harunobu-4

HIDEO MORI

9.1 Introduction

Many research groups have developed outdoor mobile robots and have tested them on roads. Martin Marietta (Denver) has developed "ALVin," an autonomous mobile robot of 2.7 m in width, 4.2 m in length, 3.1 m in height, and 6 tons in weight. It takes about 2 s to process an image. ALVin has reached about 20 km/h on an unobstructed road [15]. Carnegie–Mellon University (MU) has developed "NAVLAB," which is based on a commercial van chassis of 4 tons in weight. NAVLAB has reached 1.6 km/h on driveways on the CMU campus. It takes about 20 s to process an image [16].

Universitat der Bundeswehr Munchen has developed "VaMoRs," which is based on a commercial 5-ton van. It has reached 96 km/h on an unopened highway. The multiprocessor architectures BVV1 and, recently, BVV2 have been equipped in VaMoRs. The feature detection repetition is done 60 times per second [4].

The Harunobu project for mobile robot development has been ongoing since 1982 in Yamanashi University. This project has two aims. The first aim is the development of small outdoor robots that have the following advantages.

1. The intelligent robot is applicable to the common road environment. The robot can move on a variety of road environments. In other words, not only on an asphalt paved road but also on a color-tile paved sidewalk. It can detect and avoid obstacles and distinguish them from building and tree shadows.
2. The town robot moves on roads or sidewalks at almost the same speed that humans walk. It performs simple tasks, such as carrying goods from home to home, scavenging roads, painting lane markers, and guiding visually handicapped persons.
3. The robot is economical. The cost is less than two or three years' salary of a laborer between skilled and unskilled, which is about $35,000 to $50,000 a year.

TABLE 9.1. Characteristics of two visions.

Vision analyzing system	Visuo-motor system
Focal vision	Ambient vision
Visual cortex	Superior colliculus
Cone system	Rod system
Day eye	Night eye
Central field	Peripheral field
What	Where
Analysis	Localize
Goal object	Cue object
Intensive	Extensive
Object holding	Field holding

4. The small size and light weight of the robot allow it to move on sidewalks, and also it is light enough not to injure accidentally humans and the environment.
5. The robot is low in electricity consumption; it consumes less than 400 W in electric power, 200 W for motors and 200 W for computer systems, including the vision system.

The second aim of the Harunobu project is the investigation of human visual processing.

In the 1960s and 1970s, from experiments on animal brain ablations, two visual systems, the *visuo-motor system* and the *vision analyzing system*, were proposed to explain human behavior. The former is concerned with visual orientation, and the latter is concerned with visual discrimination. In a series of studies on the golden hamster, Schneider drew a distinction between visual orientation and visual discrimination [10, 11, 13, 14]. He showed that lesions to the superior colliculus selectively destroy the hamster's ability for visual orientation, and lesions of the visual cortex selectively disturb pattern discrimination. Trevarthen concluded that the visual system is decomposed into two systems, which was originally written by Trevarthen [13, 14] and slightly modified for this paper, the focal vision system and the ambient vision system. Table 9.1 shows a list of characteristics of the two vision systems.

The vision analyzing system, which lies on the retina-geniculo-striate cortex pathway, processes stimuli projected on the central field (fovea), and the visuo-motor system, which lies on the retino-superior colliculus pathway, processes stimuli projected on the peripheral field. The rolls of the vision analyzing system are the identification of objects and the 3-D perception of the environment. On the other hand, the roll of the visuo-motor system is to guide behavior, such as head motion, eye movement, handling, and walking.

Lower animals have considerably rich visuo-motor systems and very poor, almost negligible vision analyzing systems. For many years, biologists and psychologists have been interested in the homing behavior of insects, especially the honey bee. The bee learns to find its hive in a certain place on coming home from the field and goes straight to the opening of the hive without any hesitation. Its life span is usually about a month or so. In the first half of its life, it works in the hive for repairing the nest and nursing infants, and in the second half of its life, it goes out to get honey and pollen. When it changes its work from indoor to outdoor, it flies out of the hive with its fellows of the same age and floats about half an hour above the hive, orientating its head to the hive as if to remember its location. More quick localization behavior is seen when it leaves a flower after the first visit. It leaves the flower and flies up in the sky, drawing two or three circles around the flower, and goes back to the hive. All of the bees show the same orientation flight when they leave the flower after the first visit.

Specific behavior, such as orientation flight, is inherent by species and is called a fixed action pattern. Fixed action patterns that are defined by temporal and spatial contraction patterns of a group of muscles are released when a specific stimulus is given or a physiological level of the nervous system caused by hormone exceeds a certain level. Tinbergen proposed a paradigm about the mechanism of fixed action patterns [12].

It is assumed that even in higher animals voluntary and nonvoluntary visuo-motor actions are decomposed into fixed action patterns that are more flexible and rich in variety. However, concepts and functions of two visual systems have been proposed, but the models and algorithms that realize the systems have not yet been well defined. Kawato and Suzuki [5] studied by computational methods the visually guided voluntary movement control in the central nervous system and proposed an iterative learning scheme as a possible algorithm that includes transformation of the coordinates of the desired trajectory to the body coordinates and generation of motor command. Their control model is assumed to be a model of the visuo-motor system.

Brooks has been developing an autonomous robot named Seymour, which has six linear CCD sensors attached around the robot and three standard TV cameras, looks around the environment, and does reflexive behavior completely autonomously [2]. It seems that Seymour is a visuo-motor system rather than a vision analyzing system.

In the Harunobu project, the functions and rolls of the visuo-motor system in road following have been studied and implemented on a mobile robot and tested in the real world. Only a few aspects of a vision analyzing system are realized in Harunobu-4.

In this chapter, concepts of the visuo-motor system are described, and the recent progress about active sensing and shadow elimination is also described.

9.2 Visuomotor System in a Mobile Robot

Assume that a robot is going to move from the entrance of the building of Yamanashi University to Kofu station. At the entrance, the robot will orient itself to the direction of an edge of a garden in front of the building. Then, it will move along the edge of the garden until it reaches a corner, and then it will turn to the left and move along another edge of the garden until it reaches the gate of university. The robot will look right and left to see whether or not cars are coming. After cars pass by, the robot will leave from the gate and move along the road boundary until it reaches a crossing.

As shown in this example, the robot can move more than 90% of the course by simple visuo-motor actions that are *moving along* a certain visual guide and *moving toward* a visual target. When the road is crowded with people, the robot will not be able to see the visual guide or target and it will do a motion called *following a person*, which occurs when a person walks in the same direction as the robot's target. When the robot is going to turn a corner, it will go forward until it reaches the turning point where the next visual guide is in view. This motion is called *moving for sighting*. The italicized motions are called *vision-based stereotyped motions*, and the visual guide or target along which or toward which the robot does a streotyped motion is called a *sign pattern* [7].

Only at a corner or crossing will the robot focus its TV camera to identify a landmark for verification with a cognitive map. These focusing tasks are performed by a *vision analyzing system* that is based on models of landmarks represented by their size, color, shape, and location. The vision analyzing system is intelligent in function but takes a long time to process images. When the vision analyzing system is activated, the robot must slow down its speed or stop until the system finishes analyzing. The vision analyzing system has not yet been implemented on Harunobu-4.

As with vision-based stereotyped motions, sonar-based stereotyped motions, touch-sensor-based stereotyped motions, and internal sensor-based stereotyped motions have been defined and implemented on Harunobu-4 [8]. In Figure 9.1, the system configuration of Harunobu-4 is shown.

In Harunobu-4, data from different kinds of sensors are not fused but are used separately in proper stereotyped motion, because the receptive field and precision of a sensor are quite different depending on the kind of sensor.

The vision sensor is a middle range sensor and useful for stereotyped motion of visual guidance, the ultrasonic sensor is a short-range sensor and useful for obstacle avoidance, and the touch sensor is a very short range sensor that is useful for collision protection.

The advantages of a stereotyped motion-based robot are as follows.

1. It is adaptable to various environments. Stereotyped motions identify only sign patterns and ignore all the other visual patterns; therefore, they are adaptable to various environments.
2. It is easy to implement. A sign pattern is specified by a long edge lying on the road, a long edge standing on the road, or a moving object on the road. The robot detects only the road and sign pattern and ignores all the other patterns, such as buildings, trees, and the sky. Detection algorithms of road and sign patterns are very simple and easy to implement.
3. It is quick in the processing cycle. Very simple algorithms result in a quick

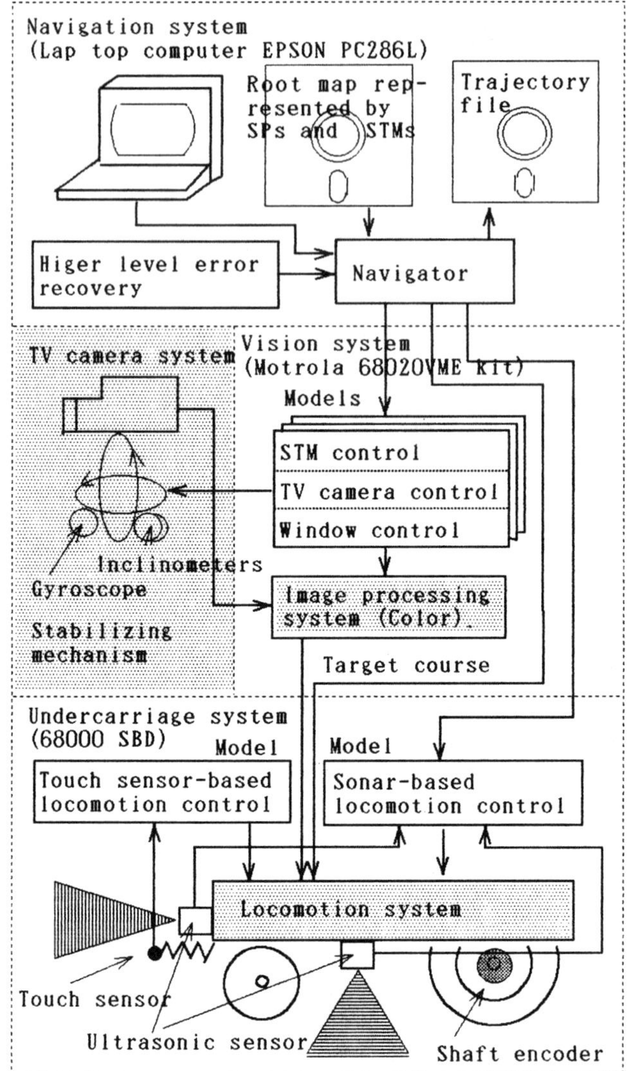

FIGURE 9.1. System configuration of Harunobu-4 (STM, stereotyped motion; SP, sign pattern).

response to visual input and dynamic vision. In dynamic vision, an image of a road scene is not so different from the last image and locations and attributes of the road and sign pattern are predictable [4].

4. Its cost is low. To implement these simple algorithms, a distributed microcomputer system, TV camera system, ultrasonic sensors, and touch sensors are required. Expensive computers or additional devices are not required.

5. It is useful in practical applications. In practical usage, such as construction applications, the robot moves along the same course repeatedly. The movement along this course can be specified by a sequence of sign-pattern-based stereotyped motions.

9.3 Active Sensing in Stereotyped Motion

Active sensing defined in this chapter is an improvement of the animate vision proposed by Ballard [1]. In the active sensing proposed by the author [9, 17] TV camera control patterns are fixed in each stereotyped motion.

Advantages of active sensing in stereotyped motions are as follows.

1. There is a decrease in searching failure when the TV camera scanning pattern and the locomotion control pattern are fixed.
2. The image processing time is shortened by the addition of a dynamic window for sign pattern searching.
3. By adding observation motion to locomotive motion there is a decrease in localization error.

In Figure 9.2, the control structure of stereotyped motion with active sens-

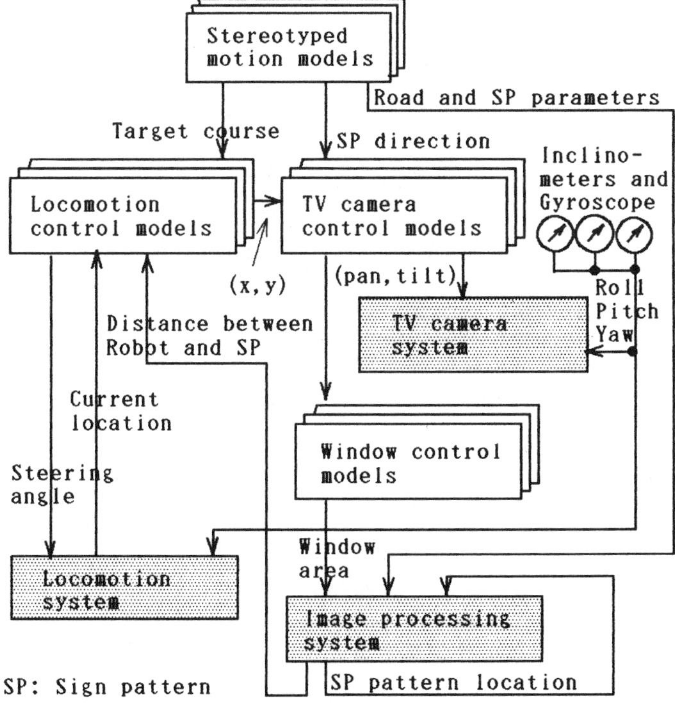

FIGURE 9.2. Control structure of stereotyped motions (SP, sign pattern).

ing is shown. A stereotyped motion model is decomposed into three control models, the locomotion control model, the TV camera control model, and the window control model. Each of the control models has a very simple parameter, as shown in Figure 9.2. The essential techniques of the active sensing in stereotyped motion are as follows.

9.3.1 Moving for Sighting

To search for a new sign pattern and place the robot at the starting point for the new stereotyped motion based on the new sign pattern, a stereotyped motion named *moving for sighting* is defined. It is used when the robot is going to turn a corner, as shown in Figure 9.3, or when avoiding an obstacle. The robot keeps its motion along or toward an old sign pattern, while the TV camera is directed to a new sign pattern until the new sign pattern is caught in the center of TV camera.

9.3.2 Dynamic Window

In active sensing, relevant regions of an image [6] are specified by dynamic windows. Dynamic windows are useful not only to speed up the image pro-

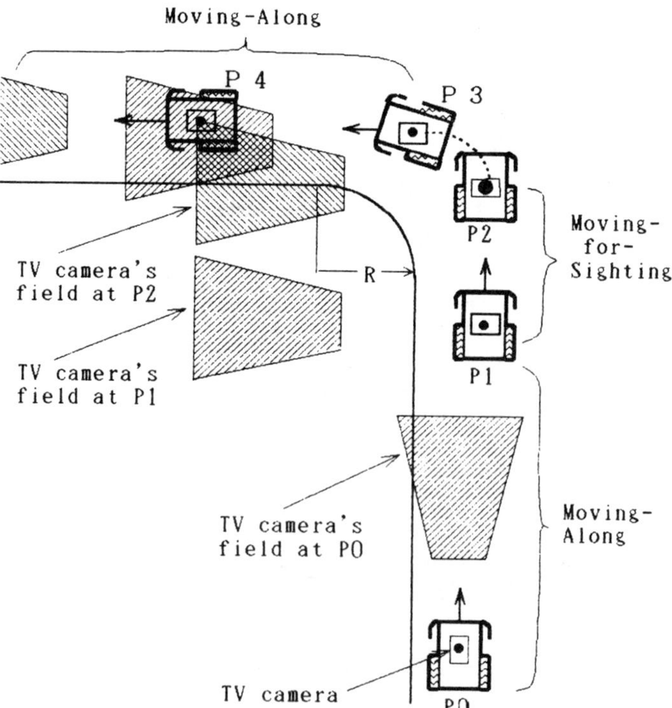

FIGURE 9.3. Moving for sighting in corner turn.

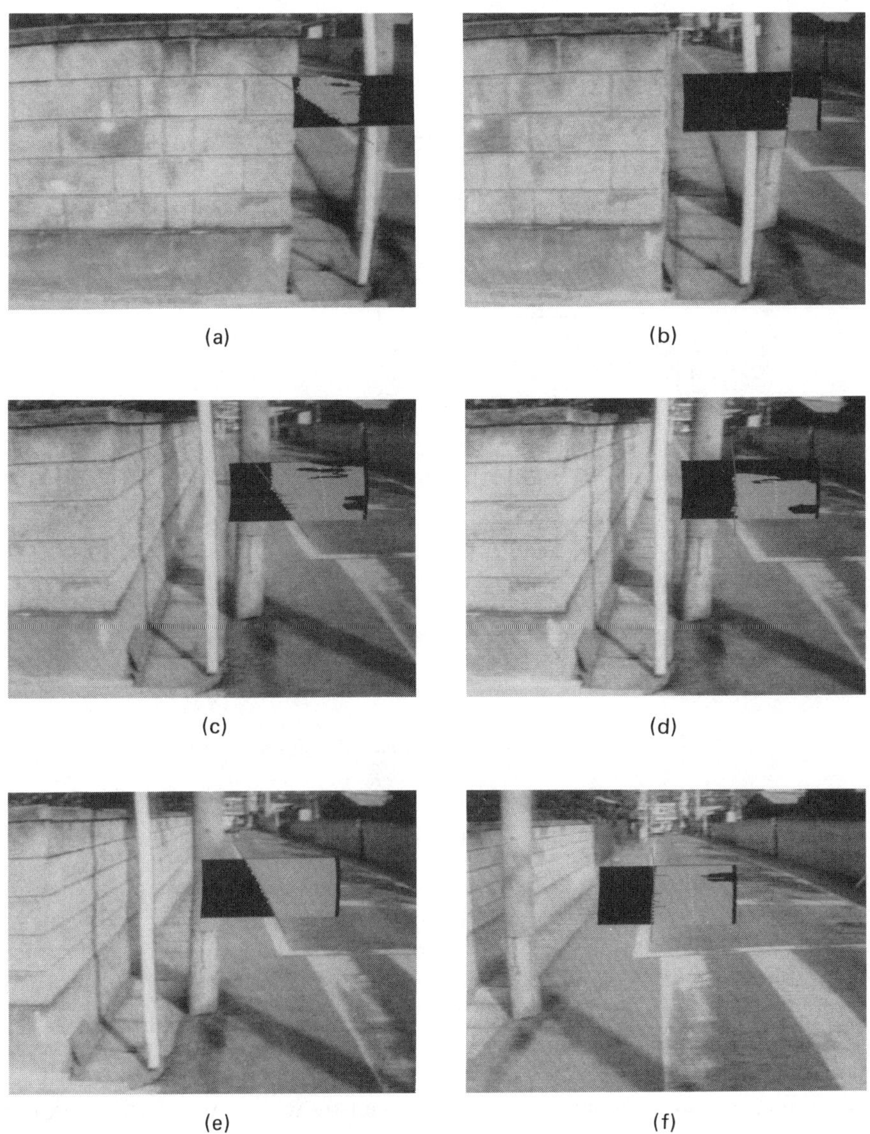

FIGURE 9.4. Examples of dynamic windows.

cessing time but also for focusing on search areas and simplifying sign pattern detection rules. Examples of a dynamic window in moving for sighting are shown in Figure 9.3. At first, a window is set up on the right side of an image frame, waiting for a new sign to appear in the frame as shown in Figure 9.4(a). The sign pattern detection method is applied to the image in the windows [7]. A sign pattern is represented by

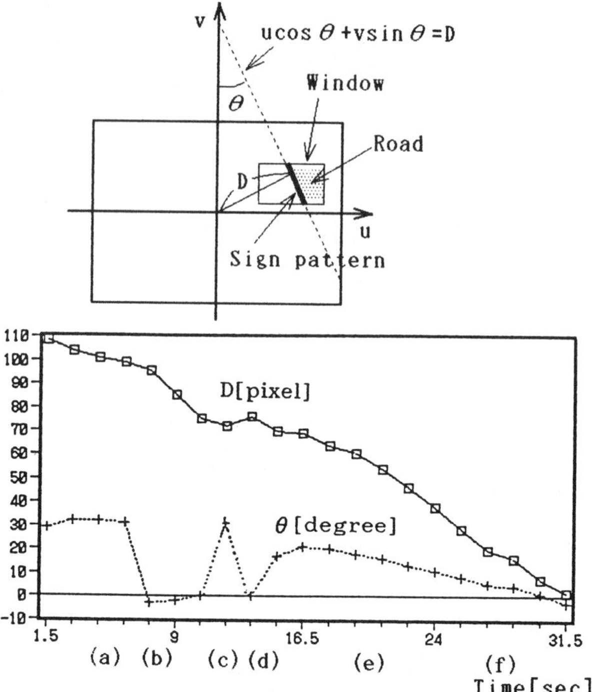

FIGURE 9.5. D and θ curves in moving for sighting.

$$U \cos \theta + V \sin \theta = D.$$

In Figure 9.5, values of θ and D are plotted along robot movement. In Figures 9.4(a) and 9.4(c), edges of the wall and pole are detected as sign patterns at first but are then identified as non-sign patterns by their angle θ and distance D. In Figures 9.4(d), (e), and (f), θ and D of the new sign pattern lying on road gradually decrease to zero when the robot approaches the corner.

Another example of a dynamic window is sign pattern detection in color-tile-paved sidewalks. Figure 9.6 shows a sidewalk that is decomposed into two different geometric-shaped regions paved by white- and brown-colored tiles. As in Gestalt psychology, one region is identified by its color as the *figure* region and the other regions are disregarded as *ground* regions. In general, the smaller region, the brown-colored region in this case, is specified as figure one. Boundaries between figure and ground regions are specified as a sign pattern. Windows should be located on the relevant part of the boundary between the figure and ground regions. Windows are set up as shown in Figure 9.6.

Vertical line scans are done to search for figure and ground mixing zones, as shown in Figure 9.7. A filter, represented by Equation (9.1), is operated on images of vertical scans to detect boundaries between the figure and ground regions.

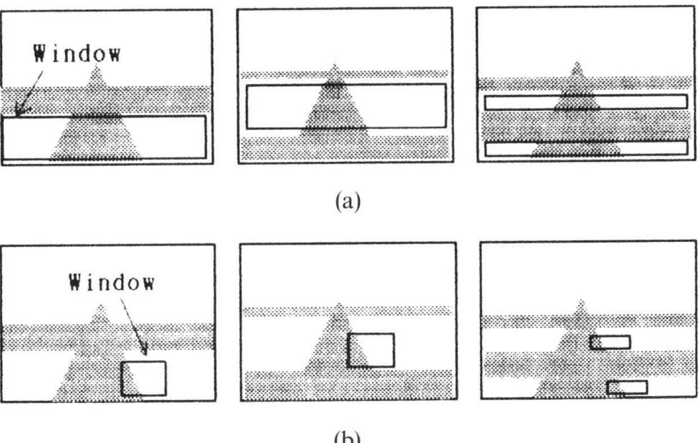

(a)

(b)

FIGURE 9.6. Window patterns: (a) windows in finding mode; (b) windows in following mode.

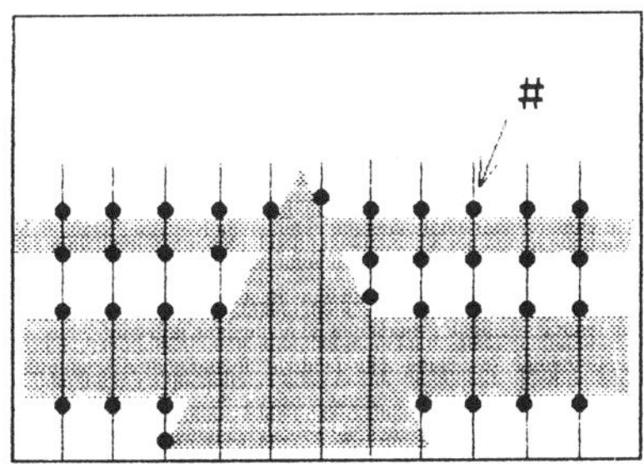

FIGURE 9.7. Vertical scans.

$$F(u, v) = \text{abs}[DR(u, v)] + \text{abs}[DG(u, v)] + \text{abs}[DB(u, v)], \qquad (9.1)$$

where

$$DR(u, v) = \sum_{j=1}^{3} R(u, v + j) - \sum_{j=2}^{4} R(u, v - j),$$

$$DG(u, v) = \sum G(u, v + j) - \sum G(u, v - j),$$

$$DB(u, v) = \sum_{j=1}^{3} B(u, v + j) - \sum_{j=2}^{4} B(u, v - j).$$

Figure 9.8 shows the $R(u, v)$, $G(u, v)$, $B(u, v)$, and $F(u, v)$ curves on the vertical scan, which is denoted by # in Figure 9.7. Figure 9.9 shows the dynamic windows obtained by the above method. Figure 9.10 shows failure examples of sign pattern detection. Table 9.2 shows the processing time; and Tables 9.3 and 9.4 show the success and error rates in sign pattern detection, respectively.

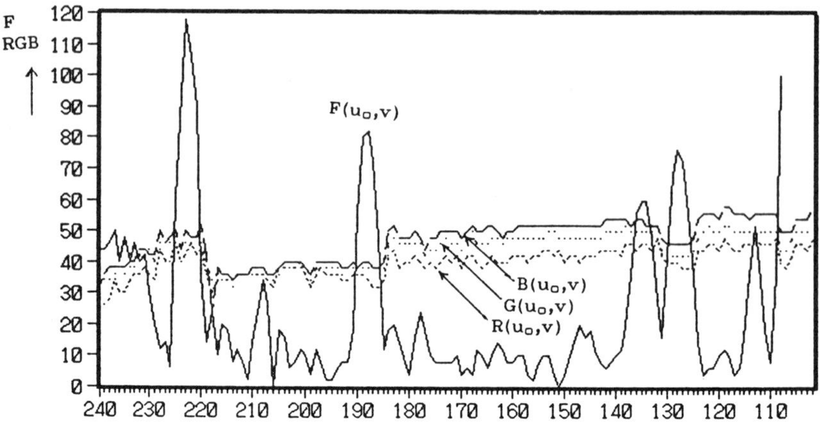

FIGURE 9.8. $F(u, v)$ on a vertical scan.

9.3.3 TV Camera Stabilization

A TV camera is mounted on a platform that is driven by two servo motors for pan and tilt. Two inclinometers (Sshaevitz Co. AccuStar) are equipped to sense the pitching and rolling of the robot, and a gyroscope (Hitatch Cable Ltd. Optical Fiber Gyroscope OFG2) is equipped to sense the yawing of the robot. Rolling and yawing angles are fed to the locomotion system, and rolling, pitching, and yawing are fed to the TV camera system, respectively, as shown in Figure 9.2. In the test course at Yamanashi University, the rolling angle reaches 5° by the traversal slope for drainage or the asphalt patch.

9.4 Shadow Elimination

In road following of a vision-based mobile robot, the problems of shadow are very important and difficult. The first problem is that it is difficult to detect accurately sign patterns on shaded roads. The second problem is that the robot may misidentify a shadow as an obstacle and avoid it, or it may not be able to find a path and stop moving.

Several research groups have reported color-vision-based approaches for shadow problems. In the robot ALVin, a shadow detection method called "shadow boxing" has been developed to solve the first problem [15]. In this method, the red and blue axes of a scatter diagram are rotated by performing two tricolor operations, and a shadow region in the scatter diagram is bounded by a rectangle. A pixel whose red and blue components are located in this rectangle is identified as a shadow element.

In NAVLAB, a road-following algorithm deals with the first problem [16]. This algorithm classifies each pixel of the image into road and nonroad cate-

FIGURE 9.9. Successful examples of dynamic windows.

(a)

(b)

FIGURE 9.10. False examples of sign pattern detection.

TABLE 9.2. Window setup and sign pattern detection time.

Mode	Window size	One window (s)	2 windows (s)
Finding	$100 * 300$	1.4	1.7
Following	$100 * 42$	0.80	0.95

TABLE 9.3. Success rate in sign pattern detection.

Mode	Success	False	Total
Finding	195 (82%)	43 (18%)	238 (100%)
Following	171 (84%)	33 (16%)	204 (100%)

TABLE 9.4. Causes of sign pattern detection errors.

Window setup, 68%	Segmentation, 26%	Figure selection, 3%	Others, 3%

gories by its color models and uses the results of classification to select the best-fit road position, and then it updates the color models based on the detected road and nonroad regions. In order to solve the second problem, both of these two groups utilize range sensors to detect obstacles.

In Harunobu-4, the color vision system is utilized to process the shadow image [3].

On a sunny day, shadows on a road cause edges between sunny and shaded regions, as shown in Figure 9.11(a). An edge-detecting filter is applied to the image, as shown in Figure 9.11(b). After the edge-based segmentation, the road will be separated into several different regions, as shown in Figure 9.11(c). The largest region in area is selected as a road candidate region. Holes in the candidate region are filled for noise elimination. Sign pattern is detected by line scanning in the road region, as shown in Figure 9.11(d). The details are as follows. The sign pattern detected in the last scene is used for predicting

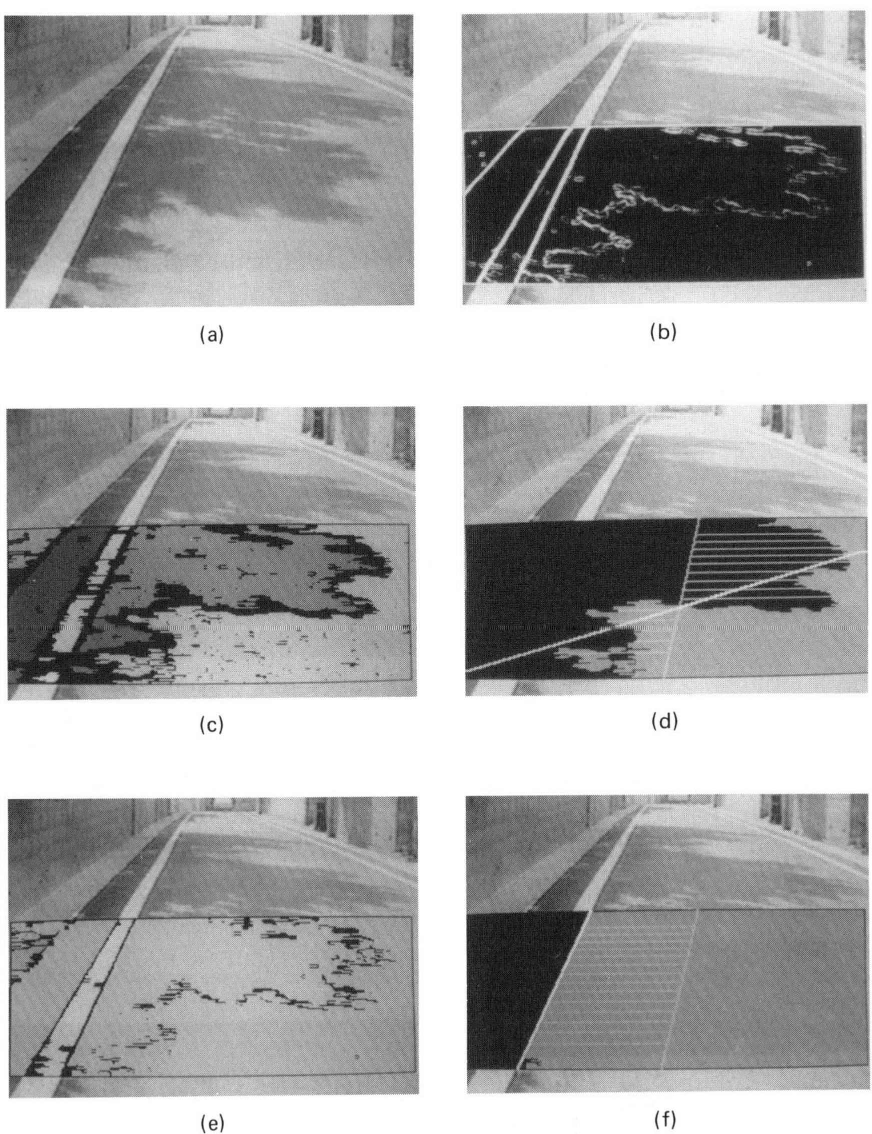

FIGURE 9.11. An example of a shadow elimination sequence.

future sign patterns, for which the road candidate region is scanned horizontally up to its boundary. Boundary points obtained by horizontal scanning are approximated by a line that is a new sign pattern candidate. In this example, the sign pattern candidate is judged to be a false sign pattern, because θ and D of the sign pattern are different from the predicted one.

To detect the real sign pattern, a shadow elimination method was developed. In this method, each pixel is classified into shadow or nonshadow

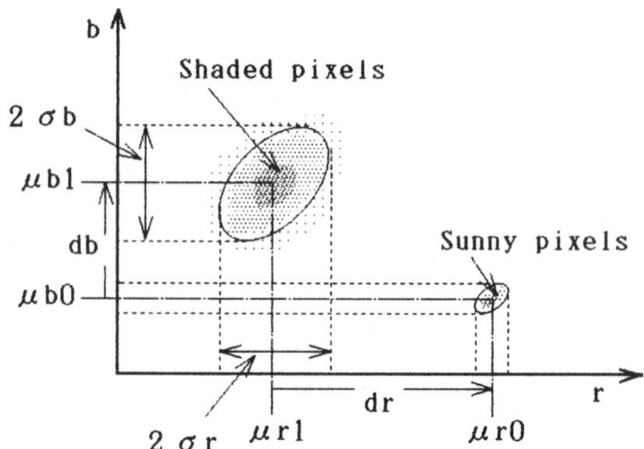

FIGURE 9.12. An r–b scattering diagram.

groups by its brightness ($Y = R + G + B$) and normalized red ($r = R/Y$) and blue ($b = B/Y$) components. The pixels that are classified as the shadow group are changed in RGB components from the original one to the average RGB components of a sunny road region. Thus, shadows on the road can be eliminated, as shown in Figure, 9.11(e). After this shadow elimination, a sign pattern is detected successfully, as shown in Figure 9.11(f).

9.4.1 Shadow Elimination Algorithm

Sunny roads are illuminated mainly by sunlight of about 5,000° K. On the other hand, shaded roads are illuminated by blue sky of about 14,000° K. The hue of a shaded road slightly shifts to blue. In other words, the average b component $\mu b1$ of the shaded region is greater in its value than the μ_{b0} of the sunny region, whereas the average r component μ_{r1} of the shaded region is smaller in value than the μ_{r0} of the sunny region, as shown in Figure 9.12. Caused by poor S/N ratio of the video signal, standard deviations σ_{r1} and σ_{b1} of the shaded region have much larger values than σ_{r0} and σ_{b0} of the sunny region. If a pixel falls into the box in which the center is fixed at (μ_{r1}, μ_{b1}), the pixel is judged to be a shaded pixel.

Figure 9.13 shows $d_r(=\mu_{r0} - \mu_{r1})$ and $d_b(=\mu_{b1} - \mu_{b0})$ and σ_{r1} and σ_{b1} for a scene of a road shaded by trees; σ_{r1} and σ_{b1} remain almost constant, while d_r and d_b show slight changes. In general, a sunny region of a new image is detected successfully; therefore, (μ_{r1}, μ_{b1}) of shaded a region are predicted by (μ_{r0}, μ_{b0}) and (d_r, d_b).

9.4.2 Implementation and Results

To test the performance of the elimination algorithm, the shadow elimination method is applied on three shadow scenes of a white car, a red car, and a dark car. Results are shown in Figure 9.14 and 9.15. The shaded part beneath the

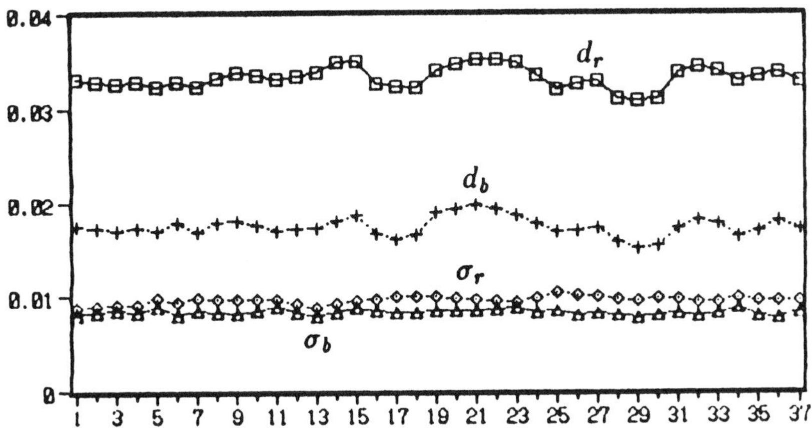

FIGURE 9.13. Here, d_r and d_b and σ_{r1} and σ_{b1} are the changes in a 9.13 moving-along shaded road.

(a)   (b)

FIGURE 9.14. White car: (a) before shadow elimination; (b) after elimination.

(a)   (b)

FIGURE 9.15. Red car: (a) before shadow elimination; (b) after elimination.

TABLE 9.5. Sign pattern detection time.

Mode, windows (pixels)	Finding, 300(H) × 250(V)	Following, 100(H) × 100(V) × 2
Image transfer	0.30 sec	0.10 sec
Edge detection	1.65	0.50
Edge-based segmentation	0.65	0.25
Road detection	0.55	0.35
Sign pattern detection	0.35	0.15
Total	3.50	1.35

TABLE 9.6. Shadow elimination time.

Mode	Finding	Following
Windows (pixels)	300(H) × 250(V)	82(H) × 150(V) × 3
Shadow elimination	1.7–0.6 s	0.8–0.3 s

car is not eliminated because this part is darker than the other shaded part. This remaining shaded part reinforces the obstacle region. The images of the dark car are eliminated in part but not as a whole, because the car is composed of many differently oriented surfaces that reflect the incident light in different ways. Therefore, the brightness of the car image will be different in different places. For these reasons, the dark obstacle is able to be detected under the shadow elimination method.

Table 9.5 shows the sign pattern detection times spent on the VME 68020 (16 MHz) system in finding and following modes that are different in window size. Table 9.6 shows the shadow elimination time table, in which time is maximum when all of the pixels in the window(s) are considered to be points in the shaded region.

9.5 Concluding Remarks

Through experimental tests the following results were obtained:

1. The Dynamic window was useful for sign pattern detection of a sidewalk that was decomposed into two different geometric-shaped regions of different color.
2. Moving-for-sighting was shown to be useful for localization of a new sign pattern when the robot went to turn a corner or avoid an obstacle.
3. The shadow elimination method was useful in sign pattern detection of a shaded asphalt road with lane mark.

To promote the study of vision guided robots at Japanese universities, one of the most important things is to find a useful application, because most of studies are done as personal projects and need financial and technical support from companies. Since 1989 two application projects of a guide dog robot and a lane mark drawing robot have been started.

References

[1] Ballard, D. H. (1989). "Reference Frames for Animate Vision, *Proc. 11th Int'l. Conf. Artificial Intelligence*, 1635–1641.

[2] Brooks, R. A., and Flynn, A. M. (1989). "Robot Beings." *Proc. IEEE Int'l. Workshop on Intelligent Robots and Systems '89*, 2–10.

[3] Chen, H., and Mori, H. (1989). "Sign Pattern Detection on Shaded Road." *Proc. IEEE Int'l. Workshop on Intelligent Robots and Systems '89*, 350–357.

[4] Graefe, V. (1989). "Dynamic Vision Systems for Autonomous Mobile Robots." *Proc. Int'l. Workshop on Intelligent Robots and Systems '89*, 12–23.

[5] Kawato, M., Isobe, M., Maeda, Y., and Suzuki, R. (1988). "Coordinates Transformation and Learning Control for Visually Guided Voluntary Movement with Iteration." *Biological Cybernetics* **59**, 161–177.

[6] Kunert, K.-D. and Graefe, V. (1988). "Vision Systems for Autonomous Mobility." *Proc. IEEE Int'l. Workshop on Intelligent Robotics and Systems '88*, 477–482.

[7] Mori, H., Ishiguro, H., Kotani, S., Yasutomi, S., and Chino, Y. (1988) "A Mobile Robot Strategy Applied to Harunobu-4." *Proc. 9th Int'l. Conf. on Pattern Recognition*, 525–530.

[8] Mori, H., Nakai, M., Chen, H., Nakayama, S., and Yuguchi, T. (1989). "A Mobile Robot Strategy—Stereotype Motion by Sign Pattern." *Proc. 5th Int'l. Symposium of Robotics Research*, 225–236.

[9] Mori, H. Nakai, M., Chen, H., Yuguchi, T., and Nakayama, S. (1989). "Steering Panning Control in Stereotyped Motion." *Proc. IEEE Int'l Workshop on Intelligent Robots and Systems '89*, 155–162.

[10] Schneider, G. E. (1967). "Contrasting Visuomotor Functions of Tectum and Cortex in the Golden Hamster." *Psychol. Forsch.* **31**, 52–62.

[11] Schneider, G. E. (1969). "Two Visual Systems." *Science* **163**, 895–902.

[12] Tinbergen, N., (1951). *The Study of Instinct.* Oxford Univ. Press, London and New York.

[13] Trevarthen, C. B. (1968). "Two Mechanisms of Vision in Primates." *Psychol. Forsch.* **31**, 299–337.

[14] Trevarthen, C. B., and Sperry, R. W. (1973). "Perceptual Unity of the Ambient Visual Field in Human Commissurotomy Patients." *Brain* **96**, 547–570.

[15] Turk, M. A., Morgenthaler, D. G., Gremban, K. D., and Marr, M. (1988). "VITS—A Vision System for Autonormous Land Vehicle Navigation." *IEEE Trans. on PAMI* **10** (3), 342–361.

[16] Thorpe, C., Herbert, M. H., Kanade, T., and Shafer, S. A. (1988). "Vision and Navigation for the Carnegie–Mellon Navlab. *IEEE Trans on PAMI* **10** (3), 362–373.

[17] Mori, H. (1990). "Active Sensing in Stereotyped Motion." *Proc. IEEE Int'l Workshop on Intelligent Robots and Systems '90*, 167–174.

10
Visual Navigation of an Autonomous On-Road Vehicle: Autonomous Cruising on Highways

AKIHIRO OKUNO, KENJI FUJITA, AND ATSUSHI KUTAMI

10.1 Abstract

This chapter describes issues relating to vehicle autonomy on highways. First, the needs for vehicle autonomy are reviewed, and an autonomous vehicle concept to meet those needs is presented. Then, a framework for the autonomous control system is proposed based on the knowledge and skills of a human driver. A prototype of the knowledge base for highway cruising was developed, and its workability was confirmed by computer simulation. A lane tracking test using an actual vehicle was conducted to develop driving skills for curved courses. Last, future research themes to realize vehicle autonomy on highways are described.

10.2 Introduction

With increasing expectations toward highway automation, the vehicle autonomy issue has gained much attention [1, 2]. In the background is the need to improve the safety and comfort of highway driving. From this viewpoint, the needs for vehicle autonomy are reviewed first.

10.3 Vehicle Autonomy

10.3.1 Human Aspects of Highway Driving

As highway speed limits tend to be raised and the average age of drivers increases yearly, the ability of people to drive safely on highways is becoming a matter of serious concern. The first problem is that of human fatigue. Although driving itself is not a difficult job, driving on highways for a long period of time is strenuous. The reasons for this are well understood; drivers on highways must, for example, always keep their eyes on lane markers to keep their cars on course. Such a monotonous, continuous stimulus fatigues the drivers' nerves and causes drowsiness. In addition, drivers in a traffic flow on highways need always to be alert to any hazard that may arise at any time. Since there are a tremendous number of possible hazards that may cause an accident, drivers are subjected to continual fear of accidents. This is also a great cause of mental fatigue for drivers.

The second problem is the matter of human error. It is well known that more than 90% of traffic accidents are caused by such human errors as lack of attention, misjudgment, and poor maneuvering. Among these, drivers' inattention is the most serious problem, since just one second of inattention may cause an accident. Given the mental stress mentioned above, it is too demanding to expect a human to drive for hours without one second of inattention. Further, in high-speed driving, drivers' inability in emergency handling also becomes a serious problem.

10.3.2 Needs for Vehicle Autonomy

To realize safer and more comfortable highway transportation, it is first necessary to relieve drivers from their fatigue and stress. For this purpose, it is effective to reduce the cognitive load of the driver and also to relieve the driver from vehicle control operations. Further, if a driver could rely on a system that recognizes safety margins at any instance of driving, mental tension would be relieved substantially. Therefore, a system to monitor and warn of any potential dangers is also considered to be effective. Moreover, it is necessary to contrive a system that can assist drivers to escape from dangerous emergency situations.

In the distant future, a vehicle in which a driver will be able to reach his or her destination by pressing only buttons may be realized, but this would require completely new traffic infrastructures. In the near future, the need for vehicle autonomy will be satisfied by a sophisticated driving support system which covers the weak points of human drivers.

10.3.3 Autonomous Highway Vehicle Concept

The autonomous highway vehicle may be defined as a car equipped with an intelligent driving support system called an autonomous control system, as shown in Figure 10.1. Here, the autonomous control system has capabilities of automatically navigating the vehicle on highways just as a human would. The driver can either control the vehicle by himself or herself, or leave the driving operation to the autonomous control system.

The autonomous control system consists of four subsystems: a perception system to perceive outside environments, a cruise planner to understand driving situations and make judgments, a vehicle control system to maneuver the vehicle as a human does, and a human interface system to communicate with the human driver. These subsystems correspond to human cognition, reasoning, and action functions.

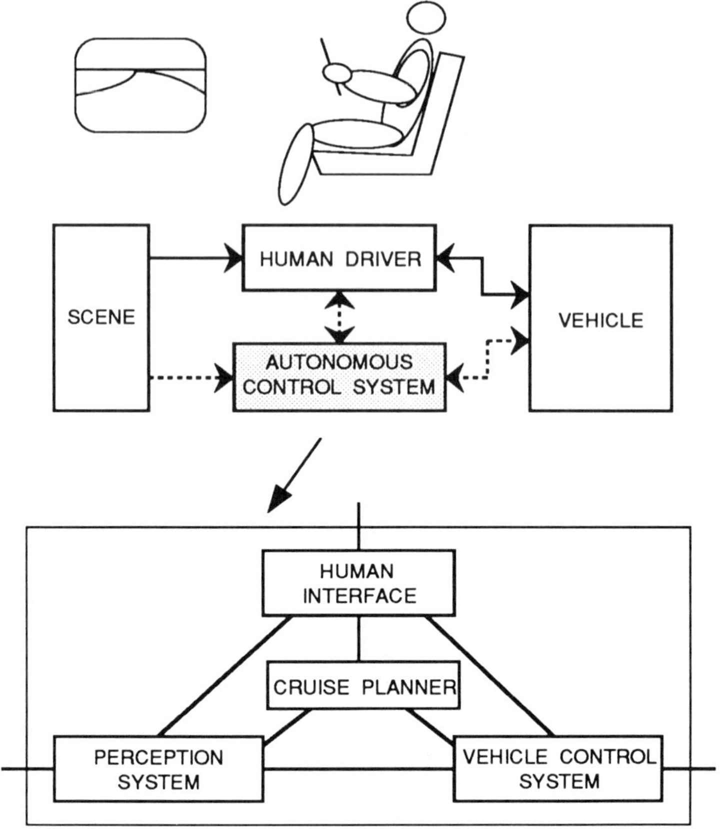

FIGURE 10.1. Autonomous vehicle concept.

The modes of operation of the autonomous control system are shown in Figure 10.2. When the driver wants to control the vehicle himself or herself, for example, this system functions as an intelligent drive-by-wire system. The driver is released from such low-level control operations as fine steering and speed adjustments to compensate for various environmental changes with the help of the autonomous control system. The system can also function as a monitoring and warning system. In this case, the driver does all the control operations, but the autonomous control system informs the driver of the safety margin and warns of any improper situation, incorrect maneuvering, etc. To detect danger, the system uses information on the driver's operation and vehicle conditions, as well as on the outside environment. The driver can also select the autonomous cruising mode. In this mode, the driver will be fully released from driving operations and will be able to devote himself or herself to higher level activities, such as monitoring overall traffic conditions.

As described, the autonomous control system functions as an intelligent driving assistant to the human driver. Such a basic concept was introduced in

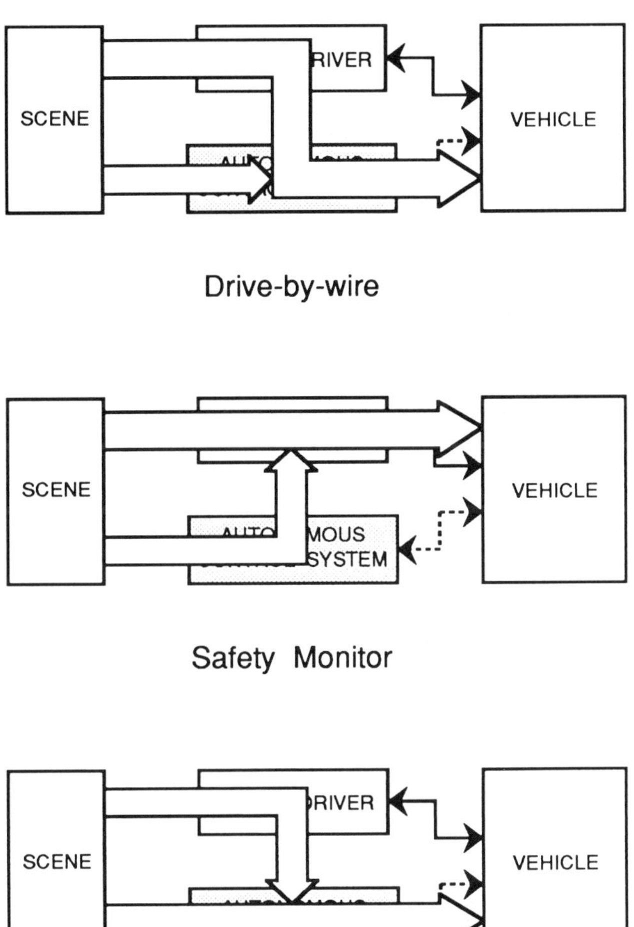

FIGURE 10.2. Operation of autonomous control system: (a) drive-by-wire, (b) safety monitor, and (c) autonomous cruising.

the Prometheus Project in Europe; and this chapter considers a concrete design for such an autonomous system.

10.4 Framework of the Autonomous Control System

In this section, the basic concepts in the design of the autonomous control system are described. It is considered to be the best approach to use the knowledge of a human driver to construct an autonomous control system,

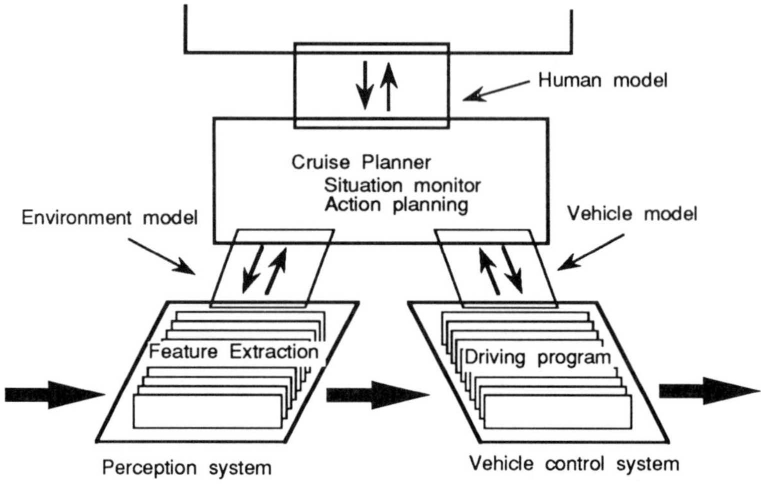

FIGURE 10.3. Autonomous control system.

since humans can drive safely and smoothly in highway traffic flow. The authors have therefore been developing an autonomous control system based on a human driver model[3]. Figure 10.3. shows the structure of such an autonomous control system. The vehicle control system contains many driving skills for different driving situations, and the cruise planner has a knowledge base for highway cruising. Details on the system are described in the rest of this section.

10.4.1 Driving Skills

It takes a long time for a human to learn how to drive a car properly. At first, it is even difficult for a human to drive the car in a straight manner, but, gradually, he or she masters the skill of compensatory steering to keep the car on the desired course. Next, a driver masters the skills of steering and speed control on curves. Further, a human driver learns to control the car in a traffic flow, where more complicated control actions are required. It is considered that a human driver can execute such complicated controls by combining basic driving skills he or she has learned from experience.

Such driving skills, or driving patterns, are executed unconsciously by reflexive controls using the feedback or feedforward of visual information. The input data at this level of control are signals; therefore, the control operation of a human can be modeled as a transfer function. This type of model is called a "driver model," and various driver models have been proposed by researchers. Thus, driving skills can be prepared as a number of driving programs, which will be detailed in Section 4.

10.4.2 Cruise Planning

The action of selecting an appropriate driving pattern in a variety of driving situations to reach a specified destination is termed "cruise planning." For example, suppose that a vehicle is on a road with a single driving lane. In this situation, the most basic driving pattern is a lane-keeping, pattern, which includes steering and speed control to negotiate curves. But, if other vehicles appear in front of or behind the vehicle, different speed control patterns are needed to adjust the distances between the vehicles. When it comes to a road with multiple driving lanes, the situations become much more complicated because the existence of vehicles in neighboring lanes will also affect the vehicle control. For example, it is not natural to drive a vehicle exactly parallel with another vehicle, and there should be an appropriate distance between them. If another vehicle cuts in front of the vehicle from the neighboring lane, speed must be adjusted to avoid possible collision. Further, if the vehicle in front is driving too slow, it should be overtaken; this overtaking action requires both lane-change and passing maneuvers. All these situations require different driving patterns. The cruise planner must be able to understand driving situations and select an appropriate driving pattern in a dynamically changing environment.

To recognize the driving situation, the following information is necessary:

1. the environmental condition, for example, the road structure, traffic conditions, traffic regulations, and weather conditions;
2. the vehicle's driving condition, for example, the dynamic states of the vehicle and the operating conditions; and
3. the driver's condition, for example, his or her purpose, motivation, and mental state.

The selection rules of driving patterns depending on the driving situations are described as a rule base that can be extracted from the empirical knowledge of a human driver. To develop such a rule base, the driving conditions need to be formally recognized; this is done through the models for the environment, vehicle, and driver, as illustrated in Figure 10.3. Of the driving situations, the driver's condition is the most difficult to quantify. Motivation, for example, could be "as safe as possible," or "as fast as possible." In addition, the mental state of the driver affects the selection of the driving pattern.

Anticipation of future events is another important function of the cruise planner. This is primarily necessary for predicting possible dangers and taking actions to prevent them. If a vehicle is in front of the autonomous vehicle in a neighboring lane, for example, the driver would anticipate that it may move into his or her lane. Such capabilities of anticipating future events can be used to prevent the vehicle from falling into situations in which the human driver feels uneasy or unsafe.

10.4.3 Perception System

The perception system is responsible for detecting the parameters of road structures and the location of other vehicles in the surrounding area as required by the driving programs. Such low-level information must be processed at an updating cycle of 0.15 or less, because it is used for reflexive control in the driving programs. On the other hand, cruise planning requires a higher level of information, that is, relative locations of surrounding vehicles, a wider range of road structures, traffic signs, etc. The updating cycle of the information at this level may be assumed to be slower by one order than that of the low-level information processing.

In order to recognize environmental conditions, the parameters defined in the environmental model need to be identified from the sensor data. In a well-structured environment, such as highways, restrictive knowledge on the environment, such as road design standards, etc., can be used effectively. In this method, all other information not set forth by the environmental model is ignored.

As for the sensors in the perception system, image sensors are indispensable for acquiring low-level information. Particularly, driving lanes can be detected only from visual information since they are marked with paint on most roads. Higher level information, such as traffic conditions, are assumed to be available through future communication systems, such as car-to-car communication systems and car-to-roadside communication systems.

However, it will be in the distant future that all the roads and cars are suitably equipped; in the meantime, it is more realistic to consider a sharing of tasks of environmental perception between the on-board sensor and the human driver.

10.5 Autonomous Cruise Simulation

The authors have developed a cruise simulator to examine the effectiveness of the autonomous control system. The present simulator, as shown in Figure 10.4, consists of three modules: a road environment simulation module; a knowledge-base module, which corresponds to the cruise planner; and a driving program module. The road environment simulation module consists of a road model and a vehicle dynamics model. The road environment simulation module generates scene data and sends the perception parameters obtained from the scene data to both a selected driving program and the cruise planner. The driving program outputs a steering angle, throttle value, and brake value based on the perception data. These output values are then fed back to the vehicle dynamics model to calculate the vehicle's motion. New scene data are generated based on the calculated motion of the vehicle. A two-wheel vehicle model was used for the vehicle dynamics model [3]. The knowledge-base module stores the knowledge for cruising to select a driving program with reference to the environmental data.

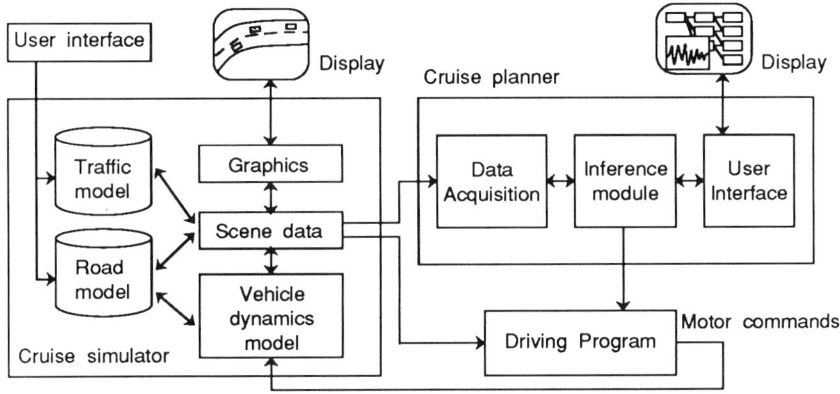

FIGURE 10.4. Autonomous cruising simulator.

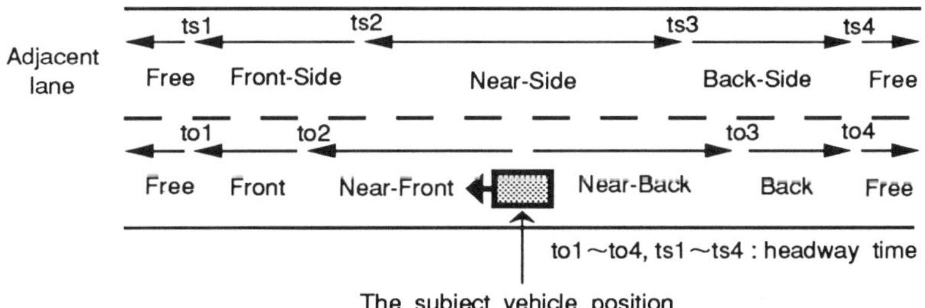

FIGURE 10.5. Road and traffic model.

Prototypes of the cruise planner and the vehicle control system were developed and tested on the simulator. Perfect perception was assumed in the simulation.

10.5.1 Environmental Model

A flat, one-way road with two lanes is considered in the simulation. Structural parameters, such as average curvatures at near, intermediate, and far distances are fed into the cruise planner. A total of four vehicles (excluding the autonomous vehicle) are considered: one vehicle in front and one behind in both the driving and adjacent lanes. The model parameters used are the location and the speeds of the surrounding vehicles; those parameters can be preprogrammed in a simulation.

In order to formally express the relative positions of the vehicles, the traffic model shown in Figure 10.5 is introduced. To represent vehicle-to-vehicle distances, a headway time instead of an actual distance is used. The headway time, which is a scale to express time allowance resembling human perception, is defined by the formula.

$$Ht = Slf/Vf,$$

where Ht is the headway time, Slf is the vehicle-to-vehicle distance, and Vf is the speed of the following vehicle.

The road environment around the autonomous vehicle is divided into 12 zones by use of the headway time, as shown in Figure 10.5. The speeds of the surrounding vehicles are also expressed by symbolic terms, such as fast, slow, and moderate, thus allowing a logical expression of the driving situation.

10.5.2 Driving Programs

Nine types of driving programs are presently available for use. There are two types of driving programs for steering control: a lane-following driving program and a lane-change driving program. There are also seven types of driving programs for speed control: a car-following driving program, a passing-car driving program, a leading-car driving program, etc. [4]. All these driving programs were developed from the basic three control algorithms described in the following subsections.

STEERING CONTROL ALGORITHM

A driver model called the "preview tracking model" [7], shown in Figure 10.6, is adopted for the lane-keeping driving program. First, the future position P of the vehicle, after an elapsed time τ seconds, is predicated assuming the current steering angle and speed remain the same. For the prediction of the future position, the steering angle θ, speed V, and acceleration α of the vehicle are used. Then, the future deviation ε_ϕ of the vehicle from the desired course is calculated, and the deviation is compensated by a discrete integral control action, that is, an increment in proportion to the future deviation is added to the steering angle.

TURNING SPEED CONTROL ALGORITHM

An incremental value in proportion to the difference between a desired speed V_d and a predicted speed V_p after an elapsed time τ is added to either the acceleration (throttle) value (ACC) or the brake value (BRK), as shown in Figure 10.7. Here, the desired speed must be set not to exceed the allowable lateral acceleration αt. Operation must be mutually exclusive for the throttle and brake pedals, that is, for acceleration, if the brake pedal is being depressed, first the brake pedal is released before the throttle pedal is depressed. If the brake pedal is not being depressed, the throttle pedal may be depressed directly. Deceleration works in a similar fashion.

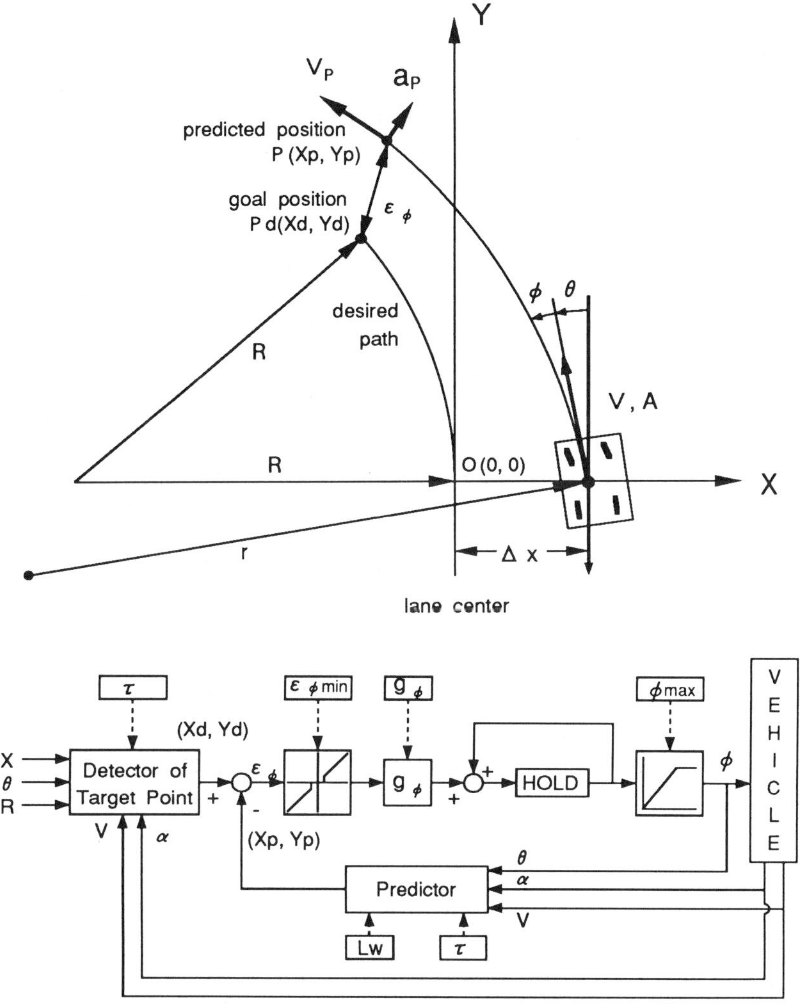

FIGURE 10.6. Prediction of feature position and preview tracking control.

VEHICLE-TO-VEHICLE DISTANCE CONTROL ALGORITHM

A model based on future deviation prediction, shown in Figure 10.8, is also adopted for controlling the vehicle-to-vehicle distances. Assuming that the current speed and acceleration of the vehicle remain the same, a future head-way time T_p for a targeted vehicle is predicted. A desired headway time T_d is given to the control system as an inherent value. Given the desired headway time, an incremental value in proportion to the deviation εT between the desired headway time and the predicted headway time is added to either the throttle or brake value. If the future deviation is less than a specified minimum

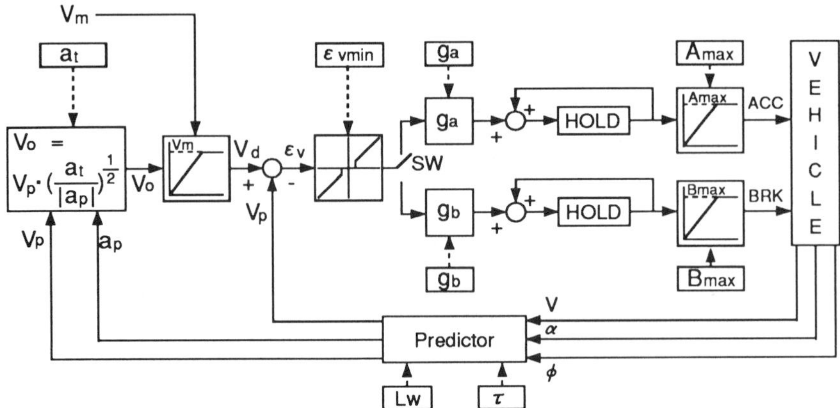

FIGURE 10.7. Turning speed control.

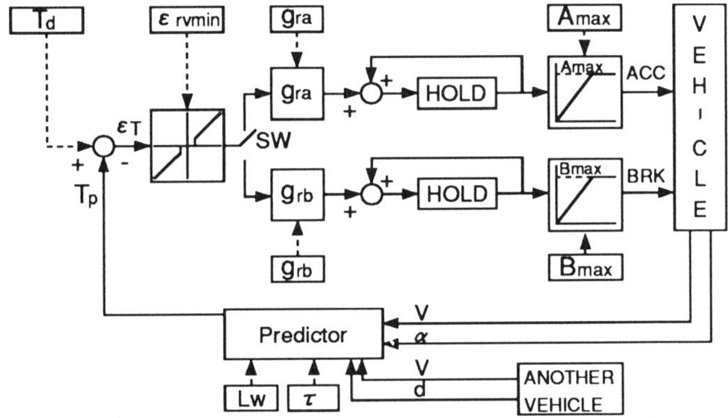

FIGURE 10.8. Vehicle-to-vehicle distance control.

value, no correction is made. Also, if the throttle and brake values exceed their specified maximum values, the maximum values are used.

10.5.3 Cruise Planner

In the prototype of the cruise planner, the knowledge for highway cruising was stored as three types of rule bases: a situation-monitoring rule base, a decision-making rule base, and an action-execution rule base. The situation-monitoring rule base gives the rules to understand the driving situations from the sensor data. The details on the cruise planner can be found in Fujita et al., 1990 [4].

The decision-making rule base selects one of the following actions in order to reach a desired condition in a variety of driving situations.

1. Free run: This is driving by maintaining a desired speed without interaction with other vehicles.
2. Follow: This is following a preceding vehicle keeping a proper spacing.
3. Overtake: This is getting ahead of a front-running vehicle.

For the selection of the actions, various information, such as the positions and speeds of other vehicles and road structures ahead, is referenced.

The action-execution rule base activates proper driving programs to execute a selected action. In overtaking a vehicle, for example, the lane-change and passing driving programs need to be executed in a proper sequence. For the execution of the selected action, a mechanism to monitor the status of achieving the goals is needed because the situation may change during the execution of the action. When a vehicle that is to be passed suddenly increases its speed and moves away, for example, this action program must be cancelled. Thus, the cruise planner must have a capability of dynamic planning; this was made possible by use of a multiple state transition network. Details are given in Fujita et al., 1990 [4].

10.5.4 Simulation Results and Discussion

The simulator is free to set any speed for the four vehicles surrounding the autonomous vehicle. Simulating various driving situations, it was confirmed on the graphic display that the autonomous vehicle could cruise on the two-lane road following a vehicle, adjusting vehicle-to-vehicle distances, overtaking slower vehicles, etc. The results proved the feasibility of the knowledge-based cruise planner.

In the current simulation, any behaviors during emergencies are not considered. For obstacle avoidance and emergency handling operations, more precise environmental models and more expert knowledge are required.

It must be mentioned that complete external perception was assumed in the simulation. There is no need to say that in reality the perception system is a technical bottleneck.

10.6 Visual Navigation Experiment

The actual vehicle tests are intended to develop visual navigation algorithms for the driving programs in various situations. Therefore, the test environment was simplified using white lane-markers to avoid the complexity of environmental perceptions. Here, a lane tracking test, which confirmed the effectiveness of the lane-keeping driving program, is described.

10.6.1 Experimental Vehicle

The experimental vehicle used was a commercial van (about 1.5 t in weight), as shown in Figure 10.9, equipped with a fixed CCD (charge coupled device)

FIGURE 10.9. Experimental vehicle MOVER-2.

camera as the vision sensor. The camera was mounted at a height of 152 cm, and its detection range was about 3 to 30 m. The main computer was a NEC PC9801 LS(80386 CPU) with an image processing board equipped with four ImPPs (Image Pipelined Processors). As a steering actuator, a stepping motor was attached to the steering shaft. The actuator torque on the shaft was 300 kg cm, and its maximum turning speed was 90°/s. The steering system itself was assisted by hydraulic power steering.

For speed control, a DC servo motor was used to control the throttle opening. The speed control was effected using a subcomputer, NEC PC9801 LX(80286 CPU), which was connected to the main computer. The main computer sent speed commands to the subcomputer, and actual speed was controlled by a feedback of the measured speed of the wheels. Details can be found in Maruya et al., 1990 [5].

10.6.2 Test Results and Discussion

The navigation tests were conducted on a straight course, circular courses, courses entering a circular section after a straight section, and S-shaped courses. The arc radii ranged form 30 to 50 m. The test speeds were a maximum 50 km/h on the straight course, a maximum 30 km/h on the circular courses, and about 15 km/h on the S-shaped courses.

The preview tracking algorithm was used to control the steering for all the courses. The look-ahead distance or preview time was adjusted depending on the speed of the vehicle. Although it was expected that there might be a steering delay upon entering a circular arc from a straight way, after adjusting control parameters, it was found that the algorithm worked well within the range of the experiments. It was also expected that there would be a considerable steering delay due to the reversal in curvature in the S-curved courses,

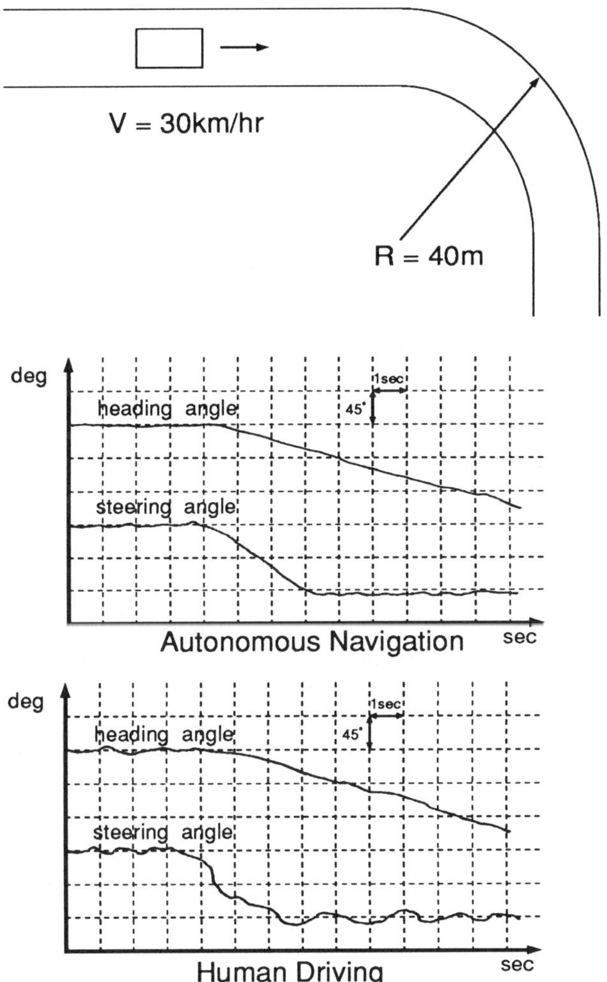

FIGURE 10.10. Experimental result.

but it was also found that the deviation from the course was within an allowable range.

An example of the test results is given in Figure 10.10, which shows an experiment of entering a circular arc section from a straight section. In the figure, measured heading angles and steering angles are shown, comparing human driving with the result of the autonomous navigation. It was confirmed that by the preview tracking algorithm the vehicle could be guided smoothly along the course.

So far, the function of tracking continuous lane markers at relatively low speeds was realized in the actual vehicle test. As the next development step, a method of tracking fixed points has been studied to realize lane-change and obstacle-avoidance driving programs [6].

10.7 Summary and Discussion

An autonomous vehicle control system that has a knowledge-based cruise planner and various driving patterns was proposed. By computer simulation, it was shown that the knowledge base could navigate the vehicle in a highway environment by switching the driving patterns depending on the situation. It was also shown that the driving programs could be realized using human driver models.

This approach is characterized by the use of the explicit knowledge of a human driver. The advantages of such a knowledge-based control system are that a complex control system can be built without difficulty by adding human knowledge to the system and prototyping can be done quickly by using the explicit knowledge of a human. Conversely, one disadvantage seems to be the lack of robustness in the system in ambiguous or uncertain situations. This is because the system relies on explicit knowledge, making it difficult to describe such situations.

Therefore, when the problem-solving algorithm and the framework of the total control system are clearly defined, it would be better to replace the part of the knowledge base by more robust control methods. For instance, the selection of the driving patterns would be done better either by neural networks or a subsumption architecture.

The vision system is a tough bottleneck in realizing an autonomous highway vehicle. The authors have been studying lane recognition of actual roads using video road images, but so far with no satisfactory results. Even in perceiving the lane markers on actual roads, which is the most basic function for the vision system, the system could not perceive the lane in about 5% of the image frames because of a variety of external disturbances. Although the deficiency of the image-sensing devices is one reason, the software technology for image understanding must be improved before the device can be improved.

The modern automobile is already a sophisticated electronically controlled system. Therefore, with a complete perception system, it is not difficult to navigate them autonomously. However, considering the navigation in actual traffic flow, certain knowledge-based and logical reasoning functions seem to be indispensable for autonomous navigation. Clarifying these will be an important research topic for the future.

References

[1] Dickmanns, E. D., and Zapp, A. (1987). "Autonomous High Speed Road Vehicle Guidance By Computer Vision." *IFAC 10th Triennial World Congress.*

[2] Bridges, G. S. (1989). "Advanced Highway/Vehicle Programs—A Texas View." *SAE Future Transportation Technology Conference and Exposition, August 1989.*

[3] Kagawa, Y., et al. (1988). "An Architecture for Intelligent Control Of Autonomous Vehicle Based on Human Driver Model." *Proc. IROS'88,* 349–354.

[4] Fujita, K., et al. (1990). "A Knowledge-Based System for Autonomous Highway Cruising." *Proc. of the Japan–U.S.A. Symposium on Flexible Automation.*

[5] Maruya, Y., et al. (1990). "Development of an Experimental Autonomous Vehicle." *Proc. of 5th Symposium on Intelligent Mobile Robot* (in Japanese).

[6] Kutami, A., et al. (1990). "Visual Navigation of Autonomous On-Road Vehicle." *Proc. of IROS'90.*

[7] Yoshimoto, K. (1981). "A Self-Paced Preview Tracking Control Model of an Automobile Driver," Preprints of I.F.A.C. 8th Triennial World Congress, XV.

11
Finding Road Lane Boundaries for Vision-guided Vehicle Navigation

L.T. SCHAASER AND B.T. THOMAS

11.1 Abstract

A method for finding road lanes using white lane markings is described. It is based on general techniques of object recognition. The lane markings are extracted as objects from the image using knowledge of their size, shape, and brightness. All possible lane boundaries that fit the markings based on the constraints of a road model are found. The correct set of lane boundaries is obtained by further application of the road model constraints. The technique is shown to be robust and can cope with markings that are intermittent, missing, or false.

11.2 Introduction

The past decade has been considerable interest in the application of computer vision techniques to the guidance of autonomous road vehicles. Initial research in the United States has demonstrated the automatic navigation of vehicles on single track roads with some capability of negotiating junctions and avoiding obstacles [1, 2, 3, 4]. Modest speeds of up to 30 km/h were achieved. Research in Germany has concentrated on navigating along more well-defined road, that is, motorways, at higher speeds of up to 100 km/h [5]. However, the fast speed was obtained at the expense of the flexibility to deal with junctions and obstacles. It is clear that a more competent visual system would need to deal with a greater variety of situations than covered to date *and* at the same time achieve reasonable driving speeds.

Research at Bristol University on road navigation has been ongoing for the past four years and ultimately aims to produce a vehicle that can cope with a variety of situations in real time. To date, Bristol has demonstrated road edge following at speeds to 30 km/h using video images for the analysis [7, 8]. A small test vehicle has also proved the algorithms in closed-loop operation by successfully navigating itself along the university garden paths. However, the algorithms have been designed to locate the extreme road boundaries and would have difficulties with multilane roads. They would also have difficulties with the intermittent nature of dashed lane markings since the algorithms are tuned to work best on continuous edges. For this reason, it has been decided to extend the work and detect the markings of multilane roads. Also, lane markings, when present, are often better guides for road navigation than edges because they usually have higher contrast.

This chapter presents a method for finding road lanes using white lane markings. The approach to this problem has been to use general techniques of object recognition that are robust rather than fast and can also be extended, in the future, to recognize other objects in the road scene. It is expected that once robust, universal techniques have been found, speed improvements will be obtained by parallel processing using our transputer-based image analysis system. First, white lane markings are extracted from the road scene based on a knowledge of their size, shape, and gray-level intensity characteristics. The markings can be dashed or continuous. A road model is then used to fit lane boundaries to the markings, and finally the lane boundaries are grouped into road lanes. This method differs from previous methods of finding road lanes because it treats the white lane markings as explicit, whole objects to be recognized rather than just edges to a road lane. Thus, once a technique has been established to extract white lane markings, alternative models can be used to find different lane marking groupings, such as junction markings and striped no-go areas.

Section 2 describes the road model used, which is in terms of the real-world coordinate system because of its invariance in the real world. Section 3 presents the "image to real-world" transformation equations that are necessary for applying the road model. Section 4 describes how the white lane markings are extracted from the input image. Section 5 details how the lane boundaries are fitted to the markings using the road model and Section 6 covers the grouping of the lane boundaries into road lanes. Finally, Section 7 presents the results of the method applied to different road scenes.

11.3 Road Model

The road is modeled in terms of the real-world coordinate system so that perspective distortion in the image does not have to be taken into account. The model has to make some assumption about the road's structure in the real world in order to fully recover 3-D information from the 2-D static image. Other researchers have made quite complex assumptions. Waxman and LeMoigne [3] modeled the road as a sequence of flat planes at various inclines. They used vanishing points and continuity between planes to recover the 3-D information. However, in practice, it is very difficult to divide a curved road into plane sections and, as a result, the 3-D information was incorrectly recovered. A more precise technique was presented by Sakurai and Zen [6], who modeled the road as a flat plane followed by a predefined

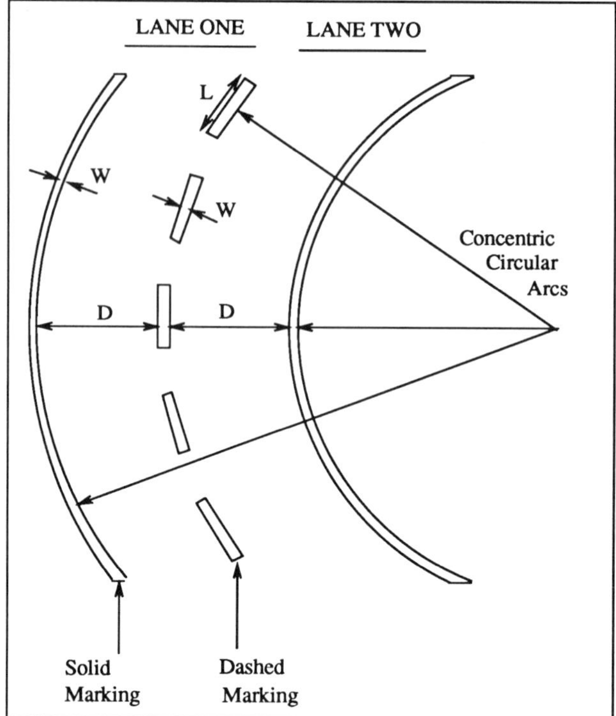

FIGURE 11.1. Road model: the road can consist of any number of lanes. Here, we show a two-lane road.

curved section and ending with another flat plane. Using vanishing points, the flat planes can be recovered in 3-D. Then, the curved section can be recovered from its endpoints, which have the same 3-D positions and tangents as the bounding flat planes. However, this method relies on all three sections being in the field of view, which rarely happens in practice.

We have chosen, at this stage, to model the road as a single flat plane. This model is then valid when the vehicle is travelling on a horizontal road or on a road of constant incline. The shape of the road in the flat plane is modeled as consisting of any number of lanes bounded by concentric circular arcs. This is shown in Figure 11.1. Straight roads are treated as circular arcs with very large radii. This model does not allow for roads with more than one bend. However, in practice, this situation does not arise very often because the field of view is limited. The road lanes have to be bounded by white lane markings that can be solid or dashed. The solid markings follow the curvature of the road and have a known width. The dashed markings are rectangular in shape and of an approximately known size. Finally, the gap between the lane boundaries has to be greater than a vehicle width.

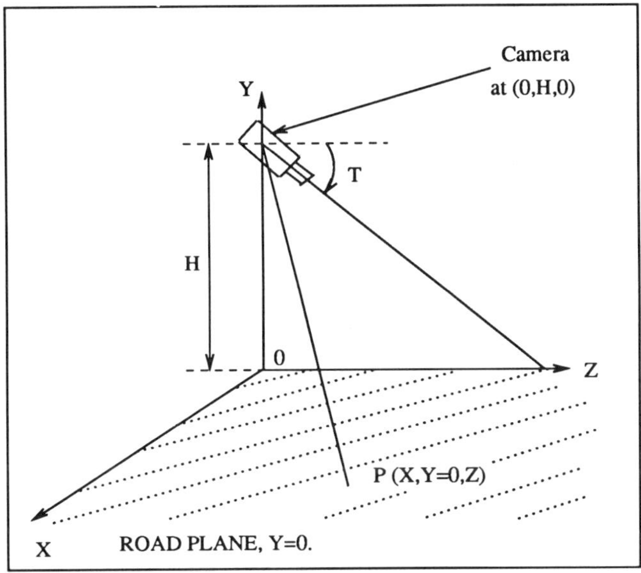

FIGURE 11.2. Camera system with world coordinates (X, Y, Z).

11.4 Transformation between Image and Real-World Coordinates

In order to apply the real-world road model to the image, it is necessary to transform image coordinates into real-world coordinates. As stated in the previous section, we make a flat road assumption that allows a point in the 2-D image plane to be fully recovered in the 3-D world. The imaging system is shown in Figure 11.2. The camera is mounted on the vehicle at a height H above the flat road surface, it is pointing in the direction of travel of the vehicle, that is, in direction Z, and it has a tilt angle τ with respect to the road plane. The road surface is the $Y = 0$ plane. The image coordinate system is shown in Figure 11.3. Using a pinhole model of the camera, the equations relating a position (x, y) in the image plane to its position in the real world $(X, Y = 0, Z)$ are

$$x = \frac{X_{size}}{2} + S_x \frac{fX}{(H \sin \tau + Z \cos \tau)}, \tag{11.1}$$

$$y = \frac{Y_{size}}{2} - S_y \frac{f(Z \sin \tau - H \cos \tau)}{(H \sin \tau + Z \cos \tau)}, \tag{11.2}$$

where (x, y) is the image coordinate position in pixel units, $(X, Y = 0, Z)$ is the real-world coordinate position in meters, Y_{size} and X_{size} are the dimensions of the image in pixel units, f is the focal length of the camera in meters, H is the

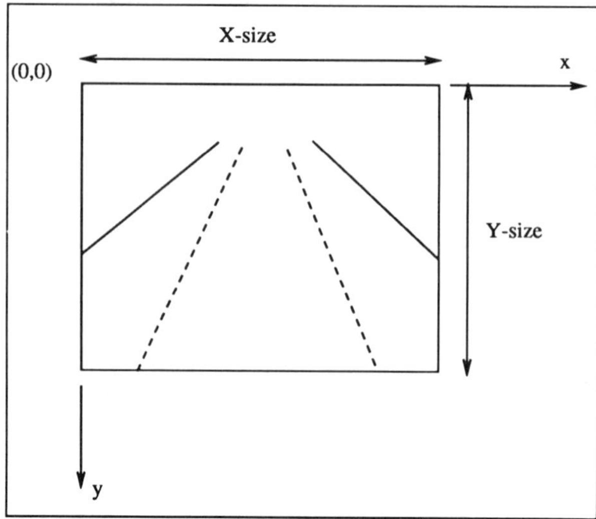

FIGURE 11.3. Image coordinate system (x, y).

height of the camera above the road surface in meters, τ is the tilt angle of the camera in radians with respect to the road plane, and S_x and S_y are factors to convert a position on the camera image plane from meters to pixels. Full details of the derivation of these equations are given in Lotufo et al., 1988 [7].

11.5 Extraction of White Lane Markings

This section describes how the white lane markings are extracted as separate objects from the static, gray-level image. The extraction of the white lane markings need not be perfect since the later use of the road model can cope with some misclassification. The method for extracting lane markings is in two stages.

First, the image is segmented into regions that are possible lane markings.
Next, feature descriptors are computed for each extracted region and used to eliminate those regions whose shapes are incompatible with the known shape of lane markings.

11.5.1 Segmentation

This first step in the processing uses knowledge about the gray-level characteristics of a white lane marking in order to segment it from the image. It is known that a white lane marking

is bright, and
has a high contrast with its surroundings.

Using these properties, the image is processed in the following way. Each pixel in the image is considered and marked as a candidate lane marking pixel if

it has a gray level intensity greater than a fixed threshold (we have taken the average of the maximum and minimum pixel intensities as the threshold, which is adequate for most situations); and

it lies between two strong intensity gradients of opposite sense within the distance of a predetermined lane marking width.

In the implementation, we assume that the lane markings are near vertical. This simplifies the processing and allows us to only compute the gradient in the horizontal direction. Also, we only scan the single row containing the pixel under consideration when looking for the required patterns in the gray-level and gradient. The gradient values are found by convolving with the x gradient Sobel filter,

$$\begin{pmatrix} -1 & 0 & 1 \\ -2 & 0 & 2 \\ -1 & 0 & 1 \end{pmatrix},$$

and the gradient threshold is set to a value equivalent to a step edge of 20 gray levels. The scanning distance for finding the two strong gradients is based on the width of a lane marking in the image and takes into account the fact that the width becomes smaller with increasing distance. From the transformation equations given in Section 3, it is known that a row in the real world, that is, Z is constant, maps onto a single row in the image, that is, y is constant. Thus, if we take two positions on the same row in the real world, (X_1, Z) and (X_2, Z), these correspond to positions (x_1, y) and (x_2, y) in the image, as shown here:

$$x_1 = \frac{X_{size}}{2} + S_x \frac{f X_1}{(H \sin \tau + Z \cos \tau)}, \tag{11.3}$$

$$x_2 = \frac{X_{size}}{2} + S_x \frac{f X_2}{(H \sin \tau + Z \cos \tau)}, \tag{11.4}$$

giving

$$(x_2 - x_1) = (X_2 - X_1) \frac{f S_x}{(H \sin \tau + Z \cos \tau)}. \tag{11.5}$$

If we assume that in the worst case the lane markings are at $45°$ in the real world to the Z axis, then their horizontal length $(X_2 - X_1)$ is $\sqrt{2} w_r$, where w_r is the lane marking width in the real world. The corresponding horizontal length in the image, $(x_2 - x_1)$, is then given by

$$(x_2 - x_1) = \sqrt{2} w_r \frac{f S_x}{(H \sin \tau + Z \cos \tau)}, \tag{11.6}$$

where

$$Z = \frac{H \tan \tau [y - (Y_{size}/2)] - S_y f H}{[(Y_{size}/2) - y] - S_y f \tan \tau}. \tag{11.7}$$

Thus, for a particular scanning row in the image y and knowing the width of a lane marking w_r, we can calculate the scanning distance $l = (x_2 - x_1)$ from Equations (11.6) and (11.7). We also use these equations to calculate a cutoff row beyond which the lane markings are too small to be processed. This is calculated using Equations (11.5) and (11.7) with $(x_2 - x_1)$ set to 1.5 pixels and $(X_2 - X_1)$ set to w_r.

Figure 11.4 shows a sample image of a road scene, and Figure 11.5 shows the results of the segmentation with the white regions as likely lane markings.

11.5.2 Feature Descriptors and Shape Analysis

Having obtained a binary image, we next obtain feature descriptors for each white region. Shape descriptors are computed to eliminate those regions whose shape is unlike the known shape of a lane marking. Position and orientation descriptors are computed to be used in the next stage of fitting lane boundaries to the markings. From the road model described in Section 2, we know the real-world shape of the markings. It is assumed that the dashed

FIGURE 11.4. Input gray-level image.

FIGURE 11.5. Results of segmentation.

markings are short and narrow enough not to be appreciably distorted in shape in the image, that is, they remain approximately rectangular in shape. The solid, continuous markings are distorted in the image because they are very long, so the assumption made is that any long object in the image is likely to be a solid marking and is therefore accepted. Thus, only the shorter objects have their shapes analyzed, and it remains to eliminate any such object that is not rectangular in shape.

The measure for rectangularity is a combination of the ratio of principal moments of inertia and a check for approximately constant width along the entire length of the object. The ratio of principal moments of inertia is calculated using the equations

$$r = I_a/I_b,$$

where I_a and I_b are the principal moments of inertia with $I_a \geq I_b$; I_a and I_b are obtained from the solutions of I in the equation:

$$(I_{xx} - I)(I_{yy} - I) - (I_{xy})(I_{xy}) = 0,$$

where I_{xx} and I_{yy} are the moments of inertia about the x axis and y axis, respectively, and I_{xy} is the product of inertia. The direction convention used for the various moments of inertia is shown in Figure 11.6. These inertias are computed for each white region:

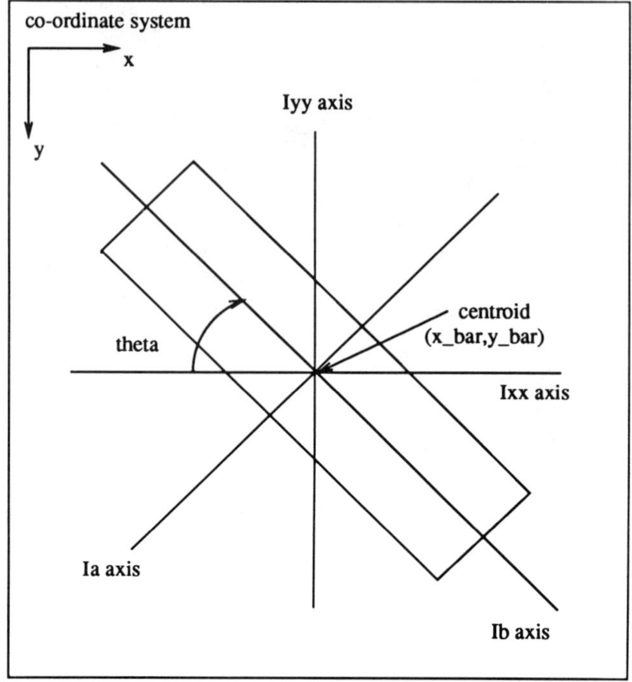

FIGURE 11.6. Direction convention for I_a, I_b, I_{xx}, I_{yy}, θ.

$$I_{xx} = \sum (y - \bar{y})^2,$$
$$I_{yy} = \sum (x - \bar{x})^2,$$
$$I_{xy} = \sum [(x - \bar{x})(y - \bar{y})],$$

and

$$\bar{x} = (\sum x)/n, \tag{11.8}$$
$$\bar{y} = (\sum y)/n, \tag{11.9}$$

where, for a white region, $(\bar{x}, \bar{y})$ is the centroid, (x, y) is a pixel position in the region, and n is the number of pixels in the region. The ratio r is related to the width w and length l of a rectangle by the formula

$$r = (l/w)^2.$$

In the implementation, we accept rectangles with (l/w) greater than two and, therefore, reject all objects whose ratio of principle moments of inertia is less than four. Objects with a greater ratio are only accepted if their width is approximately constant along their entire lengths. Finally, all objects less than five pixels in area are eliminated as they are considered to be noise.

FIGURE 11.7. Result after shape analysis.

The result of this processing is shown in Figure 11.7. We are left with lane-marking-like objects and need to provide descriptors of their positions and orientations for the next stage in the processing of fitting lane boundaries. For the short, dashed markings, we already have position information from their centroids. The orientation is calculated by the equation

$$\tan \theta = (I_a - I_{yy})/I_{xy},$$

where θ is the orientation and I_a, I_{yy}, I_{xy} are the moments of inertia of the lane boundary as shown in Figure 11.6. For the long marking, we need the position at various points along its length. These are calculated by splitting the marking into shorter, equal, sections whose centroids are individually obtained using Equations (11.8) and (11.9).

11.6 Fitting Lane Boundaries to the Lane Markings

In this stage of processing, lane boundaries are fitted to both the dashed and solid markings using the road model described in Section 2. The model is in terms of the real-world coordinate system. Therefore, an image to real-world transformation is required for the feature descriptors obtained from the previous processing stages. For the dashed markings, an approximate method is used. It is assumed that the image centroid and orientation approximately correspond to the real-world centroid and orientation when transformed by the equations given in Section 3. This assumption is valid because the lane markings are narrow. The real-world centroid $(\bar{X}, \bar{Z})$ is then obtained from the image centroid $(\bar{x}, \bar{y})$ by direct use of the equations in Section 3. The real-world orientation is obtained in a more indirect manner as follows. If θ_i is the orientation in the image, then

$$\tan \theta_i = \frac{(y_2 - y_1)}{(x_2 - x_1)},$$

where (x_1, y_1) and (x_2, y_2) are the intersection points with the image boundary of a straight line drawn from the image centroid $(\bar{x}, \bar{y})$ at an angle θ_i.

Then,

$$\theta_r = \arctan \frac{(Z_2 - Z_1)}{(X_2 - X_1)},$$

where the real-world positions (X_1, Z_1) and (X_2, Z_2) correspond to the image positions (x_1, y_1) and (x_2, y_2), and θ_r is the real-world orientation. For the solid markings, the previously obtained points along the length of the marking are transformed into real-world positions using the equations given in Section 3.

We next fit lane boundaries to the two types of lane markings based on the model that the lane boundaries are circular arcs.

11.6.1 Fitting Lane Boundaries to Solid Markings

The road model described in Section 2 requires that a circular arc is fitted to the solid marking. The center and radius of the arc are obtained by least squares fitting to the selected points along its length. The equation for a circular arc is given as

$$(X - X_0)^2 = R^2 - (Z - Z_0)^2, \tag{11.10}$$

where (X_0, Z_0) is the center of the circular arc, R is the radius of the circular arc, and (X, Z) is a point on the circular arc. This equation is not linear in the unknown parameters. Therefore, for ease of analysis, the points are actually fitted with second-order polynomials from which circular arcs are approximated. This is achieved by rearranging and expanding Equation (11.10) as a binomial series:

$$(X - X_0) = \pm \left\{ R - \frac{(Z - Z_0)^2}{2R} + \cdots \right\}.$$

The higher order terms in the expansion can be neglected since, in general, $R \gg Z$. Rearranging the last equation into a quadratic form,

$$X = \left(X_0 \pm R \mp \frac{Z_0^2}{2R} \right) \pm \frac{Z_0}{R} Z \mp \frac{1}{2R} Z^2,$$

and comparing this with $y = a + bx + cx^2$, it can be seen that the parameters X_0, Z_0, and R can be determined from the least squares fitting of a quadratic function. This method of approximation is described in more detail in Morgan et al., 1988 [8].

11.6.2 Fitting Lane Boundaries to the Dashed Markings

Fitting lane boundaries to dashed markings is harder than for the solid markings because it is not known, in advance, which lane-marking-like objects belong to the same lane boundary. A technique that could be used has been described by Vasudevan et al., 1988 [9]. Their technique searches around the endpoints of each object to find another object nearby that can be linked to it. Objects are only linked together if their position and orientation provide a smooth transition from one object to the other. This has obvious drawbacks, for example, a nearby false object can lead the linking algorithm down a false trail. For this reason, we have adopted a model-based approach that constrains the lane markings to lie on circular arcs. This provides robustness against occasional missing and false markings.

Since the parameters of the circular arcs are not known, at this stage, all possible circular arcs are found. This is achieved by taking any pair of short objects and determining if they can describe a consistent circular arc given their positions and orientations. Normals are constructed from the centroids of each of the two objects, and the intersection point is found. If the centroids are approximately equidistant from this point, a circular arc is obtained whose parameters are given by

the center (X_0, Y_0), which is the intersection point of the two normals; and
the radius R, which is the mean of the two distances from each centroid to the
   intersection point.

This process is repeated for all pairs of short objects to find all possible circular arcs. The result is a "tangle" of circular arcs, as shown in Figure 11.8. Some of the arcs describe false lane boundaries and others approximate the same true lane boundary. Since the actual road can have a varying curvature, slightly different arcs are fitted to different sections of the same lane boundary, as shown in Figure 11.8. Obviously, more processing is required to obtain the final road lane boundaries.

In preparation for subsequent processing stages, we note that the confidence in accepting a particular circular arc as a lane boundary increases with the number of objects that approximately lie on it. We already have, for each circular arc, the two objects that define it, and we then find the remaining objects that also lie on the same arc. An object is considered to "lie on" a particular circular arc if

the distance from the object centroid to the arc center is approximately that of
   the arc radius, and

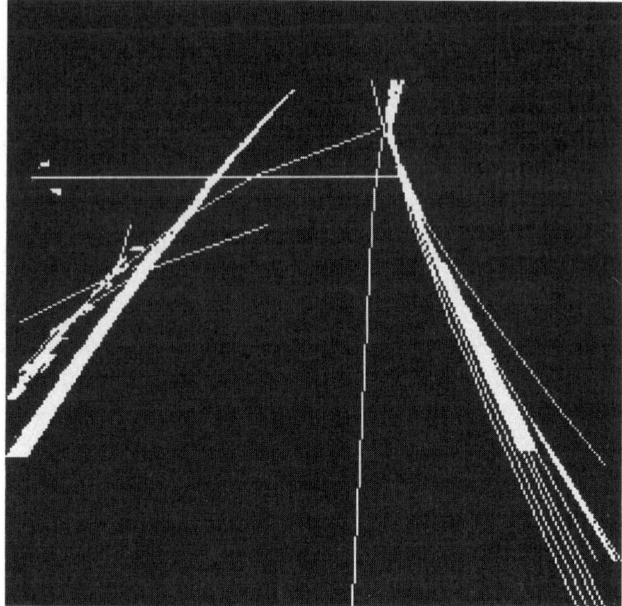

FIGURE 11.8. All possible circular arcs.

the object orientation is approximately perpendicular to the line joining the object centroid to the arc center.

The result is a list of circular arcs with their corresponding members, and this information is used in the next stage of processing.

11.7 Grouping the Lane Boundaries into Road Lanes

The previous processing stage produces a number of circular arcs, some of which are the true lane boundaries. This stage of processing finds the correct arcs based on the knowledge that they have a constant separation and are at least a vehicle width apart. We also refine the model described in Section 2 by approximating the lane boundaries to quadratic curves instead of circular arcs.

11.7.1 Finding Arcs of Constant Separation

The model of Section 2 describes the road as consisting of concentric circular arc lane boundaries. In practice, true lane boundaries are usually not concentric for the following two reasons.

For curved roads, lane boundaries can have varying curvature, in which case, different circular arcs are fitted to different sections of the same lane boundary, as shown in Figure 11.8. When pairing lane boundaries into road lanes, if the left boundary is clipped more than the right boundary at the image edge, that is, the arc lengths are different, the left boundary may be further "into the bend," which results in different arc centers for the two boundaries.

Straight roads have very large radii and thus produce large errors in the positions of their circular arc centers.

Because of these effects, it is more suitable to compare curves using the gap between their visible lengths than using the notion of "circle centers." It also turns out to be more accurate at this stage to approximate arcs, not by circles, but by quadratic curves $y = a + bx + cx^2$. We find the quadratic curves for each circular arc by least squares fitting to the centroids of the objects that belong to it. Solid markings are already described by quadratic curves, as detailed in Section 5.1. The result is a list of quadratic curves. We then find all possible sets of quadratics that have an approximately constant separation from each other as follows. Any one of the curves is taken as a "seed" and compared with all the other curves. Only those curves that have an approximately constant separation from the seed are accepted. This is repeated for all the other curves using each one as a seed in turn. The separations between the seed and a compared curve are obtained by taking a predetermined number of equally spaced points along the "length" of the seed and projecting normals from these points until they intersect the other curve. The required distances are then along the projected normals from one curve to the other.

11.7.2 Vehicle Width Constraint

This final stage processes the sets of quadratics and eliminates those curves that violate the constraint that the lane boundaries have to be at least a vehicle width apart. The curves can be too close together for any of these reasons:

one of the curves describes a false lane boundary.

different curves are fitted to different sections of the same lane boundary (shown in Figure 11.8), and

two curves describe a double marked lane boundary, in which case, the choice of curve is arbitrary.

This constraint in the road model is implemented in the following way. First, the curves that are very close together (within 30 cm) are thinned by taking the "best" curve of the group. The best curve is found using a heuristic that gives precedence to solid markings and then to curves with the most members. In the event of a tie, the curve with the longest extent in the image is taken. For a lane boundary of dashed markings, the "extent" of a curve is defined as the straight line distance in the image plane between the two members that are the furthest apart from each other. For solid markings, the extent is the straight line distance in the image plane from the start of the marking to its endpoint.

Next, curves that are less than a vehicle width apart are resolved as follows. For each set of thinned quadratic curves, all possible subsets whose curves are separated by more than a vehicle width from each other are found. This results in several sets of curves, all of which fulfill the vehicle width constraint. The number of sets if further reduced by applying the constraint that the road lanes are approximately equal in width. The final set is obtained by using a heuristic that first takes the set with the most curves and then, in the event of a tie, takes the set that has the largest sum total of curve extents.

11.8 Experimental Results

The algorithm was coded in C and PROLOG and executed on a SUN 3/50 workstation; C was used for the low-level feature extraction and PROLOG was used for the higher level fitting of lane boundaries. The algorithm has been successfully tested on a number of road images and Figures 11.9 and 11.10 show the results of two representative images. The resultant lane boundaries obtained from the processing are shown as thin white curves superimposed on the original image. The model of circular arc lane boundaries

FIGURE 11.9. Resultant lane boundaries for a curved road.

FIGURE 11.10. Resultant lane boundaries for a straight roads.

has proved to be a very good approximation for grouping dashed lane markings into their correct lane boundaries for both straight roads and roads with a single bend.

Although the aim of the algorithm is for robustness rather than speed, we give here an indication of the processing time to show that it is not excessive and that there is a possibility of implementing it in real time. The algorithm took in the order of one minute to execute. Execution time depends mainly on the number of possible circular arcs obtained, which, in turn, depends on the number of dashed markings per lane boundary and the number of lane boundaries. If the number of dashed markings per lane boundary is N and the number of lane boundaries is B, for each lane boundary there would be

$$\frac{N!}{2!(N-2)!}$$

combinations of taking two markings from a group N to produce the possible circular arcs. Therefore, the total number of possible circular arcs would be

$$\frac{N!}{2!(N-2)!} \times B.$$

The equation shows an exponential growth in processing time with N, but in practice, N is typically five or six, and B is usually no greater than four, for a three-lane highway.

11.9 Conclusions

A method has been presented in this chapter that finds road lanes using white lane markings. The markings are extracted as distinct objects from the image, and lane boundaries are fitted to them using the constraints of a road model. The model describes the road as a flat plane consisting of concentric circular arc lane boundaries. This method has been shown to work well on a variety of multilane roads. In particular, it is capable of correctly fitting lane boundaries to dashed markings and is robust against occasional missing and false markings. The model has been found to describe most road situations since the typical look-ahead distance of 30 m usually allows only one bend to be visible. In cases where this does not apply, the model can be extended.

The object recognition approach for finding lane boundaries has proved to be successful, and as a result, it is planned to find road junctions based on the same approach. In this case, the patterns of the white junction markings will be the structures to be recognized.

Acknowledgments. L. T. Schaaser is indebted to SERC and MoD RARDE (Chertsey) for financial support.

References

[1] Wallace, R., et al. (1985). "First Results in Robot Road-Following." *Proc. 1985 IJCAI, Los Angeles* 1089–1095.

[2] Wallace, R., et al. (1986). "Progress in Robot Road-Following." *Proc. 1986 IEEE Int. Conf. Robotics and Automation*, 1615–1621.

[3] Waxman, A. M., et al. (1987). "A Visual Navigation System for Autonomous Land Vehicles." *IEEE Journal of Robotics and Automation* 3, 124–141.

[4] Sharma, U. K., and Kuan, D. (1989). "Real-Time Model Based Geometric Reasoning for Vision-Guided Navigation" *Machine Vision and Applications* 2, 31–44.

[5] Dickmanns, E. D., and Zapp, A. (1986). "A Curvature-Based Scheme for Improving Road Vehicle Guidance by Computer Vision." *SPIE Conference 727 on "Mobile Robots", Cambridge, Ma., USA, October 26–31, 1986.*

[6] Sakurai, K., et al. (1987). "Analysis of a Road Image as Seen from a Vehicle." *First International Conference on Computer Vision, London*, 651–656.

[7] Lotufo, R. A., et al. (1988). "Road Edge Extraction Using a Plan-View Image Transformation." *Proceedings of Fourth Alvey Vision Conference at University of Manchester*, 185–190.

[8] Morgan, A. D., et al. (1988). "Road Edge Tracking for Robot Road Following." *Proceedings of Fourth Alvey Vision Conference at University of Manchester*, 179–184.

[9] Vasudevan, S., Cannon, R. L., and Bezdek, J. C. (1988). "Heuristics for Intermediate Level Road Finding Algorithms." *Int. Journal of Computer Vision, Graphics and Image Processing* 44, 175–190.

12
An Extracting Method of the Optical Flow for an Anticollision System

TOSHIO ITO AND SHIRO KAWAKATSU

12.1 Abstract

We examine a gradient method to detect spatial information of the targets surrounding a car using image information of the objects for the anticollision system. More pixels can be calculated and less calculation time can be expected by using a gradient method than by a correlation method or a matching method. In this chapter, we mention the algorithm used to calculate differences between two objects in images with the gradient method in one dimension. In addition, we calculate distance and velocity values of the objects in the images and examine the possibility of using the gradient method in a car application.

12.2 Introduction

We are investigating a vision system for detecting a target surrounding a car. This system aims mainly to maintain an absolute distance between the car and the target. A basic principle in extracting spatial information about objects imaged with a TV system is to estimate differences between two images of the object. In general, for shaded images obtained with a TV system (which can be simulated to a pinhole lens camera system), distance information can be extracted from differences between two images obtained from cameras that differ in spatial position. Velocity information can be extracted from differences between objects in the time sequence images. There are many methods to extract such information. Typical methods are the correlation method [1], matching method [2, 3, 4], and gradient method. This time, we examine the gradient method because it is believed that more pixels can be calculated and less calculation time can be expected by using the gradient method rather than the correlation and matching methods.

The gradient method of image understanding was developed through the research of TV image compression [5, 6]. The research derived the relation between a gradient of the brightness intensity and an optical flow. It has suggested a mathematical model from a view of psychology [7, 8]. The method to derive this optical flow directly with shaded images has been studied by Horn and Shunck, 1981 [9]. This method assumed that the brightness intensity distribution of the target would be constant whenever the relative movement between the camera and the target was very small. The gradient method was established under the this assumption and was demonstrated by actual experiments. It is possible to apply this to stereo since it can be thought of as one of the methods to extract differences between two images of the object [10].

12.3 A Gradient Method

12.3.1 A Calculation Method

In the gradient method, it is assumed that the intensity of the brightness keeps the same value at the same point of the objects in the different images. This condition can be expected when an intensity and a direction of the light source and a direction and a distance between the camera and the target are almost fixed.

Let $E(x, y)$ be the brightness intensity at the image point (x, y) of the object. Then, $E_1(x, y)$, $E_2(x, y)$ are the intensities at the point (x, y), where $E_1(x, y)$ is the intensity before the relative movement between the camera and the object will occur, and $E_2(x, y)$ is the intensity after the movement has occurred. If (u, v) are the x and y components of the movement vector at that point, we expect that the intensity will be the same at the point $(x-u, y-v)$. That is,

$$E_1(x-u, y-v) = E_2(x, y) \tag{12.1}$$

for a small movement. This single constraint is not sufficient to determine both u and v uniquely. It is also clear that we can take advantage of the fact that the motion field is continuous almost everywhere, assuming $E(x, y)$ is linear within the movement area. So, we can expand the left-hand side of Equation (12.1) in first-order terms of a Taylor series and obtain

$$E_1(x, y) - [\partial E_1(x, y)/\partial x] \times u - [\partial E_1(x, y)/\partial y] \times v = E_2(x, y). \tag{12.2}$$

Using the abbreviations $Ex = \partial E_1(x, y)/\partial x$, $Ey = \partial E_2(x, y)/\partial y$, $Et = E_2(x, y) - E_1(x, y)$, we obtain

$$(Ex \times u) + (Ey \times v) + Et = 0. \tag{12.3}$$

The derivatives Ex, Ey, and Et are estimated from the images.

Generally, the movement vector cannot be determined with only this constraint equation, so the assumption that the neighboring points of (x, y) have the same movement vector is made. Although at least two constraint equations are needed to solve the equation with respect to u and v, more equations have to be used to raise the reliability of the solutions. So, many calculations are needed.

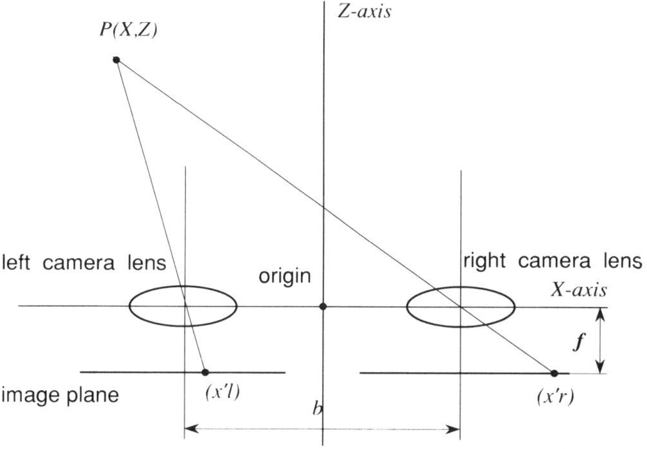

FIGURE 12.1. Camera geometry for stereo method.

In the application to cars, it is desirable to determine only the x directional components of the movement. Substituting $v = 0$ in Equation (12.3), we obtain

$$Ex \times u + Et = 0, \tag{12.4}$$

which shows that only the x directional movements occur. This equation shows that the component u is determined uniquely.

12.3.2 An Application for a Stereo Method

The most common method for extracting the distance between the camera and the target is to use the difference between the object in the stereo images. Suppose that two cameras are rigidly attached to each other so that their optical axes are parallel and separated by a distance b (Figure 12.1). Assume that the baseline (the line connecting the lens centers) is perpendicular to the optical axes, and orient the x axis so that it is parallel to the baseline. The coordinates of the point (X, Z) in the environment are measured relative to an origin midway between the lens centers. Let the image coordinates in the left and right images be $(x'l)$ and $(x'r)$, respectively. Then,

$$\frac{x'l}{f} = \frac{(X + b)/2}{Z}, \qquad \frac{x'r}{f} = \frac{(X - b)/2}{Z}, \tag{12.5}$$

where f is the distance between the lens center to the image plane in both cameras. These equations can be solved for the unknown Z:

$$Z = bf/u, \tag{12.6}$$

because the movement vector u can be regarded as the disparity $x'l - x'r$.

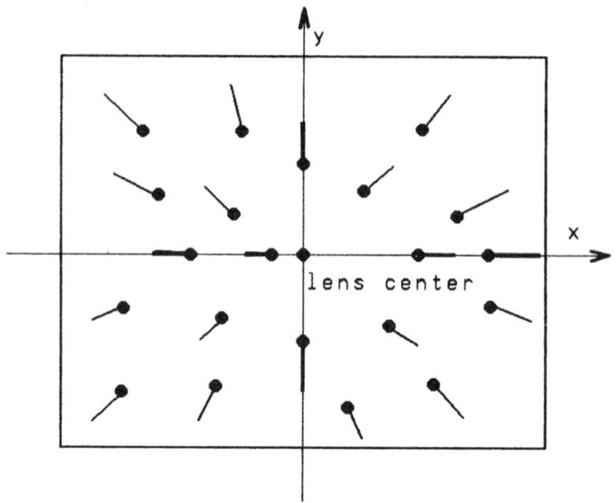

FIGURE 12.2. Optical flow field.

12.3.3 An Application for the Optical Flow

The movement vector can be regarded as the optical flow vector when the vector is extracted from time-varying images of the object obtained with one camera. When the camera is moving along its optical axis without rotation with respect to the static rigid body target, the optical flow of objects arises radially around the center of the lens in the image (Figure 12.2). The line implied by the focus of the extension on the image plane realizes Equation (12.4).

In this case, we can estimate the distance between the camera and the target. Let the coordinates of a point P in the environment be (X, Z) and the translational components of the motion of the camera be $(0, W)$ (Figure 12.3). Then, the component of the camera motion is in the form

$$dZ/dt = -W. \tag{12.7}$$

A point p on the image plane is the perspective projection from the point P; thus, the coordinate of p is

$$x = X/Z. \tag{12.8}$$

The optical flow at a point x, denoted by u, is

$$u = dx/dt. \tag{12.9}$$

Differentiating the equation for x with respect to time and using the derivatives of Z, we obtain the following equation for the optical flow:

$$u = (dX/dt)/Z - X(dZ/dt)/Z \times Z. \tag{12.10}$$

Then, the following equation holds:

$$Z = x \times W/u. \tag{12.11}$$

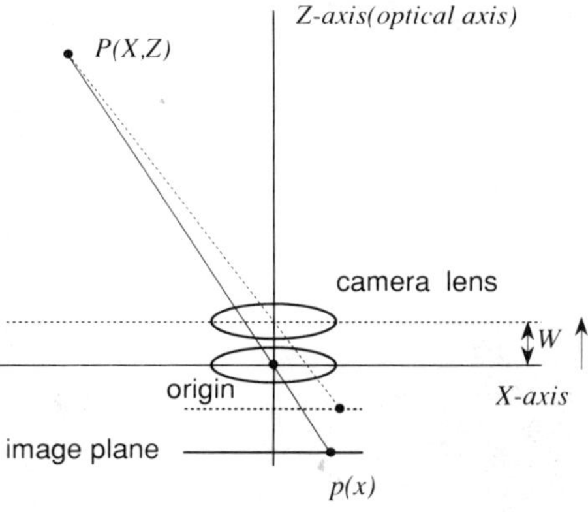

FIGURE 12.3. Coordinates when the camera moves.

12.4 A Calculation Algorithm

12.4.1 A Transformation into the One-Dimensional Image

The two-dimensional images have to be transformed into the one-dimensional case for use by our algorithms. Ideally, the object line on the image plane corresponding to the object line on the other image plane gives our one-dimensional case. However, since the images contain various errors and noise, some kinds of integration should be carried out at every pixel. The direction of the integration is perpendicular to the line. Moreover, the image intensity values of the border are multiplied by some weights because the limit of the integration has to be considered in the camera calibration. Figure 12.4 shows this algorithm.

12.4.2 A Calculation Algorithm

Figure 12.6 shows the calculation result of u by use of Equation (12.4), applied to the brightness intensity of objects shown in Figure 12.5. The one-dimensional images of objects transformed by the previous section algorithm in Figure 12.5 are obtained by use of the stereo method. The disparity u is made of as many pixels as possible under the controlled conditions in the experimental room. The dispersions of u in Figure 12.6 are caused by the difference from $E_1(x)$ to $E_2(x)$ at the calculation unit pixel. Another calculation algorithm is needed because the number of pixels that can be calculated will be decreased if the pixels are rejected where the differences are large.

The derivative of $E_2(x)$ with respect to x for u can be written

$$E_2'(x) = [E_2(x + u) - E_2(x)]/u = [E_1(x) - E_2(x)]/u. \qquad (12.12)$$

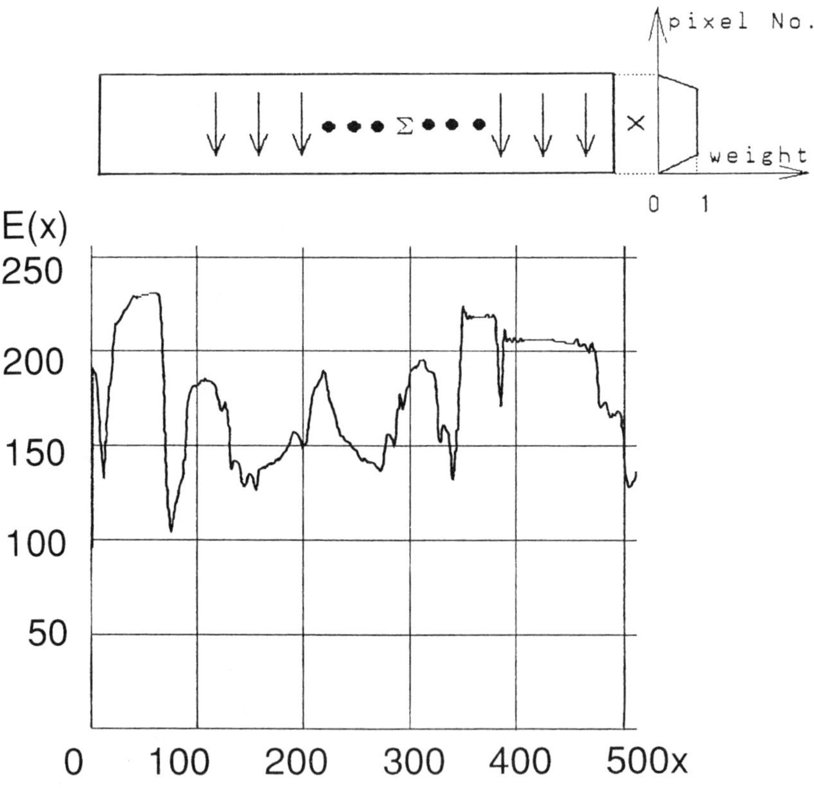

FIGURE 12.4. The algorithm of the integration.

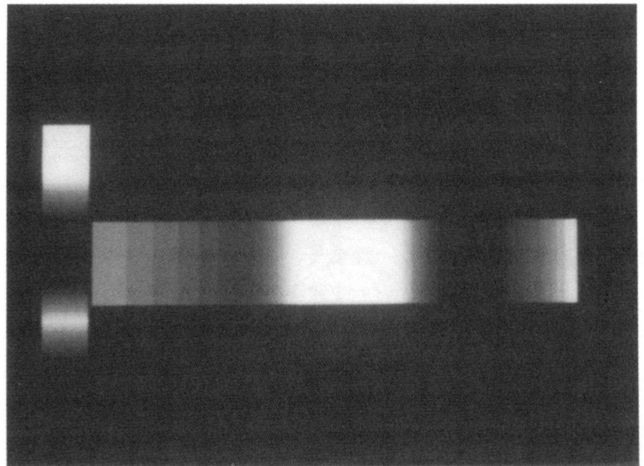

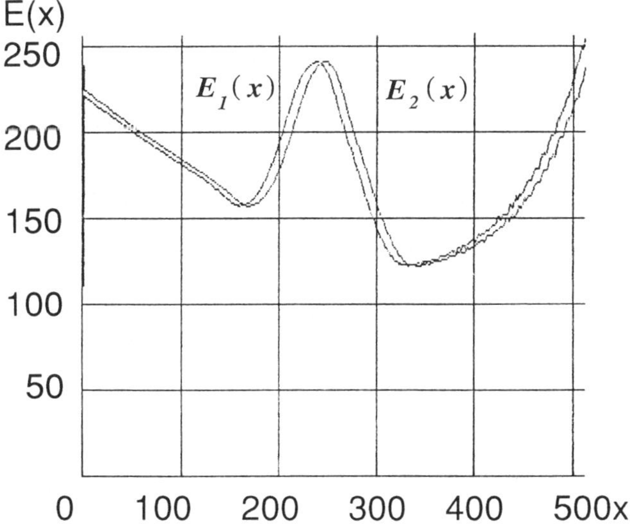

FIGURE 12.5. Brightness intensity distributions.

Therefore, we obtain

$$u = [E_1(x) - E_2(x)]/E_2'(x). \qquad (12.13)$$

This equation is equivalent to Equation (12.4). We can then move $E_2(x)$ by our estimate of u, and repeat this procedure, yielding a type of Newton–Raphson iteration. Ideally, our sequence of estimation will converge to the best u. This iteration is expressed by

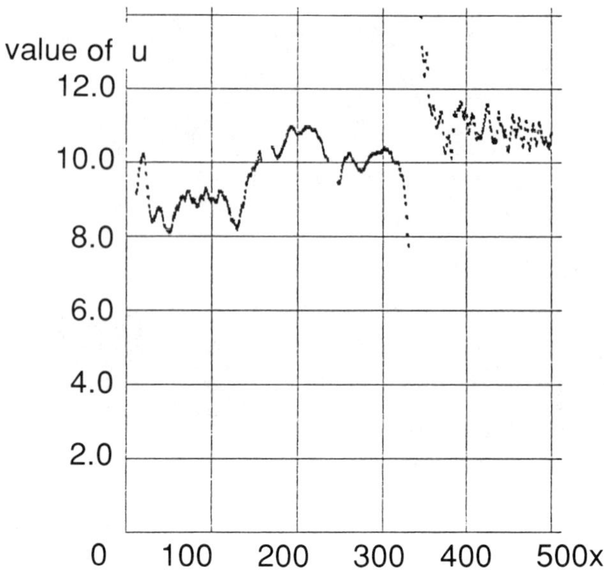

FIGURE 12.6. Values of u by use of Equation (12.4).

$$u_0 = -Et/Ex,$$
$$u_{n+1} = u_n + [E_1(x) - E_2(x + u)]/E_{2'}(x + u). \tag{12.14}$$

Figure 12.7 shows the calculation result of u by use of Equation (12.14) applied to the brightness intensity of objects shown in Figure 12.5. Values of u were obtained by two times iteration and not calculated where the coincidence of $dE_1(x)/dx$ and $dE_2(x)/dx$ were not correspondent.

12.5 Calculation Results

Figure 12.8 shows the image of the object whose distance from the camera is about 10 m. Figure 12.9 shows the brightness intensities in one dimension transformed from the two images obtained by the stereo method in which the value of b in the Equation (12.6) is 20 mm. Figure 12.10 shows the calculation result of the distance by use of Equation (12.6).

Figure 12.11 shows the optical flow, and figure 12.12 shows the calculated distance by use of Equation (12.11). In this camera measurement system, the sampling time is 1/30 mm s, and the velocity of the camera motion is 20 km/h.

12.6 Discussion

12.6.1 An Influence of the Target Reflectance Property

The image intensity of the object depends on a reflectance property of the target. Generally, the reflectance property is constituted by two components,

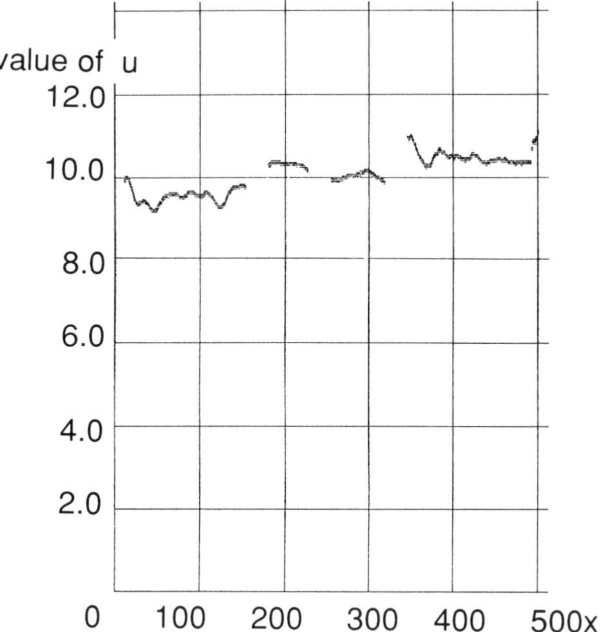

FIGURE 12.7. Values of u by use of Equation (12.14).

FIGURE 12.8. The image of the object.

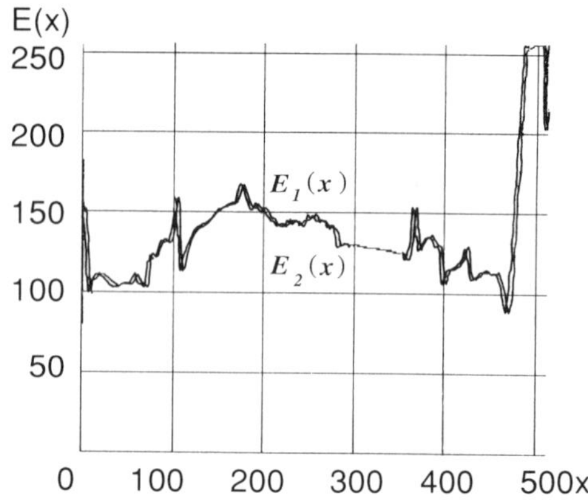

FIGURE 12.9. Brightness intensities of objects.

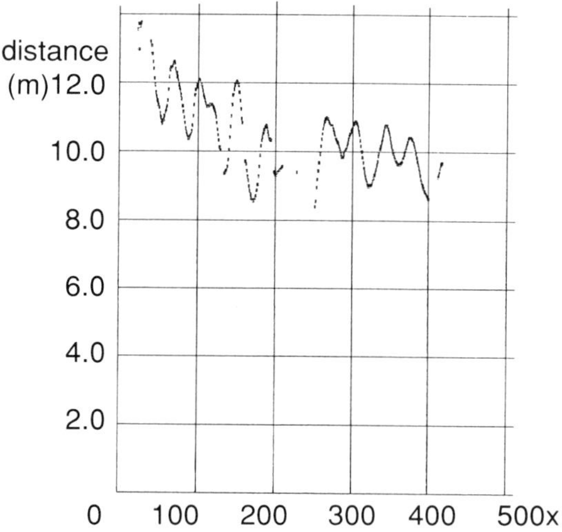

FIGURE 12.10. Calculated distance values.

a diffuse component and a specular component. The former expresses the surface property of the target. The latter mainly reflects the source of light, such as a mirror. Therefore, the extracted optical flow given by use of the gradient method in the specular area gives only the information concerning the source of light, not the object. In general, the extracted distance value between the camera and the reflecting point is larger than the actual distance

FIGURE 12.11. Calculated optical flow.

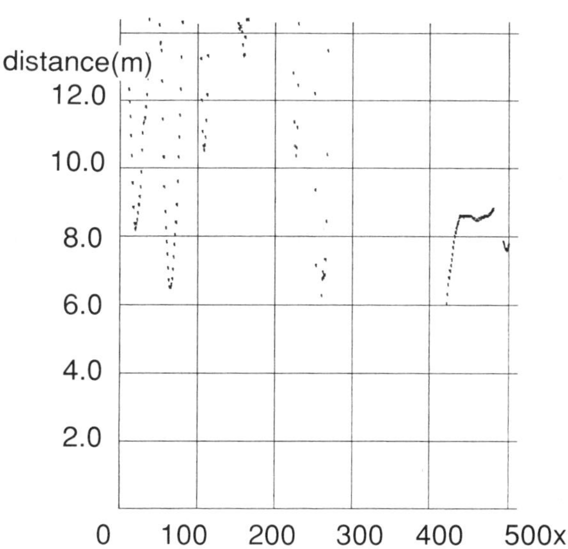

FIGURE 12.12. Calculated distance values.

between the camera and the target. So, large distance data should be omitted in case of the actual implementation.

12.6.2 An Influence of the Vision System

The image intensity also depends on the characteristic of the vision system because this transducer does not always transfer the brightness of the object. The vision system is comprised of the following components: optical lenses, a TV scanning system, an A/D converter, and some electronic circuits. Such components have many error factors, such as lens aberrations, scanning distortions, quantized errors, and heat noises. To suppress the influence of the lens aberration, we used low-pass filtered images. The scanning distortion is avoided by use of a charge coupled device (CCD). We took measures by using integrated images for the remaining errors and noises because they were random errors. The resolution of the CCD is the most important factor because the gradient method needs more correct brightness distribution functions.

12.6.3 The Problem to Be Solved in Actual Implementation

The gradient method assumes a constant brightness of the object between two images. It is hard to use the gradient method under the condition of varying illumination. We absorbed the influence of the small vibration by using the integrated images of the object for optical flow because the brightness is changed by the car vibration when the image measuring interval time is long. In the stereo images, this problem can be suppressed by shortening the image measuring interval time. In the stereo method, the calibration is another problem to be solved. Some new calibration methods and mechanisms must be introduced because the performance and specification of two cameras are not exactly the same. Generally, we need a more sophisticated vision system with an automatic gain control (AGC) and auto iris. We also need some new methods to reduce the influence of the weather conditions.

12.7 Conclusion

In this chapter from Sections 1 to 3, the outlook, the basic theorem, the way of adopting to the stereo vision system, the way of extracting optical flow, and the calculation method of the gradient method were showed. In Section 4, we showed the calculation results, and in the Section 5, we mentioned the problems in the actual implementation.

References

[1] Koga, K., and Miike, H. (1989). "Optical Flow Analysis Based on Spatio-Temporal Correlation of Dynamic Image." *Trans. of the Inst. of E.I.C.* **J72-D-II** (4), pp. 507–516.

[2] Marr, D. (1976). "Eary Processing of Visual Information." *Philosophical Trans. of the Royal Society of London* **275**, 483–524.

[3] Moravec, H. P. (1987). "Toward Automatic Visual Obstacle Avoidance." *Int. Joint Conf. on Artificial Intelligence*, 584.

[4] Enkelmann, W., Kories, R., Nagel, H. H., and Zimmermann, G. (1988). "An Experimental Investigation of Estimation Approaches for Optical Flow Fields." In W. N. Martin and J. K. Aggarwal, eds., *Motion Understanding*, Kluwer Academic Publishers, Boston, Massachusetts.

[5] Lim, J. O., and Murphy, J. A. (1975). "Estimating the Velocity of Moving Images in Television Signals." *Computer Graphics and Image Processing* **4**, 311–327.

[6] Cafforio, C., and Rocca, F. (1976). "Methods for Measuring Small Displacements of Television Images." *IEEE Trans. Information Theory* **IT-22**, 573–579.

[7] Gibson, J. J. (1950). *The Perception of the Visual World*. Riverside Press, Cambridge, Massachusetts.

[8] Gibson, J. J. (1966). *The Senses Considered as Perceptual System*. Houghton-Mifflin, Boston, Massachusetts.

[9] Horn, B. K. P., and Shunck, B. G. (1981). "Determining Optical Flow." *Artificial Intelligence* **17**, 185–203.

[10] Lucas, B. D., and Kanade, T. (1981). "An Iterative Image Registration Technique with an Application to Stereo Vision." *7th IJCAI*, 674–679.

13
Obstacle Avoidance and Trajectory Planning for an Indoor Mobile Robot Using Stereo Vision and Delaunay Triangulation

MICHEL BUFFA, OLIVIER D. FAUGERAS, AND ZHENGYOU ZHANG

13.1 Abstract

This article describes the work at INRIA on obstacle avoidance and trajectory planning for a mobile robot using stereo vision. Our mobile robot is equipped with a trinocular vision system that is being put into hardware and will be capable of delivering 3-D maps of the environment at rates between 1 and 5 Hz. The 3-D maps contain line segments extracted from the images and reconstructed in three dimensions. They are used for a variety of tasks, including obstacle avoidance and trajectory planning.

For those two tasks, we project on the ground floor the 3-D line segments to obtain a two-dimensional map, simplify the map according to some simple geometric criteria, and use the remaining 2-D segments to construct a tessellation, more precisely a triangulation, of the ground floor. This tessellation has several advantages.

It is adapted to the structure of the environment since all stereo segments are edges of triangles in the tessellation.

It can be efficiently computed (the algorithm we use has a complexity $o(n)$ if n is the number of segments used).

It is dynamic in the sense that segments can be efficiently added or subtracted from an existing triangulation.

We use this triangulation as a support for further processing. We first determine free space simply by marking those triangles that are empty, again a very simple process, and then use the graph formed by those triangles to generate collision-free trajectories. When new sensory data are acquired, the ground floor map is easily updated using the nice computational properties of the Delaunay triangulation, and the process is iterated.

We show an example in which our robot navigates freely in a real indoor environment using this system.

13.2 Introduction

Our mobile robot can use a vision machine that has been designed and built within the European Esprit project P940, also called DMA, for depth and motion analysis. Without entering into the details of this machine, it is sufficient to say that it can process three images acquired from a trinocular stereo rig and compute the position and orientation in three dimensions of a set of line segments. These line segments are produced by polygonal approximations of edges in the three images. The stereo algorithm that matches the polygonal chains has been described in [2] and implemented in parallel on a board with three Motorolla 56000 DSPs. This board, part of the DMA machine, has been designed and built jointly by INRIA and MATRA.

At the time of this writing the final version of the machine is just available in our laboratory, and we will start to use it intensively. The throughput time is between 1 and 5 Hz, depending upon the exact hardware configuration. This means that, in the best case, we will reconstruct a three-dimensional wireframe representation of the environment five times a second.

Having put things in perspective, what we want to describe in this article is a piece of work that is built on top of the DMA machine and plans to use its real time capabilities.

The task is to build local representations of the robot environment that can be used to dynamically map free space, and plan and update trajectories. We use the word trajectory and not itinerary to mark the difference in time scale and planning complexity. The representation we are going to describe is local and does not carry any semantics. It is a pure volumetric representation of the free space around the robot, as measured by the stereo system over the last few seconds. It is therefore a combination of no more than 25 three-dimensional wire frames that are used to support tracking of an itinerary that has been set up as a goal by another, slower process. The representation is thus local, both in space and time. The use of local trajectory planning is to allow this process to gather information about the actual state of the environment and in particular about events that could not be foreseen, such as the approach of a moving obstacle. This local representation is, on one hand, passed over to the higher level, slower process and, on the other hand, used immediately, in a reactive fashion, to cope with possible collisions and to perform the actual robot control operations that lead to a satisfying pattern of motions.

13.3 What Do We Do with the 3-D Wire Frames?

We assume that for the kinds of time scales we are considering, the ground on which the robot is moving can be considered flat. This local plane is known through calibration and used as a support for our representation.

We project on the ground all 3-D line segments obtained from stereo that lie between the ground and a plane parallel to it at a height equal to that of the

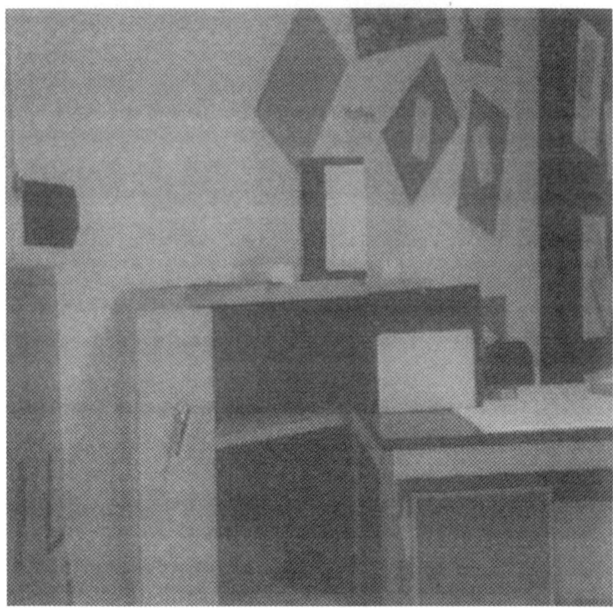

FIGURE 13.1. A typical indoor scene.

robot. We have thus reduced a three-dimensional wire frame representation to a two-dimensional one. To help guide the reader's understanding, we show in Figure 13.1 an image of a part of a typical scene with which the robot has to cope, and in Figure 13.2, we show the projection on the ground floor of the 3-D line segments reconstructed from stereo. We notice that this representation is probably not immediately useful, being still quite complex geometrically.

A further item of interest is that in the process of reconstructing 3-D line segments we also compute a measure of their uncertainty. The exact measure depends on what kind of representation we are using for the line segments [1, 5], but it always represents the amount of confidence we have in a specific segment. This confidence is a function of the reliability of the various elements in the processing chain, such as edge detection, polygonal approximation, and calibration of the trinocular stereo rig, as well as of the geometry of the scene (segments that are far are less reliable than segments that are close). It is represented by a symmetric weight matrix, also sometimes called a covariance matrix if a probabilistic interpretation is required, whose size is that of the representation and whose diagonal elements are big if the corresponding parameters in the representation are uncertain, and small otherwise.

From the 3-D representation of the line segments and the corresponding measure of uncertainty, it is possible to compute a representation for the projected segments and the corresponding uncertainty (see Section 13.4). Uncertainties are conveniently represented by ellipses in the plane, where

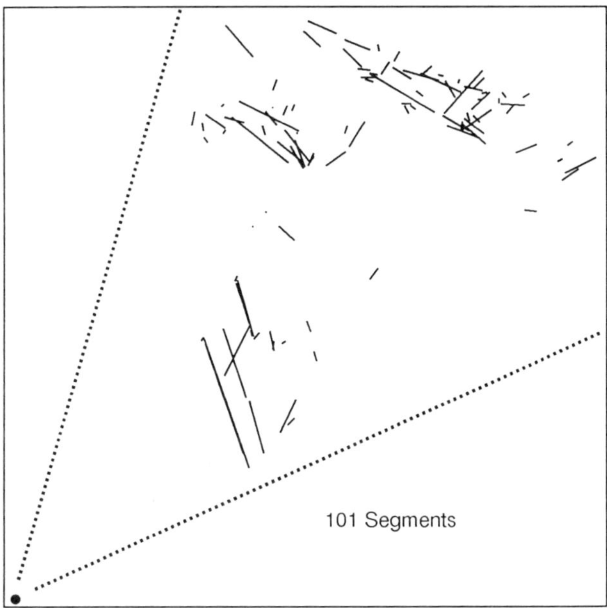

FIGURE 13.2. Projection on the ground floor of the 3-D segments produced from a stereo view of the scene of Figure 13.1.

they can be thought of as geometric upper bounds. For example, Figure 13.3 shows the ellipses representing the uncertainties of the midpoints of the projected segments of Figure 13.2. We note that segments that are far from the robot are more uncertain than those that are close.

Having such a measure is important because it allows processes that operate on the representations to give more weight to reliable measurements than to others. An example of such a process is one that estimates the robot ego-motion [1, 8, 5] and the obstacle's motion [7, 6]. In this article, we use this information to help produce a simplified version of the 2-D map in Figure 13.2. The main reason for producing such a map is that of conciseness. Since our goal is to produce a map of free space, we are uninterested in redundant information, such as too many geometrically similar segments in the same area that possibly arise from a highly textured pattern. Suppressing redundant information will allow us to keep only the information really relevant to the task at hand and will facilitate and speed up further processes. We now discuss in details how the 2-D map is simplified.

13.4 Two-Dimensional Map Simplification

We describe how line segments are represented.

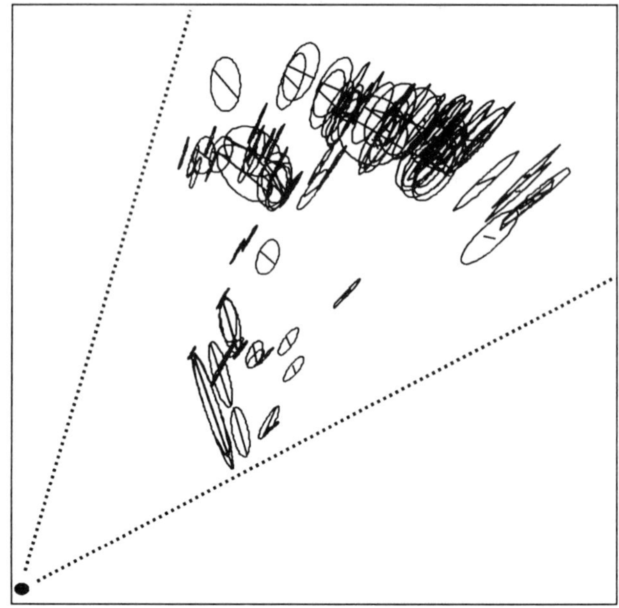

FIGURE 13.3. Ellipses of uncertainty for the midpoints of the 2-D segments of Figure 13.2.

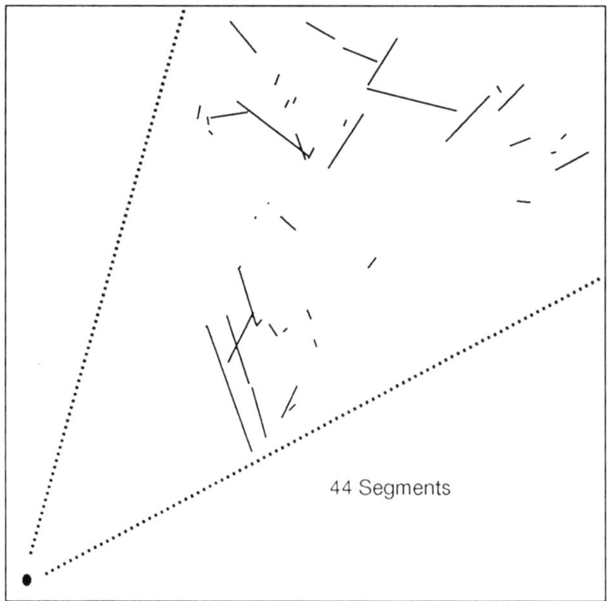

FIGURE 13.4. Simplification of the set of segments of Figure 13.2.

13.4.1 Representation of Line Segments

A 2-D or 3-D line segment is usually represented by its endpoints M_1 and M_2 and their covariance matrices Λ_1 and Λ_2. Since the endpoints of a segment are not reliable, we cannot directly use them in most cases. Considering that the longitudinal (along the segment) position of a line segment is much less precise than the transverse one, one may instead use its supporting line. However, after this abstraction, we completely lose the longitudinal information, which is useful in many cases, such as matching and fusion. Further, the uncertainty of the supporting line does not reflect that of the segment, that is, the supporting line of a very uncertain segment may be less uncertain than the supporting line of a less uncertain segment. For these reasons, we have proposed a new representation for 3-D line segments that is a trade-off between lines and segments [9, 5].

In our representation, the spherical coordinates ϕ and θ are used to represent the direction of a line segment, and the midpoint $\mathbf{m}$ is used to locate the segment. The length of the segment is denoted by l. If we denote $M_2 - M_1$ by $\mathbf{v} = [x, y, z]^t$ (the nonnormalized direction vector), the unit direction vector by $\mathbf{u} = \mathbf{v}/\|\mathbf{v}\|$, and $[\phi, \theta]^t$ by $\boldsymbol{\phi}$, we get

$$\phi = \begin{cases} \arccos \dfrac{x}{\sqrt{x^2 + y^2}} & \text{if } \quad y \geq 0 \\[4mm] 2\pi - \arccos \dfrac{x}{\sqrt{x^2 + y^2}} & \text{otherwise,} \end{cases} \tag{13.1}$$

$$\theta = \arccos \frac{z}{\sqrt{x^2 + y^2 + z^2}}.$$

The covariance matrix $\Lambda_{\boldsymbol{\phi}}$ of $\boldsymbol{\phi}$ is given to the first-order approximation by

$$\Lambda_{\boldsymbol{\phi}} = \frac{\partial \boldsymbol{\phi}}{\partial \mathbf{v}} \Lambda \mathbf{v} \frac{\partial \boldsymbol{\phi}^t}{\partial v}, \tag{13.2}$$

where $\dfrac{\partial \boldsymbol{\phi}}{\partial \mathbf{v}}$ is the Jacobian matrix of $\boldsymbol{\phi}$ with respect to $\mathbf{v}$, and $\Lambda_{\mathbf{v}} = \Lambda_1 + \Lambda_2$. Similarly, we can compute the covariance matrix $\Lambda\mathbf{u}$ of the unit direction vector $\mathbf{u}$.

The most important part of our representation is the modelization of the uncertainty of the midpoint. The midpoint of a line segment is modeled as

$$\mathbf{m} = (M_1 + M_2)/2 + n\mathbf{u}, \tag{13.3}$$

where n is a random variable. In fact, Equation (13.3) states that the midpoint of a segment may vary randomly around its position $(M_1 + M_2)/2$ along the direction of the segment. The random variable n is modeled as a Gaussian, zero mean, whose standard deviation σ_n is some positive scalar. In our implementation, σ_n is related to the length l of the segment, $\sigma_n = \kappa l$ (we choose $\kappa = 0.2$). This says that a long segment is more likely to be broken into smaller ones in other views. The covariance matrix of $\mathbf{m}$ is given by [see 5, 9]

$$\mathbf{\Lambda_m} = (\mathbf{\Lambda}_1 + \mathbf{\Lambda}_2)/4 + \sigma_n^2(\mathbf{\Lambda_u} + \mathbf{uu}^t). \tag{13.4}$$

The uncertainty in the length of a segment is not modeled because it is not required in our algorithm.

A 2-D line segment can be viewed as the orthogonal projection of a 3-D line segment on a plane. A 2×3 matrix T can be used to describe this projection, that is, we can relate a 3-D point $\mathbf{m}_3$ with its projection $\mathbf{m}_2$ (here, subscripts are used to denote the dimension, which will be omitted if no ambiguity) by

$$\mathbf{m}_2 = T\mathbf{m}_3. \tag{13.5}$$

For example, if we want to project a 3-D point on the plane $y = 0$, then

$$T = \begin{bmatrix} 1 & 0 & 0 \\ 0 & 0 & 1 \end{bmatrix}.$$

We represent 2-D line segments similarly to 3-D line segments, that is, a 2-D line segment is described by its angle θ with the axis x, its midpoint $\mathbf{m}$, and its length l. Those parameters can be easily computed based on Equation (13.5). We also compute the variance σ_θ for θ and the covariance matrix $\mathbf{\Lambda_m}$ for $\mathbf{m}$. For example, $\mathbf{\Lambda}_{\mathbf{m}_2} = T\mathbf{\Lambda}_{\mathbf{m}_3}T^t$. The uncertainty on l is not modeled.

13.4.2 Simplification

At this point, we have a set of 2-D line segments, which may contain redundant information. In the following, we describe how to intelligently merge geometrically similar segments in the same area, which are, for example, projections of some textures on a wall.

The simplification is performed as follows. All segments are sorted by a bucketing technique that allows to easily access the neighbors of a segment. The segments are also sorted according to their orientation. Now, for a segment S not yet processed, we compute a list of segments that are neighbors to it and another list of segments that have similar orientations to it. The intersection of the two lists are the first candidates. If a segment S' among those candidates is similar enough to S, we then merge S and S', yielding a new segment $\hat{S}$. We then compare $\hat{S}$ with the rest of the candidates: if a new similar segment S'' is found, $\hat{S}$ will be updated. This procedure is applied to every candidate and every unprocessed segment.

The merging technique for 2-D line segments is the adaptation of that for 3-D line segments [5, 9]. Let θ, $\mathbf{m}$, and l be the parameters of a segment S with uncertainty measurements σ_θ^2 and $\mathbf{\Lambda_m}$, and θ', $\mathbf{m}'$, and l' be that of another segment S' with uncertainty measurements $\sigma_{\theta'}^2$ and $\mathbf{\Lambda}_{\mathbf{m}'}$. Those two segments can be merged if and only if they satisfy the following relations based on the Mahalanobis distance:

$$(\theta - \theta')^2/(\sigma_\theta^2 + \sigma_{\theta'}^2) \leq \kappa_\theta, \tag{13.6}$$

$$(\mathbf{m} - \mathbf{m}')^t(\mathbf{\Lambda_m} + \mathbf{\Lambda}_{\mathbf{m}'})^{-1}(\mathbf{m} - \mathbf{m}') \leq \kappa_m, \tag{13.7}$$

where κ_θ and κ_m are thresholds. Looking up the χ^2 distribution table, we can choose $\kappa_\theta = 3.84$ for a probability of 95% with 1 degree of freedom and $\kappa_m = 5.99$ for a probability of 95% with 2 degrees of freedom. The above conditions say that we merge only segments that are the same (in the probabilistic sense).

Based on a minimum-variance estimator, we get the following parameters for the merged segment $\hat{S}$:

$$\hat{\theta} = (\sigma_{\theta'}^2 \theta + \sigma_\theta^2 \theta')/(\sigma_\theta^2 + \sigma_{\theta'}^2), \tag{13.8}$$

$$\sigma_{\hat{\theta}}^2 = \sigma_\theta^2 \sigma_{\theta'}^2/(\sigma_\theta^2 + \sigma_{\theta'}^2), \tag{13.9}$$

$$\hat{\mathbf{m}} = \boldsymbol{\Lambda}_{\mathbf{m}'}(\boldsymbol{\Lambda}_{\mathbf{m}} + \boldsymbol{\Lambda}_{\mathbf{m}'})^{-1}\mathbf{m} + \boldsymbol{\Lambda}_{\mathbf{m}}(\boldsymbol{\Lambda}_{\mathbf{m}} + \boldsymbol{\Lambda}_{\mathbf{m}'})^{-1}\mathbf{m}', \tag{13.10}$$

$$\boldsymbol{\Lambda}_{\hat{\mathbf{m}}} = \boldsymbol{\Lambda}_{\mathbf{m}}(\boldsymbol{\Lambda}_{\mathbf{m}} + \boldsymbol{\Lambda}_{\mathbf{m}'})^{-1}\boldsymbol{\Lambda}_{\mathbf{m}'}. \tag{13.11}$$

They give the orientation and the position of the merged segment. Since the two segments are considered as two instances of a single segment, their union should be its better estimate. By projecting the endpoints of the two segments on the merged segment, we choose the farthest projections as the endpoints of the merged segment, and the length can also be computed. The *real* midpoint M of the merged can be determined and can always be expressed as

$$M = \hat{\mathbf{m}} + s\hat{\mathbf{u}}, \tag{13.12}$$

where s is a scalar and $\hat{\mathbf{u}}$ is the unit direction vector of the merged segment. Equation (13.12) can be interpreted as the addition of a *biased* noise on $\hat{\mathbf{m}}$. Thus, the covariance matrix of M is given by

$$\boldsymbol{\Lambda}_M = \boldsymbol{\Lambda}_{\hat{\mathbf{m}}} + s^2(\boldsymbol{\Lambda}_{\hat{\mathbf{u}}} + \hat{\mathbf{u}}\hat{\mathbf{u}}^t). \tag{13.13}$$

For more details, the reader is referred to Zhang and Faugeras [5] and Zhang [9].

13.5 Constructing a Volume Representation of Free Space

We now have a simplified two-dimensional representation of the environment in terms of line segments in which free space is not explicitly represented. In order to obtain such a representation, we use a special data structure that has been quite thoroughly studied in *Computational Geometry* [4] and that we have been using recently to represent 3-D stereo data [3]. We use it here in two dimensions, but the basic ideas are similar. We have no space here to go into the details of the Delaunay triangulation, and we will only try to convey the main ideas and the philosophy of the technique.

13.5.1 Constructing a Triangulation of the 2-D Maps

The idea is to compute, as an intermediate representation, a triangulation of the endpoints of the 2-D segments that has the characteristic of containing the segments as edges of the triangulation. A set of planar points is said to be triangulated if its points are joined by nonintersecting straight line segments so that every region internal to the convex hull is a triangle. There exist many different triangulations of a set of points, but we have chosen the Delaunay because

it has some nice properties related to shape, and
it can be implemented efficiently.

Both reasons are well described in Faugeras et al. [3].

The shape property can be summarized in two dimensions, but it is also true in three dimensions. Suppose we measure a number of points on the boundary of an object and assume that this boundary is piecewise smooth. Then, under some fairly weak assumptions, the Delaunay triangulation of the set of measured points contains a polygon that approximates the boundary shape well. If that polygon can be easily recovered from the triangulation, then we have a nice way of representing the shape of the object. How to do this is described in the next section.

With respect to the implementation, there exist algorithms for computing the Delaunay triangulation of a set of n 2-D points with an optimal time complexity of $o(n \log n)$. The basic idea is to use the divide and conquer strategy [4]. Even though this kind of algorithm is optimal in complexity, it has one main drawback, namely, that it is not incremental; we have to wait until all data points have been collected before we can start triangulating them. This is in contrast with what is needed for our application in which data are collected sequentially and must be processed on the fly. Fortunately, there exists an incremental algorithm whose worst case time complexity is $o(n^2)$ but whose average complexity is $o(n)$.

The algorithm is described in details in Faugeras et al. [3], but the basic idea is quite simple. Suppose we have triangulated the first n points and that we have just measured the $n + 1$st point. We exploit the fundamental property of the Delaunay triangulation. If ABC is a Delaunay triangle, consider the disk it defines. That disk does not contain any other data point. This property of the disks is necessary and sufficient for a triangulation to be Delaunay. Coming back to our new data point, we only have to determine those disks of the existing triangulation it falls into (the corresponding triangles are edge connected) and retriangulate their data points together with the new point.

A further property of the Delaunay triangulation that is very relevant to our application is related to the skeleton. Indeed, if we consider the centers of the Delaunay disks and join those centers whose triangles share an edge but do not cross the boundary of free space, we obtain a set of polygonal lines that are a subset of the Voronoi diagram of the set of measured points [3]. Voronoi diagrams have been heavily used in robotics as a support for trajectory planning, and the fact that the Voronoi diagram of free space is present in our representation comes as a strong support for our approach.

In our case, we have more than points, namely, segments. If we just triangulate their endpoints, as shown in Figure 13.5, we obtain the so-called uncon-

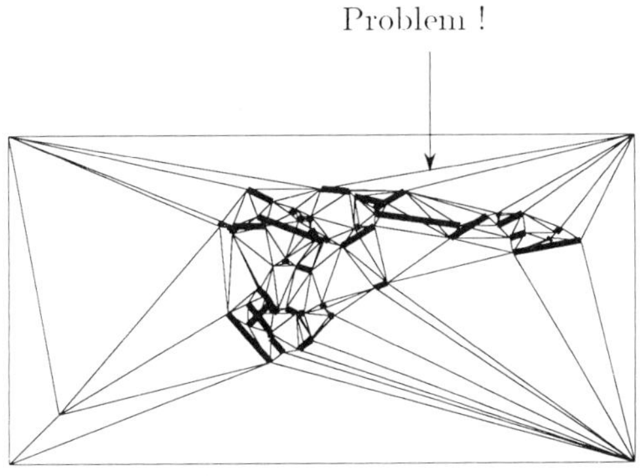

FIGURE 13.5. Unconstrained Delaunay triangulation of the set of segments of Figure 13.4.

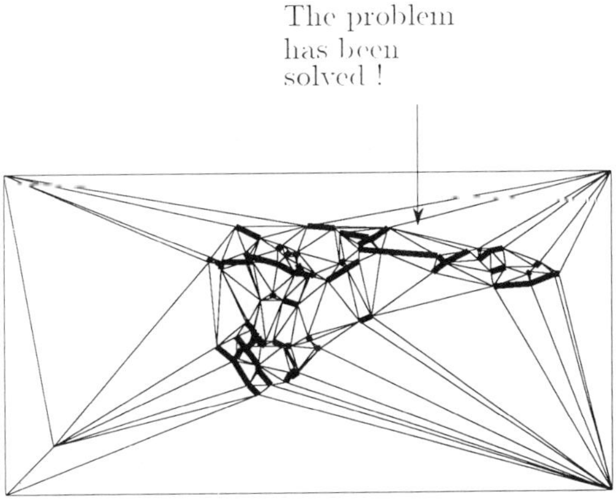

FIGURE 13.6. Constrained Delaunay triangulation of the set of segments of Figure 13.4.

strained triangulation of the set of segments of Figure 13.4. It can be seen that some stereo segments are not Delaunay edges. This is a problem because the next step described in the following section may produce a less accurate representation of free space than if all stereo segments are Delaunay edges. The problem can be eliminated by adding some more points on some of the stereo segments. The procedure is described in Faugeras et al. [3], and the results on the same scene are shown in Figure 13.6, which is the so-called constrained Delaunay triangulation of the set of stereo segments.

13.5.2 Marking Empty Triangles

Having constructed the constrained triangulation, we now wish to identify those triangles that are part of free space. In order to do this, we exploit a very

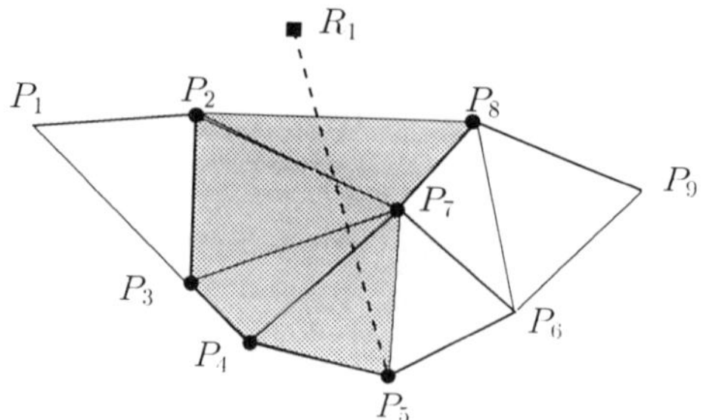

FIGURE 13.7. How to mark empty triangles.

simple visibility property that is best explained by looking at Figure 13.7. In this figure, the robot position is indicated by R_1 and a number of segments (P_1P_2 to P_8P_9) are present and represented as thick lines. Only subsets (marked with black bullets) of those segments are visible from R_1. The other Delaunay edges are represented by thin lines. Since point P_5 has been seen by the robot in position R_1, it means that all triangles crossed by the optic ray R_1P_5 are empty and should be marked as such. We have, therefore, a very simple way of constructing from the Delaunay triangulation a set of marked triangles that represent free space.

13.5.3 Taking into Account Several Viewpoints

Having discovered a portion of its environment that is free space, the robot can plan to move to a new position. Once it has moved to this position, it needs an estimate of its motion. This is provided by another process described elsewhere [1, 6], which provides at the same time an estimate of the motion and a measure of its uncertainty. Using this information, the previous set of segments can be transformed in the current coordinate system, and simplification can be attempted with the set of newly measured segments.

This previous set of segments is actually the model of the environment that the robot builds. It is the concatenation of everything it has seen in the previous instants. How this model has been built has been described in Section 13.4. The robot does not have infinite memory in the sense that segments that have not been seen for a long time have their measures of uncertainty increased. When those measures reach a certain threshold, the corresponding segments are discarded.

The model is triangulated and free space is explicitly represented in it by marked triangles. Segments in the model also "know" from which robot position they have been seen. When a new segment is fused with an old one, the old segment is erased from the triangulation and the new one inserted. This operation can be implemented quite efficiently with the incremental algorithm that has been sketched in Section 13.5.1. Once the new segments have been inserted into the model, free space can be recomputed using the triangulation and the various robot positions.

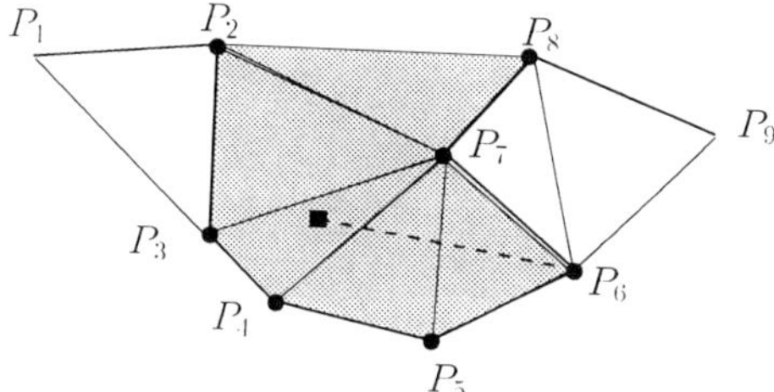

FIGURE 13.8. Taking into account a second viewpoint.

In Figure 13.8, the robot has moved from R_1 to R_2 and discovered segments $P_5 P_6$ and $P_6 P_7$. Again applying the visibility property allows it to improve its local representation of the environment.

13.5.4 Parallelization

There are two stages where parallelism can be introduced to speed up computation time. The Delaunay triangulation can be parallelized by splitting the environment into square buckets and computing independently the Delaunay triangulation of the segments within each bucket. There are, of course, side effects, such as if a segment crosses two buckets then its point of intersection with their boundary must be added. The result is that the union of the Delaunay triangulations is not globally Delaunay, only locally. This is no real problem since the Delaunay property is used only for building the triangulation but not for the representation of free space. This parallelization is being implemented on the parallel machine CAPITAN, built by MATRA, which will be the host for the DMA machine.

The other step that can be made parallel is the marking of the empty triangles. Each optical ray can be traced independently for each robot position. We have not yet implemented this idea.

13.6 Results

The constrained Delaunay triangulation (Figure 13.6) of Figure 13.4 can be used to compute free space. The result is shown in Figure 13.9. The robot then plans a motion in the free space zone and acquires a new set of 3-D data. This is used to estimate its displacement (Figure 13.10), and the previous simplified 2-D model is fused with the new data after projection on the ground plane. The result of this fusion is shown in Figure 13.11, which is to be compared with Figure 13.4. The robot displacement between the two views is 4.5° and 15 cm. The new representation of free space is shown in Figure 13.12, which is to be compared with Figure 13.9.

Finally, we show in Figure 13.13 the result of integrating 13 views obtained while the robot was moving, and in Figure 13.14, we show the resulting border of free space.

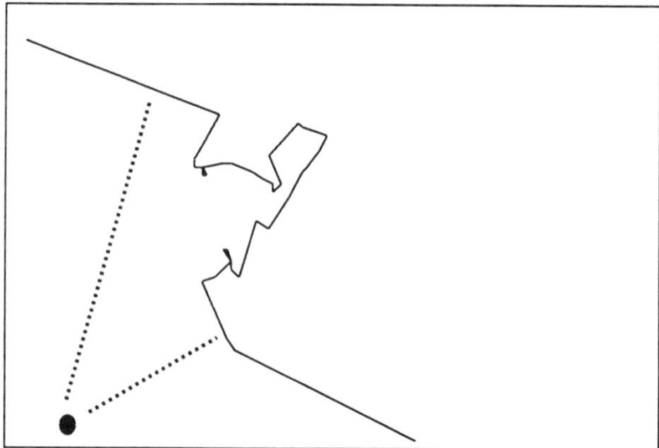

FIGURE 13.9. The border of free space for the environment of Figures 13.1. and 13.4.

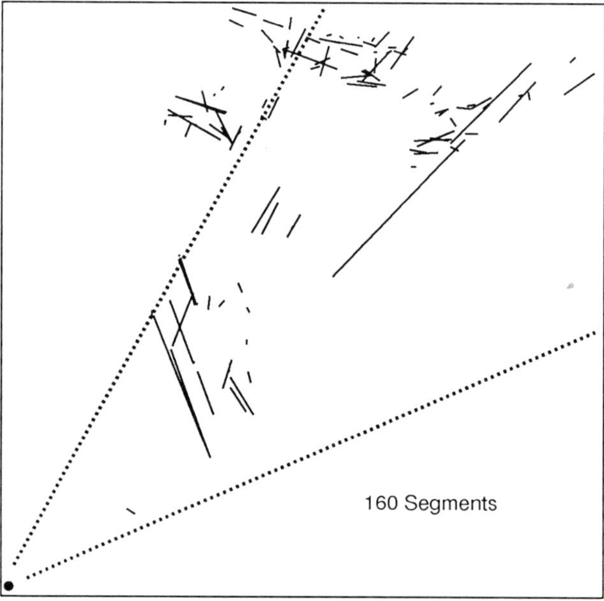

FIGURE 13.10. Before the fusion of two different viewpoints.

13.7 Conclusions

We have presented the current work at INRIA on obstacle avoidance and trajectory planning for a mobile robot using stereovision. In order to facilitate and speed up further processes, we project 3-D line segments reconstructed by our trinocular system on the ground plane. We have detailed how the re-

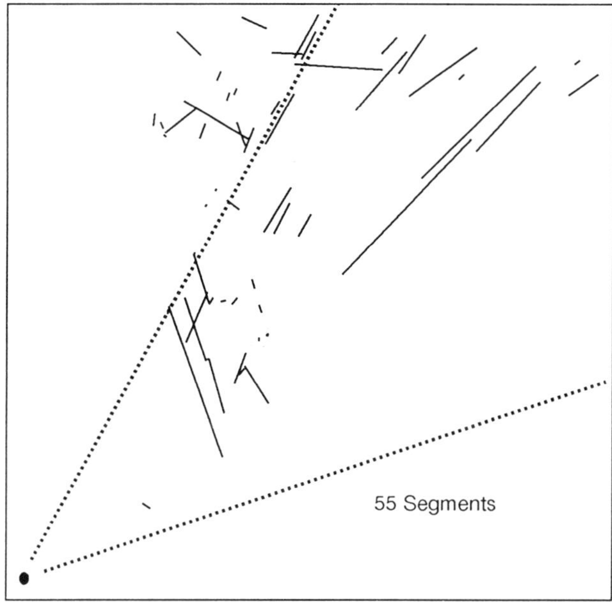

FIGURE 13.11. After the fusion of two different viewpoints

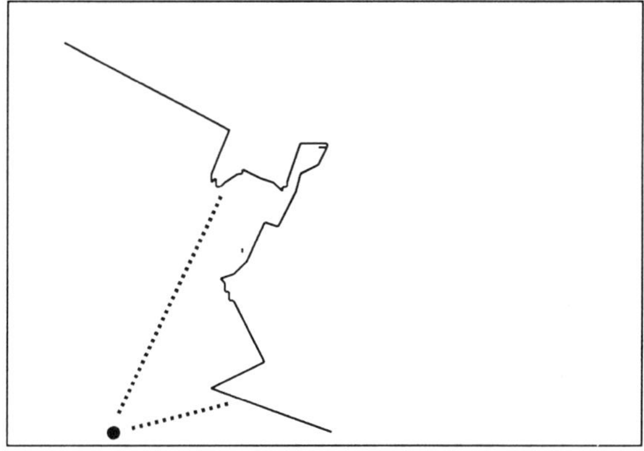

FIGURE 13.12. The border of free space for Figure 13.11.

sulting 2-D map is further simplified based on some geometric criteria. Performing the Delaunay triangulation on the remaining segments, we get a
tessellation of the ground floor. The free space can then be determined by
marking empty triangles, and it is used to generate collision-free trajectories.
After the robot arrives in a new position, the stereo system builds a new set of
3-D line segments. They are again projected on the ground plane, which is

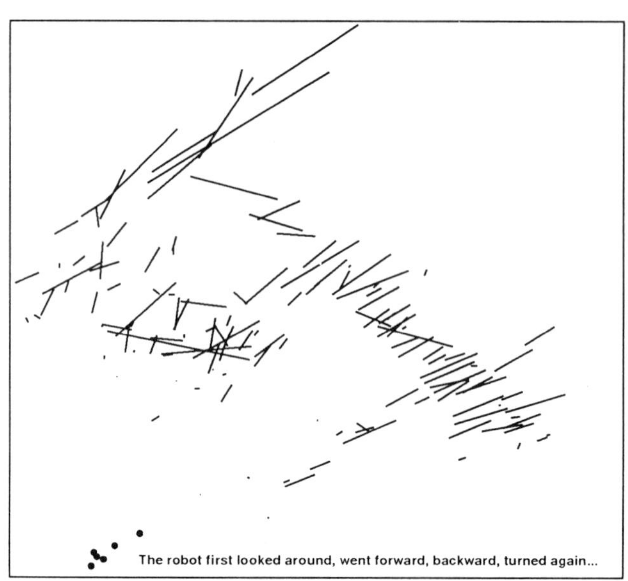

FIGURE 13.13. The fusion of 13 scenes.

FIGURE 13.14. The border after 13 scenes.

integrated into the existing 2-D map. The Delaunay triangulation can be easily updated. We have also described that most parts of our algorithms can be implemented in parallel.

References

[1] Ayache, Nicholas, and Faugeras, Olivier D. (1989). "Maintaining Representations of the Environment of a Mobile Robot." *IEEE Transactions on Robotics and Automation*, December. Also INRIA Report 789.

[2] Ayache, N., and Lustman, F. (1987). "Fast and Reliable Passive Trinocular Stereovision. In *Proceedings ICCV '87, London*, 422–427, IEEE, June.

[3] Faugeras, Olivier D., Lebras-Mehlman, Elizabeth, and Boissonnat, Jean-Daniel. (1990). Representing Stereo data with the Delaunay Triangulation. *Artificial Intelligence Journal* **44**(1–2), July (also INRIA Tech. Report 788).

[4] Preparata, F., and M. Shamos, (1985). *Computational Geometry*. Springer-Verlag, Berlin and New-York.

[5] Zhang, Zhengyou, and Faugeras, Olivier D. (1990). "Building a 3D World Model with a Mobile Robot: 3D Line Segment Representation and Integration." In *Proceedings of the 10th International Conference on Pattern Recognition*, 38–42, IEEE Computer Society, June.

[6] Zhang, Zhengyou, and Faugeras, Olivier D. (1990). "Motion Analysis of Two Stereo Views and Its Applications. In *Proceedings of the ISPRS Symposium on Close-Range Photogrammetry Meets Machine Vision*, September.

[7] Zhang, Zhengyou, and Faugeras, Olivier D. (1990). "Tracking and Motion Estimation in a Sequence of Stereo Frames." In *Proceedings of the 9th European Conference on Artificial Intelligence, ECAI, August 1990*.

[8] Zhang, Z., Faugeras, O. D., and Ayache, N. (1988). "Analysis of a Sequence of Stereo Scenes Containing Multiple Moving Objects Using Rigidity Constraints." In *Proc. the Second International Conference on Computer Vision*, 177–186, IEEE, Tampa, Florida, December.

[9] Zhang, Zhengyou (1990)."Analysis from a Sequence of Stereo Frames and Its Applications." PhD Thesis, University of Paris-Sud, Orsay, Paris, France, (in English).

14
A Parallel Architecture for Curvature-based Road Scene Classification

AMY POLK AND RAMESH JAIN

14.1 Abstract

A parallel architecture for road scene classification is presented. The system is designed to classify scenes on the basis of road curvature and operate in daytime, highway environments. The system processes input in five stages and uses the same decision-making paradigm, Hough transforms, in all three higher levels of processing. Many parallel implementations of Hough transforms have already been developed [3, 7]. In addition, we present our own implementation, which expands the algorithm to more general feature detection applications. Using only algorithms that are easily implemented in VLSI (very large scale integrated) circuitry, we develop a powerful real-time image processing module for autonomous steering control.

14.2 Introduction

14.2.1 Navigation

We present an architecture for a detachable, intelligent steering control mechanism: intelligent because it makes few assumptions about road illumination, width, or degree of curvature; detachable because it can easily be implemented in VLSI circuitry, attached to any automobile.

The Carnegie-Mellon University (CMU) NAVLAB system [18] uses color information to classify areas of the image into known road and nonroad features. Once the entire image is determined, the system uses the shape of the road region to determine road curvature. The CMU system's color and shape models were built for a simple road scene, a bicycle path near the CMU campus. Changes in illumination such as dusk, nighttime, or even momentary cloudiness will render color information different from the assumed model. The system's shape model assumes a straight road of known width and will fail for sharply curved and irregularly shaped roads.

VITS [19], developed by a team of researchers at Martin Marietta, consists of a vision module that constructs an information-rich world description, a reasoning module that uses this information to perform complex tasks, and other control modules. Like CMU's NAVLAB, the system uses color information to classify pixels and extracts a road region. For navigation, the system then determines vehicle-centered 3-D points denoting left and right road edges. A reasoning system convertes this information into a fixed world-coordinate system for navigation. In addition, the VITS system has been able to perform other tasks, such as road following, obstacle avoidance, landmark recognition, and cross-country navigation. Implicit in the navigation system is a color model for the road, which may not be accurate for environments outside the Martin Marietta test site. Kluge and Thorpe [11] also note that VITS discards valuable color sensing capability for computational efficiency.

The University of Maryland [20] has developed a road-following system that divides the navigation task into a bootstrapping stage, where the initial model is constructed, and then a feed-forward stage, where information from previous frames simplifies analysis on the current road scene. The bootstrapping stage uses edge detection and edge linking by Hough transforms to locate road edges. In successive frames, the system looks for road edges similar in orientation and position to the previous scene. This system assumes that edge information in the window of interest corresponds only to road boundaries and has been shown to fail on scenes that contain other prominent edges.

Pomerleau [15], also of the CMU NAVLAB, uses a neural network that encodes both color and laser range data as input and encodes road curvature and elevation descriptions as output. The back propagation weight update algorithm learns the transition function from the input to output of synthetically generated road scenes. Thus, the network develops its own internal representation of the navigation task. However, the lack of structure brings the traditional deficiencies of neural networks as well; the performance of the system is only as good as the quality of training examples.

Dickmanns and Zapp [5] attach special-purpose hardware to an ordinary Mercedes to control the vehicle with speeds up to 100 km/h on the German autobahn. Road geometry, vehicle turning radius and speed, and the location of visually tracked features are all fed into a single filtered state model. The system incorporates assumptions about the stable geometry of the autobahn into its control function: straight lines, constant radius curves, and smooth transitions between curved and straight sections. Therefore, this system cannot navigate roads that do not fit these geometrical constraints.

Our system extends the capability of Dickmanns and Zapp to operate in more variable road environments while using algorithms that may be easily implemented into VLSI hardware. What makes this possible is that we use a multilayered modular architecture, in which the same decision-making paradigm, Hough transforms, is used for all higher levels of processing: edge linking, vanishing point location, and vanishing point analysis. Because we use vanishing point analysis, not explicit shape models, in determining road curvature, the classification algorithm is not limited to roads of a fixed width or curvature.

14.2.2 Hough Transforms

An algorithm that combines parameter transformation and a voting mechanism, Hough transforms have traditionally been a successful feature detector. Originally developed to find equations of lines [8, 6], Hough transforms have been extended to more general problems in feature detection [2, 10, 21, 4]. They are more robust in the presence of noise than numerical analysis methods, such as least squares. Serial implementations, however, have proven much too slow for use in real-time vision systems, especially when applied to large input and parameter spaces. However, with the development of specialized hardware, we can use Hough transforms to design a sophisticated, real-time road scene classification system.

The parallel nature of Hough transforms has lead researchers to consider parallel implementations as early as 1975 [14], and interest continues to this day [17, 2]. Recently, wafer-scale VLSI implementations have been developed [7, 16, 22]. With the exception of the design by Rhodes et al. [16], implementations consider only the straight-line detection function of Hough transforms, not the more, general applications of parameter-based feature detection that we require in the latter stages of our road navigation system. Therefore, our proposed parallel Hough transform considers the implementation in a more general context. We have taken characteristics from artificial neural networks to design a general-purpose feature detection algorithm.

14.2.3 Vanishing Points

Various techniques have been developed for determining the vanishing points of an image: mapping line segments with a common vanishing point of circles that pass through the origin [Kender, 9], transforming lines with the same vanishing point into a common line in the inverted space [Kender, 9], and using cross products of line segments [Badler, 1]. Magee and Aggarwal [23] present a method similar to Badler's with improved computational efficiency. Liou and Jain [12] start with a limited set of candidate vanishing points and have edge points give supportive evidence to each candidate. They use this technique in the feedforward phase of a static-scene road following system.

Like Kender, we use an accumulator array to collect votes for candidate vanishing points. Here, the accumulator array is explicitly described as a Hough transform transition from input to parameter space. We use our own computationally inexpensive algorithm for determining the support a certain line gives a candidate vanishing point.

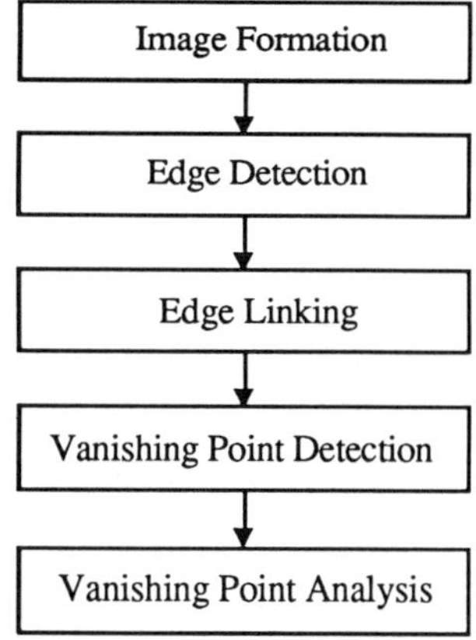

FIGURE 14.1. System schematic.

14.3 System Outline

Step 1: Image Formation. A CCD camera (charge coupled device, solid state image sensor) is mounted on the front of a car, sending a continuous stream of 480×480, 8-bit gray-scale images to successive stages of processing. We assume that road boundary lines will be detected and both the curvature and elevation of the road vary gradually.

Step 2: Edge Detection. A Sobel edge detector is used to measure the local image intensity gradient at each pixel. Thresholding is then used to distinguish edge points from the rest of the image.

Step 3: Edge Linking. Individual edge points are linked together to form oriented line segments. The image is divided into 15×15 tiles. A parallel implementation of Hough transforms are used to find dominant line segments in each tile.

Step 4: Vanishing Point Detection. The effects of perspective cause any set of parallel lines that are not parallel to the image's plane to converge to a vanishing point. Here, since road boundaries are considered to extend into parallel lines, we look for the vanishing at the intersection of right and left road boundary lines. Edge data are taken from a window of interest in which edge points are assumed to correspond only to road boundaries. The window of interest is divided into six horizontal bands. Assuming road boundaries are straight within these bands, each of the oriented line segments indicates a possible set of locations for the road boundary intersection. Again, a parallel implementation of Hough transforms is used to find the location of the vanishing point in each band.

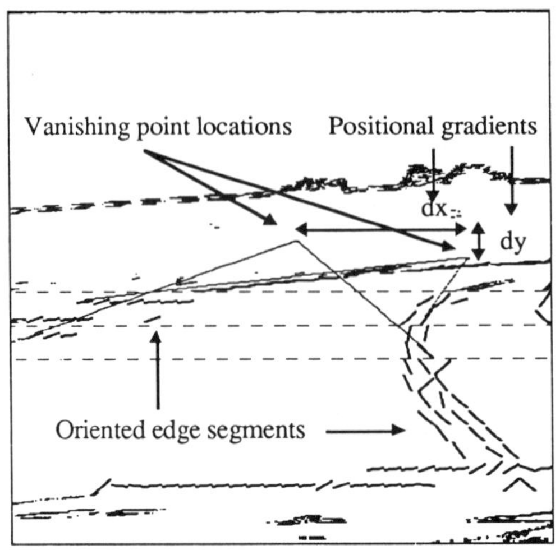

FIGURE 14.2. Results of steps 1 through 5. Oriented edge segments appear within window of interest. These segments point to possible locations for the vanishing point. Vanishing points in two horizontal bands are detected. Finally, positional gradient between the two vanishing points forms a model of road curvature. Here, the wide horizontal displacement between vanishing points indicates sharp curvature to the right.

Step 5: Vanishing Point Analysis. Gradient displacement between vanishing points of adjacent bands are used to estimate road curvature. Hough transforms are used once more to convert gradient values into a signal to the steering control mechanism. The steering control interface requires that the road scene be placed into one of five classes: straight, curve left, sharp curve left, curve right, sharp curve right.

14.4 Parallel Implementation of System Algorithms

14.4.1 Processing Equations

In most artificial neural network algorithms, the activation of a unit is some function of the weighted sum of all units connected to it. These equations have been developed from studying activation behavior of neurons in the brain [13]. Precisely, a unit's activation is

$$o_j = f(\sum w_{ij} o_i),$$

where o_i are the activations of all units connected to o_j, w_{ij} is the strength of the connection between units o_i and o_j, and is f an activation function, usually performing some thresholding capability. Typically, units are organized into ordered layers, where every unit in the lower layer is connected to every unit in the upper layer. In each stage of our system, connection weights and activation functions have been selected to perform the desired functions.

14.4.2 Sobel Edge Detector

The Sobel edge detector measures the edginess of a pixel by summing the gray-level values within a small (typically 3×3) neighborhood and multiplying by negative and positive values on either side of the pixel. Weights from the input image to the edge image have been hard wired to perform this multiplication, and the activation function has been selected to perform the thresholding mentioned previously. Since the Sobel operator measures edginess in the vertical and horizontal directions separately, using two different weight masks, the activation equation is modified slightly:

$$wx = \begin{array}{|c|c|c|} \hline -1 & 0 & +1 \\ \hline -2 & 0 & +2 \\ \hline -1 & 0 & +1 \\ \hline \end{array} \qquad wy = \begin{array}{|c|c|c|} \hline +1 & +2 & +1 \\ \hline 0 & 0 & 0 \\ \hline -1 & -2 & -1 \\ \hline \end{array}$$

$$o_j = f(|\sum wx_{ij}o_i| + |\sum wy_{ij}o_i|),$$

$$f(x) = \begin{cases} 255 & \text{if} \quad x > \text{threshold} \\ 0 & \text{if not.} \end{cases}$$

14.4.3 Hough Transforms

In Hough transforms, one point in the input space corresponds to a set of points in the parameter space. The transformations are applied to each point in the input space, accumulating votes in the parameter space. Areas of density in the parameter space correspond to parameter values that best describe the input. All that are needed to apply Hough transforms to a feature detection problem are the transformation equations used to map points in the input space to the parameter space.

Our parallel implementation of Hough transforms uses processing equations similar to other neural network architectures, such as multilayer perceptron. In contrast to neural networks that learn by updating their weights, the weights of this network are hard wired to perform the transformation function. Afterward, multiple iterations in the parameter space, using lateral inhibition, converge to determine the unit with maximum activation.

The lateral weights in the parameter space are such that a unit strengthens itself and weakens all others. Through successive iterations, weaker units will fall to negative values. Once below this threshold, they are considered inactive and are assigned zero value:

$$w_{ij} = \begin{cases} 1 + k*i & \text{if} \quad i = j \\ -2/\text{size} & \text{if not,} \end{cases}$$

$$f(x) = \begin{cases} x & \text{if} \quad x > 0 \\ 0 & \text{if not,} \end{cases}$$

where size is the number of units in the parameter space, and k is a small constant on the order of $1/\sqrt{\text{size}}$..

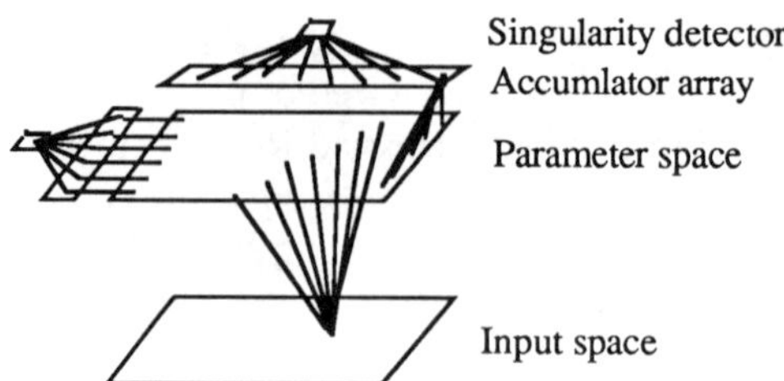

FIGURE 14.3. Parallel architecture for Hough transform. Here, the dimensionality of both input space and parameter space is 2.

Accumulator arrays count the number of active units in each dimension of the parameter space. (All points in the parameter space for which $r = 3$ are connected to the third unit in the accumulator array.) These accumulator units, in turn, are connected to units called singularity detectors. As always, the activation unit is the weighted sum of all units connected to it. The system detects the presence of a unique active unit in the parameter space when each singularity detector unit equals one.

It is possible that two units in the parameter space will have exactly the same value and that this is the maximum value in the parameter space. To avoid the two units canceling each other through successive iterations, a small adjustment value, different for each unit in the parameter space, is added to each lateral excitatory weight. After a few iterations, there will always be a unique maximum value.

In some implementations of Hough transforms, particularly those with large parameter spaces, activity is averaged over a local neighborhood. This function can also be implemented in massively parallel form, passing an averaging mask over the parameter space. Note that we do not use this averaging function in our road classification architecture because, in each stage of Hough transform processing, the parameter space tends to be small. Hence, a unit's activation does not need the support of its neighbors.

Our application required that the system detect one unique global maximum in the parameter space, for example, in the edge linking domain, we wanted to detect the *most prominent* line in the image. However, some applications require the detection of all local maxima, for example, *all* lines in the image. This modification is also realizable in parallel by subtracting each unit in the parameter space with a threshold of a likely high value for that point. Any point with activation greater than zero is a local maximum. Thus, singularity detector units would be removed, and the maxima would be detected in a one-pass, not iterative, algorithm.

14.4.4 Edge Linking

Weights implement finding the slope-offset line equation that best represents the points. Both the input space and parameter space are 2-dimensional 15×15 matrices. Offset r ranges from -7 to $+7$; possible slope values ($-\pi$

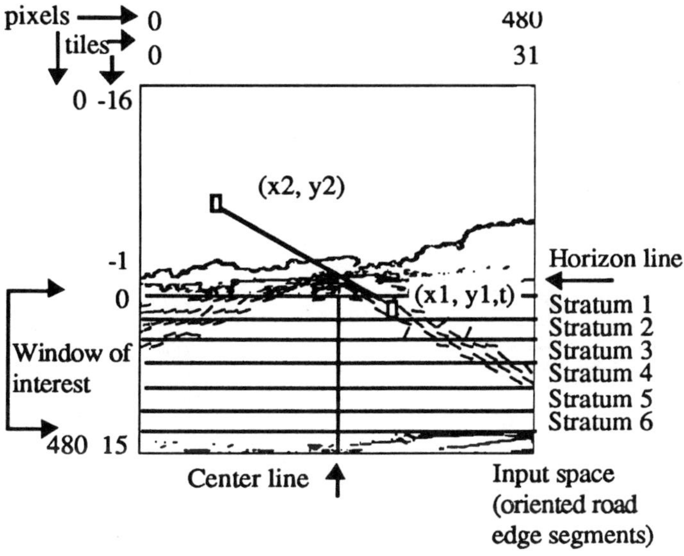

FIGURE 14.4. Vanishing point geometry. Input space unit (x_1, y_1, t) is connected to parameter space unit (x_2, y_2) if $x_1 \cos(t) + y_1 \cos(t) = x_2 \cos(t) + y_2 \sin(t)$.

to π) are partitioned into 15 intervals of 12°ds each. A unit in the parameter space is connected to a unit in the input space if its parameters are plausible values for a slope-offset equation involving the $x–y$ values in the input space:

$$W_{(x,y)(r,t)} = \begin{cases} +1 & \text{if} \quad r = x\cos(t) + y\sin(t) \\ 0 & \text{if not,} \end{cases}$$

where both x and y range from -7 to $+7$.

14.4.5 Vanishing Point Detection

The orientation of segment (t), the location of the tile (x, y), and the number of edge points in each tile (n) are passed from the edge linking module for each 15×15 tile. As stated, each of the oriented line segments indicates a possible set of locations for the road boundary intersection. The input space here is three dimensional, where the orientation of each segment forms the third dimension. The possible vanishing point locations, from a 480×270 range, are partitioned into intervals of 15×15 pixels each, forming a 2-D 32×18 parameter space. If a certain orientation is found, by the edge linking module, to be present in that tile, its activation is set to the number of edge points (n) in that tile. Otherwise, its activation is zero. A unit in the input space is connected to a unit in the parameter space if the extension that oriented the line segment upward intersects with the parameter segment:

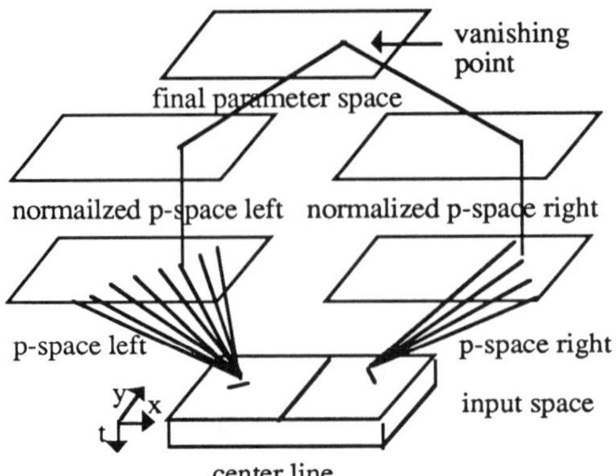

FIGURE 14.5. Normalization of parameter spaces corresponding to left and right road boundaries.

$$w(x_1, y_1, t)(x_2, y_2) = \begin{cases} +1 & \text{if} \quad [x_1 \cos(t) + y_1 \sin(t) = x_2 \cos(t) + y_2 \sin(t)] \\ 0 & \text{if not} \end{cases}$$

where x_1 and x_2 range from 0 to 31, y_1 ranges from -16 to -1, and y_2 ranges from 0 to 15.

Specifics of road imaging geometry require a modification to the data-space-to-parameter-space transformation. In a double-lane highway situation, one boundary (road edge) has dense edge data; while edge data for the other boundary (dotted road line) are sparse. Therefore, each oriented edge segment is classified as a left or right road edge, and parameter spaces are calculated for each side. The left and right parameter spaces are then normalized, and the resulting parameter spaces are then summed. Thus, both left and right edge orientations have equal votes. Lateral inhibitive weights in the final parameter space iteratively locate the best guess for the vanishing point.

Currently, left–right classification is based on the position of the edge in the image; edges appearing to the left of center of the image are classified as left. However, in sharply curving road situations, this is not the case. In the future, we hope that this center point will be selected automatically.

14.4.6 Vanishing Point Analysis

The positional gradient of vanishing points between successive strata model road curvature. Large displacements between vanishing points signify sharp road curvature, and the direction of horizontal displacement signifies the direction of the curve (left or right), whereas in straight road scenes, road edges from successive strata point to the same local neighborhood of points. This simple metric is an integer that indicates the degree of road curvature:

$$\text{metric} = (|\Delta x| + |\Delta y|)^* \, \text{sgn}(\Delta x).$$

Experimental values of this metric range of -8 to $+8$, with partitions into 5 curvature classes as shown.

Sharp curve left	Smooth curve left	Straight		Smooth curve right	Sharp curve right
-8 -7 -6	-5 -4 -3	-2 -1 0 1 2		3 4 5	6 7 8

The input space for the final Hough transform module is a two-dimensional gradient space, and the parameter space is a one-dimensional space with five units, one unit for each road class. A unit in the input space (dx, dy) is connected to a unit in the parameter space (c) if the metric described previously falls into the partition of this curvature class.

The number of classes is determined by the requirements of the interface to the steering control device. Here, we use 5 classes, but the vanishing point analysis module can easily be reconfigured to place roads into one of 7, 9, or any number of curvature classes.

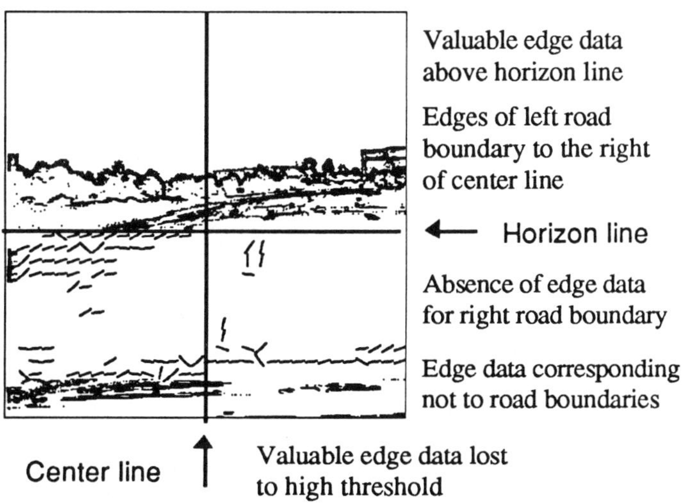

FIGURE 14.6. A road scene that violates every constraint. (1) The window of interest contains additional objects (2) Edge data for the right road boundary are absent; the vanishing point analysis module will not be able to detect an intersection. (3) Much valuable edge data in the lower right corner of the image are lost because an inaccurate threshold value. (4) The road curvature is so steep that left road edges are to the right of the centerline and will be classified as right edges in later stages of processing. Valuable edge data are lost because they appear above the horizon line.

14.5 Implementation Results

The road classification algorithm was tested on 10 digitized videotape images of actual road scenes of highways I-94, M-14, and I-275 in the southeastern Michigan area. Results for some of the images are shown in Figure 14.7. Images were selected on the quality of the edge data. (Implications for improving the quality of edge data for each image are discussed later.) Parallel algorithms were simulated by C language programs on an Apollo 4000 workstation. Because these algorithms are intended for parallel circuitry but implemented on a serial computer, they are considerably slower than our real-time processing goals. Complete road classification requires about 1.5 h for each image, an indication of the drastic speed up expected when we move to parallel hardware. Note that system performance did not suffer when analyzing roads of sharp curvature, as do other road navigation systems [18, 19]. In fact, the system performed best under these conditions.

14.6 Directions for Future Research

Clearly, in order for this system to be robust enough to operate correctly under a variety of highway environments, much work needs to be done. Accurate road classification is dependent on accurate edge data, which performs well on only a subset of the images that we wish to analyze. Our objectives in future research are to remove these constraints.

Constraint 1. For the empty highway, edge data within the window correspond only to road boundaries. In the current system, we assume that edge data inside the window of interest correspond only to road boundaries. In actuality, many other edge points are present: other vehicles, potholes, the road shoulder, etc. In highway environments, road boundaries are denoted by white lines on a dark road. In the future, we hope to incorporate this domain of knowledge into our massively parallel system by including the original intensity and gradient direction of edge point, as well as its gradient magnitude, in the calculation of the edginess value.

Constraint 2. For the well-bounded highway information on the presence of edge data for both left and right road boundaries for each strata is needed. Since the vanishing point is defined as the intersection of left and right road lines, detection requires the presence of edge data on both left and right road boundaries in at least two adjacent strata. In actual highway environments, however, the lane is bounded on one or both sides by discontinuous edges (dotted lines). This constraint prevented us from performing complete analysis on many of our highway images. The most significant improvement to the system will be to supplement edge data by extending these lines to all horizontal bands of the image.

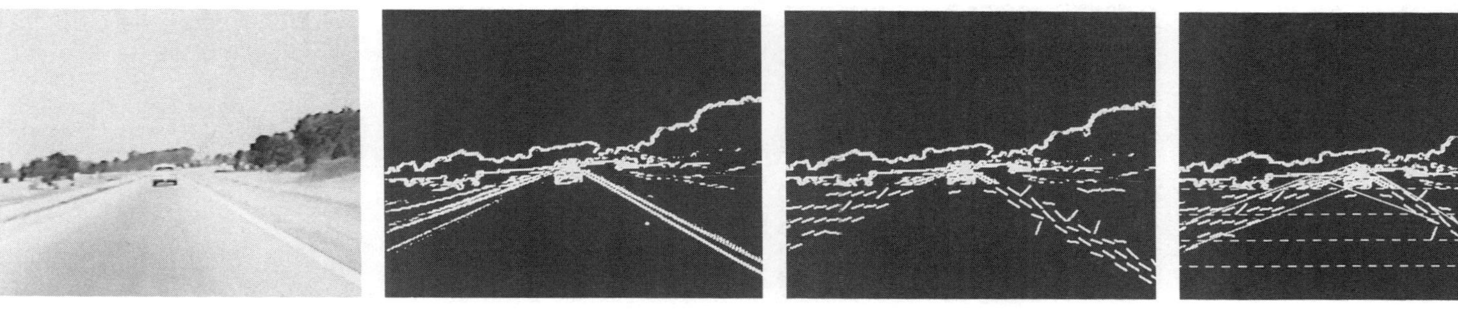
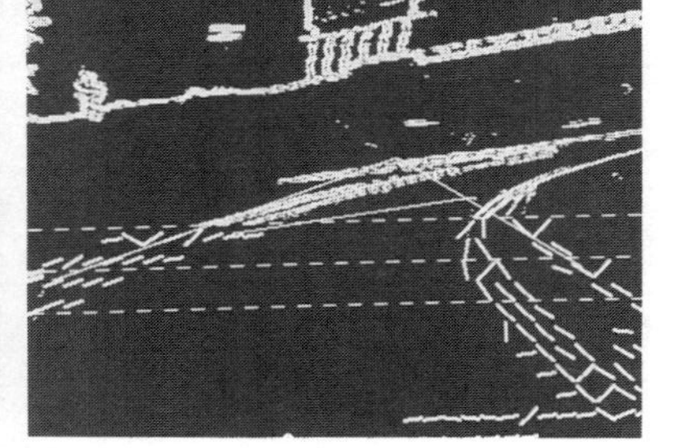

FIGURE 14.7. Results of image classification for five types of road curvatures. Intensity image is original digitized image. Edge Image is the result of the Sobel edge detecter. Oriented Edge Segment image is the result of Hough transforms performed on the window of interest, Vanishing Points are detected in two horizontal bands in each image. Curvature classification is shown next to the image name.

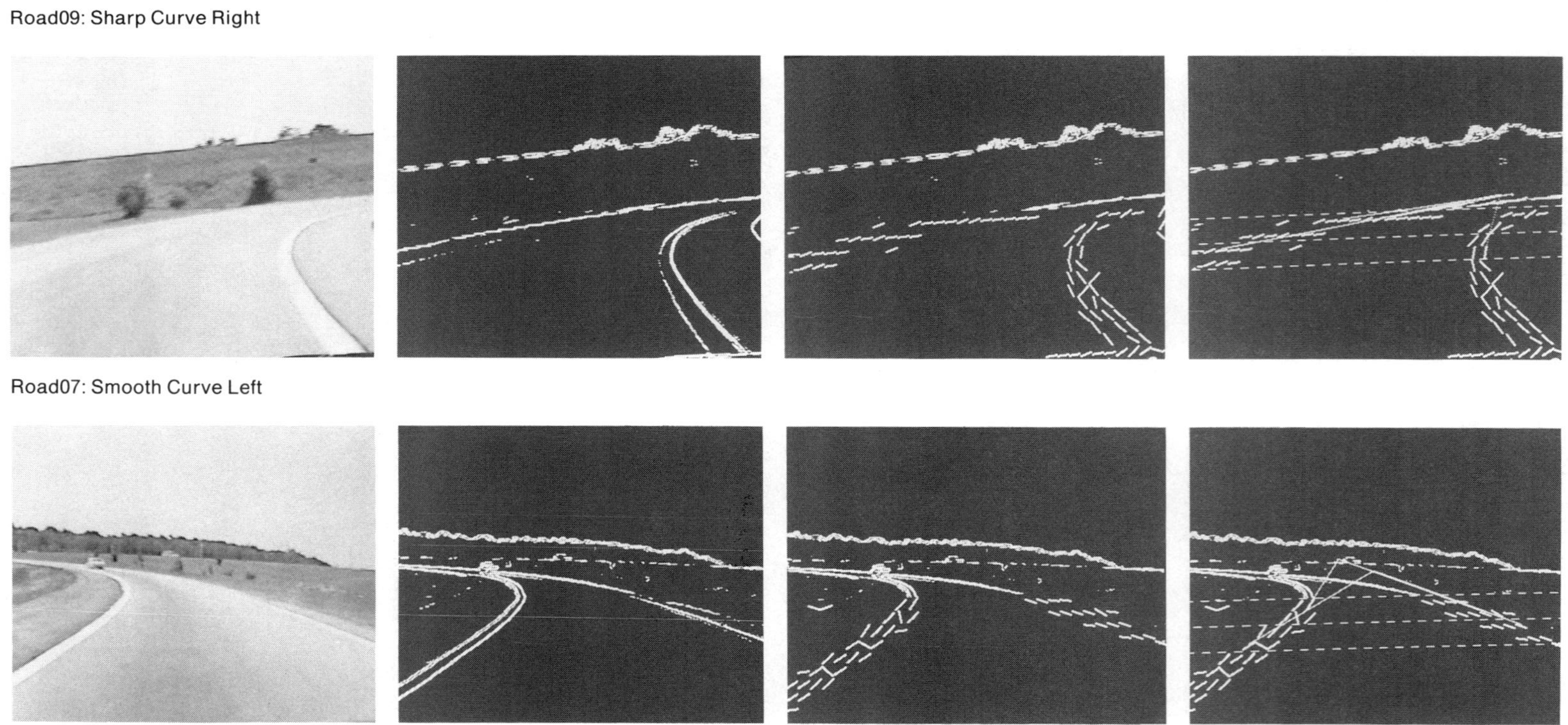

FIGURE 14.7 (*cont.*)

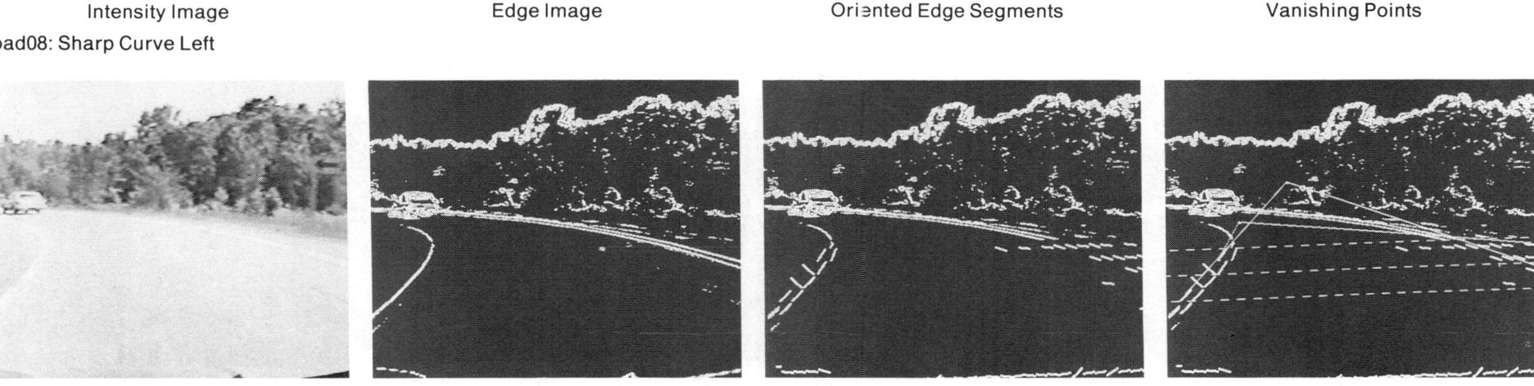

FIGURE 14.7 (cont.)

Constraint 3. For the daytime environment a constant edge threshold is needed. The Sobel edge detection process returns an edginess value for each point, and the threshold differentiates between valuable edge data points and noisy intensity variations. This threshold was selected by trial and error because it returned the best results for an initial set of images. A new edginess threshold is needed to extend the system's capabilities to dusk nighttime environments. In the future, we hope to select the edge threshold automatically by a histogram analysis of the window.

Constraint 4. Both constant elevation and gradual curves, that is, constant horizon and centerline parameters, are needed. Currently, the centerline and horizon line are static parameters, and any images for which these parameters do not fit are rejected. In the future, the initial centerline and horizon line will be detected on a bootstrapping phase by analyzing a straight road scene. On the assumption that road elevation and road curvature vary gradually, the system will continually update the new values.

14.7 Conclusion

A vision-based vehicle navigation system need not consist of a specialized vehicle equipped with on-board computing facilities. Instead, by carefully analyzing the road following problem, we have encoded algorithms and domain knowledge into simple mathematical transformations that can easily be implemented in VLSI circuitry. Although the need for a high quality of edge data severely restricts the number images our system can analyze correctly, we have presented an architecture with which one may perform sophisticated image processing tasks necessary for autonomous navigation, in real time, using small detachable devices.

References

[1] Badler, N. (1974). "Three-Dimensional Motion from Two-Dimensional Picture Sequence." *Proceedings of the International Joint Conference on Pattern Recognition, [IJCPR] August*, 157–161.

[2] Ballard, D. H. (1981). "Generalizing the Hough Transform To Detect Arbitrary Shapes." *Pattern Recognition*, **13** (2), 111–122.

[3] Ben Tzvi, D., and Sandler, M. (1989). "Efficient Parallel Implementation of the Hough Transform on a Distributed Memory System." *Microprocessing and Microprogramming (Netherlands)*, August, 147–152.

[4] Davies, E. R. (1988). "Application of Generalized Hough Transform to Corner Detection." *IEEE Proceedings Part E on Computers and Digital Techniques*, January, 49–54.

[5] Dickmanns, E., and Zapp, A. (1986). "A Curvature Based Scheme for Improving Road Vehicle Guidance in Computer Vision." *Proceedings of SPIE Conference on Mobile Robots*, Cambridge, Massachusetts.

[6] Duda, R. O., and Hart, P. E. (1972). "Use of Hough Transforms to Detect Lines and Curves in Pictures." *Commutations of the ACM* **15** (1), 11–15.

[7] Hanahara, K., Maruyama, T. and Uchiyama, T. (1988). "A Real-Time Processor for the Hough Transform." *IEEE Transactions on Pattern Analysis and Machine Intelligence*, PAMI **10** (1), 121–125.

[8] Hough, P. V. C. (1962). "Method and Means for Recognizing Complex Patterns." U.S. Patent 3,069,654.

[9] Kender, J. (1979). "Shape from Texture: An Aggregation Transform That Maps a Class of Textures into Surface Orientation." *Proceedings of the International Joint Conference on Artificial Intelligence*, August 1979 [IJCAI 79], 335–337.

[10] Kimme, C., Ballard, D., and Sklansky, J. (1975). "Finding Circles by an Array of Accumulators." *Communications of the ACM* **18** (2), 120–122.

[11] Kluge, K., and Thorpe, C. (1989). "Explicit Models for Robot Road Following." *IEEE Conference on Robotics and Automation*, May, 1148–1154.

[12] Liou, S., and Jain, R. (1987). "Road Following Using Vanishing Points." *Computer Vision, Graphics and Image Processing* **39** (CVGIP 87), 116–130.

[13] Lippman, R. P. (1987). "An Introduction to Computing with Neural Nets." *IEEE ASSP Magazine*, April, 4–22.

[14] Merlin, P., and Faber, D. (1975). "A Parallel Mechanism for Detecting Curves in Pictures." *IEEE Transactions on Computers* **C-24** (1), 96–98.

[15] Pomerleau, D. A. (1989). "ALVINN: An Autonomous Land Vehicle in a Neural Network." CMU Technical Report, CMU-CS-89-107.

[16] Rhodes, F. M., Dituri, J. J., Glenn, H. C., Emerson, B. E., Soares, A. M., and Raffel, J. I. (1988). "*A Monolithic Hough Transform* Processor Based on Restructurable VLSI." *IEEE Transactions on Pattern Analysis and Machine Intelligence*, *PAMI* **10** (1), 106–110.

[17] Stockman, G. C., and Argawala, A. K. (1989). "Equivalence of Hough Curve Detection to Template Matching." *Commutations of the ACM* **20** (11), 821–822.

[18] Thorpe, C., Herbert, M., Kanade, T., and Shafer, S. (1988). "Vision and Navigation at Carnegie–Mellon Navlab. *IEEE Transactions on Pattern Analysis and Machine Intelligence*, *PAMI* **10** (3), 362–373.

[19] Turk, M., Morgenthaler, D., Gremban, K, and Marra, M. (1988). "VITS- A Vision System for Autonomous Land Vehicle Navigation." *IEEE Transactions on Pattern Analysis and Machine Intelligence*, *PAMI* **10** (3), 342–361.

[20] Waxman, A., LeMoigne, J., Davis, L., Srinivasan, B., Kushner, T., Liang, E., and Siddalingaiah, T. (1987). "A Visual Navigation System for Autonomous Land Vehicles," *IEEE Journal on Robotics and Automation*, 124–141, vol. RA-3, No. 2, April.

[21] Wechsler, H., and Sklansky, J. (1977). "Finding the Rib Cage in Chest Radiographs. *Pattern Recognition* **9**, 21–30.

[22] Wilson, R. (1989). "Monolithic Processor Performs Histogram, Hough Transforms at 20 MHz." *Computer Design*, March, p. 91.

[23] Magee, M. J., and Aggarwal, J. K. (1984). "Determining Vanishing Points from Perspective Images." *Computer Vision, Graphics and Image Processing* **26**, (CVGIP 26), 256–267.

15
Mobile Robot Perception Using Vertical Line Stereo

JAMES L. CROWLEY, PHILIPPE BOBET, AND KAREN SARACHIK

15.1 Abstract

This chapter describes a real-time 3-D vision system that uses stereo matching of vertical edge segments. The system is designed to permit a mobile robot to avoid obstacles and position itself within an indoor environment. The system uses real-time edge tracking to lock onto stereo matches. Stereo matching is performed using a global version of dynamic programming for matching stereo segments.

The chapter starts by describing the hypotheses that make possible an inexpensive real-time stereo system. The process for detecting vertical edge points and describing these points as vertical edge segments is then described. This is followed by a brief description of the edge tracking process. Stereo matching by global dynamic programming is then described.

Stereo matching makes it possible to recover the 3-D position of vertical lines. The mathematics for projecting 3-D vertical segments from stereo matches is presented. A parametric representation for 3-D vertical segments is then described. This representation describes 3-D segments as a vector of parameters with their uncertainty.

15.2 Introduction

Indoor man-made environments contain many vertical contours. Such contours correspond to environmental structures that a mobile robot may perceive to position itself and navigate. This chapter describes a stereo vision system that detects and matches vertical edges in real time. This system uses a pair of cameras mounted on an manipulator arm on a mobile robot. The system currently operates in software on a workstation with cycle times of 2

This work was sponsored by the Société ITMI and Project EUREKA EU 110: MITHRA.

300

min per stereo pair. A hardware implementation of the modules of this system is under construction and should bring the cycle time to 10 images per second.

15.2.1 Design Principles

In order to achieve real-time stereo at 10 images per second, a number of design decisions were necessary. These design decisions, which result in a number of simplifications in the design of the system, are possible because of the operating conditions of the system. This system was inspired by the system of Kriegman, Treindl, and Binford [8].

The first design decision is that the camera *optical axes* and *baselines* are to be constrained to be *parallel to the floor* (the $x-y$ plane). The camera mount is considered fixed with respect to the robot vehicle. The only source of camera motion is motion by the mobile robot. Vibrations and minor changes in camera tilt and roll are modeled as an uncertainty during the 3-D reconstruction process.

This decision permits us to define the translation from a camera coordinate system to the global coordinate system using:

1. constant values for the 3-D position and orientation of the cameras with respect to the robot and
2. the position and orientation of the robot vehicle with respect to the world coordinate system.

The position and orientation of the robot vehicle, as well as covariance estimates of their uncertainties, are available from the robot vehicle controller [5].

The second design decision is that only *vertical edges* ($90° \pm 45°$) are to be detected and matched. As a result, the system is capable of detecting and reconstructing only 3-D segments that are within $45°$ of the Z (or vertical) axis. As we shall see, this decision permits us to simplify greatly the segment extraction process by limiting the process to segments that are within $45°$ of the vertical in each image. It further permits the 3-D segments that are recovered to be represented in a unique coordinate system. Thus, this decision amounts to restricting the system to those segments from which it can reasonably obtain 3-D information.

The third design decision is that the cameras are to have coplanar retinas. It is well known that the equations for recovering depth from disparity are uniform for all pixels for coplaner stereo cameras. Experiments, however, show that it is nearly impossible to rigidly mount two cameras with coplaner retinas, and the error in depth is very sensitive to the error in coplanarity. Thus, we are obliged to rectify the position of edge lines by projecting them onto virtual coplaner retinas. This rectification is performed on matched edge lines. The parameters of the rectification transformation are obtained from the calibration of the stereo camera configuration, described later. These design decisions make possible a relatively simple real-time stereo system, as described in the following section.

15.2.2 System Organization

The stereo system is organized as a pipeline of relatively simple modules, as illustrated in Figure 15.1. The first module in the system is concerned with detecting vertical edges. A cascade of simple filters is used to first smooth the image and then approximate a first vertical derivative. Filters in this cascade

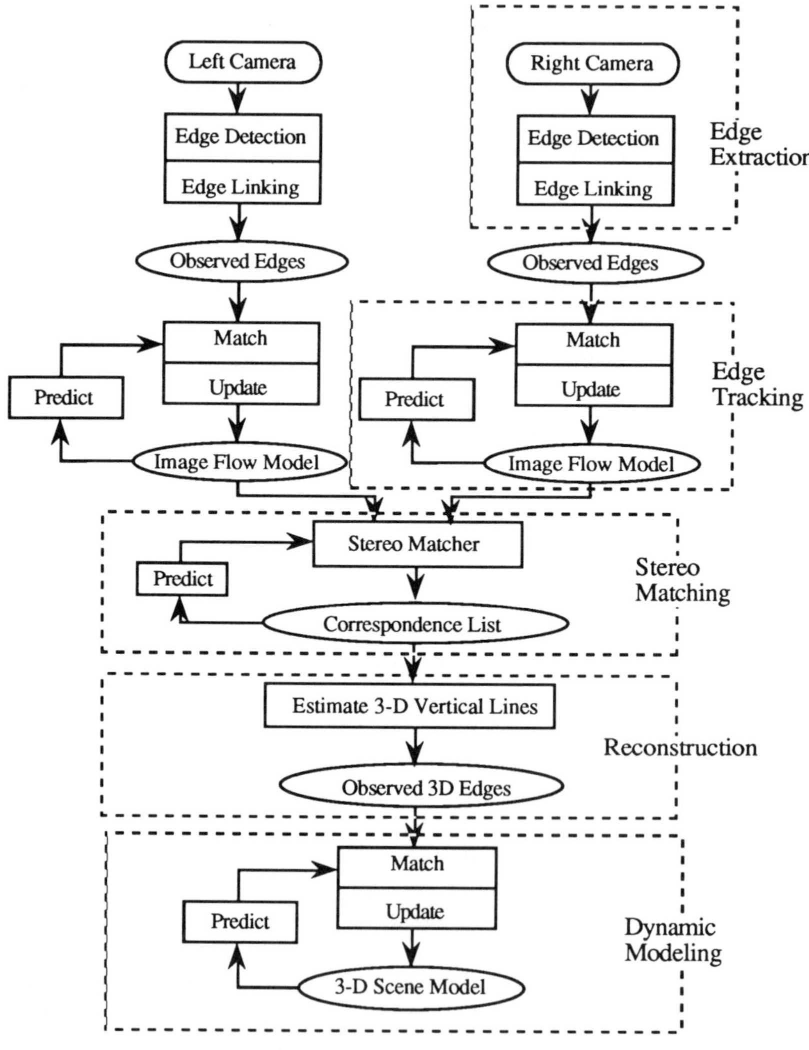

FIGURE 15.1. The system organization.

are composed of a binomial kernel for smoothing and a first difference kernel for calculating the derivative. The design of the vertical derivative filters is described in Section 2.

The second module in the system is responsible for edge chaining and straightline approximations. Raster-scan-based chaining algorithms are well suited to real-time implementation. However, such algorithms are normally greatly complicated by the presence of near horizontal line segments. The restriction to detecting vertical edges yields a very simple one-pass chaining algorithm. The resulting chains are expressed as segments by a recursive line splitting algorithm.

The third module in the system describes the maintenance of an image flow model by tracking edge lines. As this tracking process has recently been described at the 1988 ICCV conference [4], this section is limited to a brief description of the flow model and the tracking process. Tracking allows us to preserve the correspondence between an observed edge and information in the 3-D scene model.

Stereo matching is performed by a single pass of a dynamic programming algorithm over the entire image. Dynamic programming is based on matching the ordered sequence of edge lines in the left and right flow models. The algorithm determines all possible costs for matching the two image sequences in order. A cost measure is used based on the Mahalanobis distance for edge direction, overlap, and distance from a default disparity. Matches are saved in a match list and used to generate the default disparity for subsequent cycles. This process is described in Section 4.

The result of matching is a vertical 3-D segment. A representation for vertical segments and the inference of their 3-D parameters is described in Section 5. Section 6 describes the transformation from a camera-centered coordinate system to a scene-based coordinate system. The chapter concludes with sample results from 3-D reconstructions.

15.3 Detecting and Linking Vertical Edges

This section describes a system for detection of vertical edges. This section begins with a description of a filter for detecting first derivatives in the vertical direction. It then describes the detection and linking of edge points by a raster-scan linking algorithm. We conclude with the extraction and parametric representation of edge segments.

15.3.1 The Filter for First Vertical Derivatives

The basic components of our derivative filter are the binomial low-pass filter and a first difference filter. Let us define the family of binomial filters as b_k. Each filter in this family can be expressed as k "autoconvolutions" of the kernel filter $b_1 = [1, 1]$. We denote this operation by the superscript $*k$.

$$b_k = b_1^{*k} = [1, 1]^{*k}.$$

The set of such filters is given by Pascal's pyramid. Filters from this family are the best integer coefficient, finite impulse response approximation to the Gaussian low-pass filter. We are interested in the low-pass filters that have an odd number of coefficients, and so we restrict our filter to values of k that are even integers. Thus, our low-pass filter will be composed of cascades of the filter

$$b_2 = [1, 2, 1].$$

A circularly symmetric filter, $b_k(i, j)$, may be defined by a cascade of a filter in the row direction, $b_k(i)$, and a filter in the column direction, $b_k(j)$. That is, $b_k(i, j) = b_k(i) * b_k(j)$. For $k = 2$, this filter has the form

$$b_2(i, j) = \begin{bmatrix} 1 & 2 & 1 \\ 2 & 4 & 2 \\ 1 & 2 & 1 \end{bmatrix}.$$

For a first difference filter, we employ the filter

$$d(i) = [1, 0, -1]$$

We wish to be able to tune selectivity the orientation of our filter. This can be accomplished by cascading an additional vertical low-pass filter, also using the binomial kernel. Let us refer to the number of vertical cascades as the parameter m. The vertical low-pass filter is thus $b_m(j)$. We observe that when $m = k$, the resulting filter has a 3-dB cutoff for edges at approximately $90° \pm 45°$.

Thus, our vertical edge filter has the form

$$f_{km}(i, j) = d(i) * b_k(i, j) * b_m(j).$$

The results presented in the following are based on a filter of $k = 4$ and $m = 2$. These values were obtained empirically.

This filter will give a negative response for transitions from dark to light and a positive response for transitions from light to dark. The sign of the response is a useful feature for matching. The sign is preserved in the output of the filtering step by encoding the values from -127 to 128 for each pixel.

15.3.2 Detecting Vertical Edge Points

A prefectly sharp change in gray level, oriented in exactly the vertical direction, will result in a peak whose width is blurred to $k + 1$ pixels by the edge detection filter. The cross section of this blurring resembles the binomial low-pass filter. However, true edges are rarely perfectly sharp, even when the camera is in perfect focus. One of the benefits of the low-pass filter in the the edge detector is to smooth edges to give a single local extremum, even when the edge is noisy and spread over several pixels.

For edge directions that are not perfectly vertical, the vertical component of the edge filter introduces an additional blurring. In particular, for a perfect edge at 45°, the edge filter will blur the peak to $m + k$ pixels. In order to detect points that belong to vertical edges, each row of the filtered image is scanned for extrema. An extremum, or edge point, is any pixel $e(i, j)$ that is a local maximum and has more than twice the absolute value of neighbors three pixels away. That is,

$$|e(i - 1, j)| \leq |e(i, j)| > |e(i + 1, j)|,$$

$$|e(i, j)| \geq 2|e(i - 3, j)|,$$

and

$$|e(i, j)| \geq 2|e(i + 3, j)|.$$

15.3.3 Raster-Scan Edge Chaining

A vertical edge chain is a list of adjacent pairs of image coordinates, (i, j). Edge chaining during a single raster scan makes possible a chaining process that can be implemented in real time, with minimal hardware support. Raster-scan edge chaining is greatly complicated by the presence of edges near the scan direction [6]. By restricting edges to directions perpendicular to the scan direction, the process becomes quite simple. This process is illustrated in Figure 15.2.

As the image is scanned by row, the process maintains two lists of edge chains: an open list and a closed list. The lists are maintained in order of their position in the scan. A chain for the last column is maintained at the tail of the open list. At each step, the algorithm designates a "current" edge chain that is expected to be found next. Let us call the column number of the last pixel in the current chain by the letter c.

Each time an edge pixel is detected, its column number, i, is compared to the column number of the current edge chain. Three cases are possible.

$$\text{Case 1:} \qquad i < c - 2.$$

In this case, a new edge has been detected before the designated "current" edge. A new chain is created and inserted in the list before the current chain.

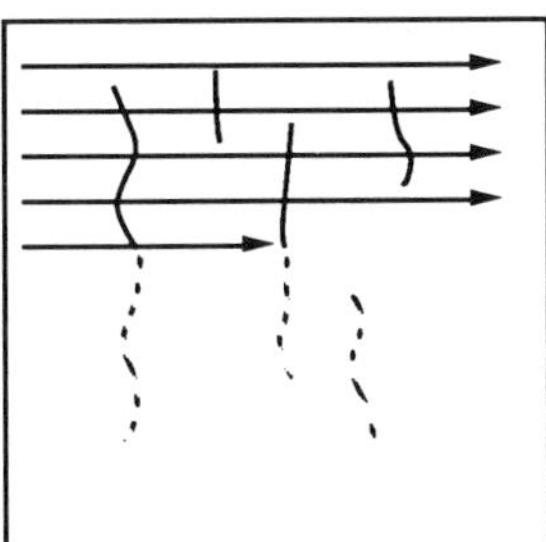

FIGURE 15.2. The raster-scan chaining algorithm maintains a list of open chains.

$$\text{Case 2:} \qquad c - 2 \le i \le c + 2.$$

The detected edge point is within a pixel of the current chain. In this case, the pixel is added to the current chain, and the next chain in the list is selected as the new current chain.

$$\text{Case 3:} \qquad c + 2 < i.$$

In this case, the current chain was not detected. If the row number of the current chain is more than three rows from the the current row, the chain is extracted from the open list and added to the closed list. The next chain in the list is selected as the current chain and the test is repeated.

Edge chains are converted to edge segments by the well-known "recursive line splitting" process [7]. This algorithm is known to exhibit instabilities when representing curved edges. In experiments with curved objects, these instabilities have not proved a problem for subsequent stages in our system.

15.3.4 The MDL Edge Segment Representation

As edge segments are detected, they are transformed to a parametric representation [3] composed of the midpoint, direction, and length. We refer to this representation as the MDL. This representation is designed to facilitate the matching step in the segment tracking phase.

The MDL parametric representation for segments is illustrated in Figure 15.3. A segment is represented by a vector $\mathbf{S} = \{c, d, \theta, h\}$ of parameters. Here, c is the perpendicular distance of the segment from the origin, d is the distance from the perpendicular intercept of the origin to the midpoint of the segment, θ is the orientation of the segment, and h is the half-length of the segment.

For each segment, we also save the midpoint P_m, expressed in image coordinates (i, j) as well as the endpoints P_1 and P_2. The sign of segments is encoded by adding $180°$ to the angle θ of negative segments (segments that mark transitions from dark to light).

15.3.5 Example of Detected Edge Segments

Figure 15.4 shows the edge segments detected in a typical laboratory scene. The top shows the original stereo images. The bottom shows the results of

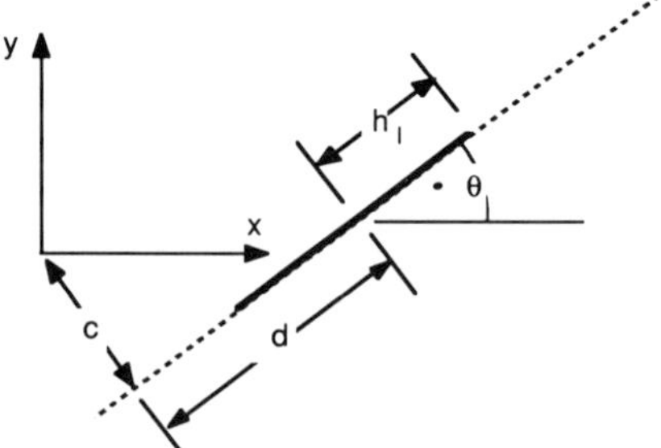

FIGURE 15.3. The MDL parametric representation for line segments.

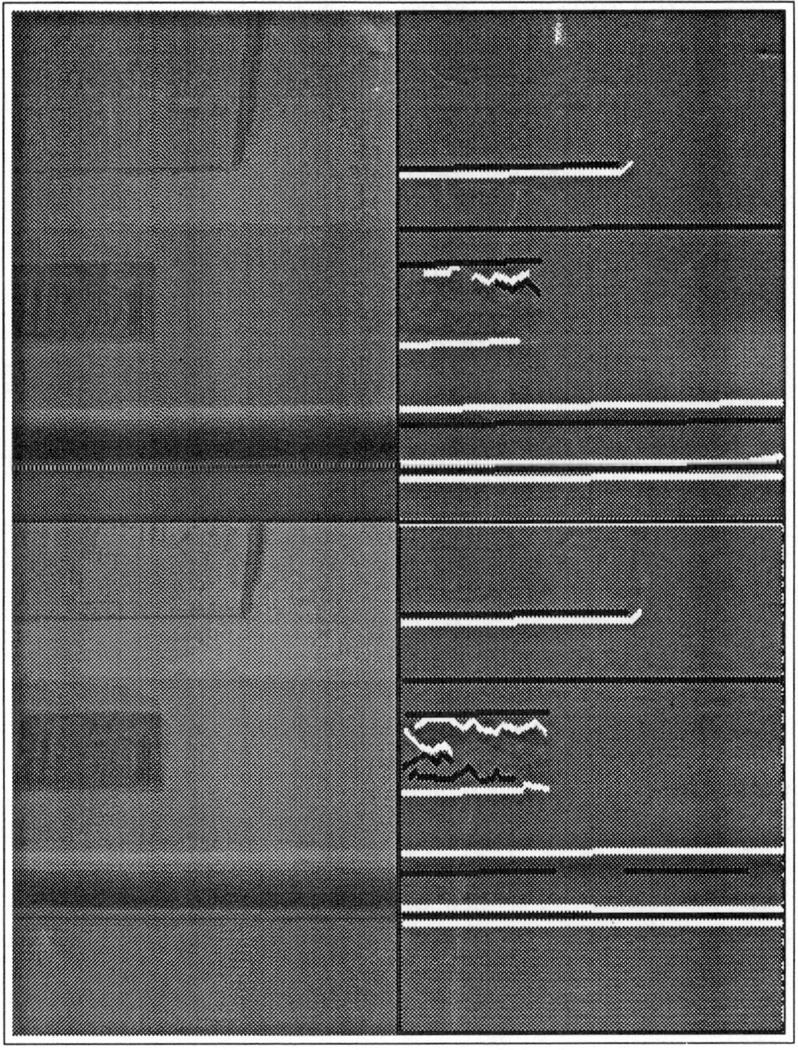

FIGURE 15.4. An example of the vertical edge segments detected in a laboratory scene.

filtering for vertical edges. The detected edges are superimposed on the filtered images. White segments are transitions from light to dark, while dark segments are transitions from dark to light. This is one of a sequence of stereo images for which 3-D data are reconstructed in the following.

15.4 Measuring Image Flow by Tracking Edge Segments

Correspondence of edges in a dense temporal sequence of images can be maintained by a very simple tracking process based on a Kalman filter. Such a process has been described in Crowley et al., 1988 [4]. This tracking process is well debugged and has been used in a number of our projects. Real-time hardware has recently been constructed using this algorithm [23]. We will restrict our presentation to an overview of this process. Interested readers are referred to the presentation in Crowley et. al., 1988 [4] for details.

15.4.1 Representation for the Image Flow

The tracking process maintains a list of "active" edge segments composed of the parameter vector, $\mathbf{S} = \{c, d, \theta, h\}$. However, for each attribute A of $\mathbf{S}$, the vector includes the temporal derivative, a', and the covariance between the attribute and its temporal derivative. That is, for each A of $\mathbf{S}$,

$$A = \begin{bmatrix} a \\ a' \end{bmatrix} \qquad C_A = \begin{bmatrix} \sigma_a^2 & \sigma_{aa'} \\ \sigma_{aa'} & \sigma_{a'}^2 \end{bmatrix},$$

where

$$a' = \partial a/\partial t.$$

The flow model also contains a confidence factor, CF, represented by a state from the set $\{1, 2, 3, 4, 5\}$ and a unique identity, ID. The identity of a segment permits the process to preserve the association between a segment, its corresponding segment in the other image, and the resulting 3-D segment.

In summary, the flow model segments are represented by the following parameters.

Perpendicular position	$\{c, c', \sigma_c^2, \sigma_{cc'}, \sigma_{c'}^2\}$
Tangential position	$\{d, d', \sigma_d^2, \sigma_{dd'}, \sigma_{d'}^2\}$
Orientation	$\{\theta, \theta', \sigma_\theta^2, \sigma_{\theta\theta'}, \sigma_{\theta'}^2\}$
Half-length	$\{h, h', \sigma_h^2, \sigma_{hh'}, \sigma_{h'}^2\}$
Confidence factor	CF
Identity	ID

15.4.2 Maintenance of a Dynamic Flow Model

The segments in the flow model are tracked by a three-phase process illustrated in Figure 15.5.

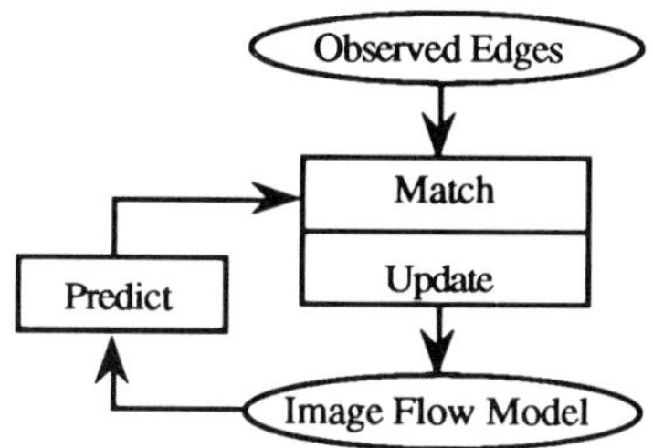

FIGURE 15.5. The flow measurement process.

The phases of this cyclic process are

1. prediction,
2. correspondence, and
3. model update.

The prediction phase uses the Kalman filter prediction equation to predict the position and uncertainty for each of the parameters in each of the segments in the image flow model. Correspondence is performed by matching differences of the segment parameters c, d, h, and θ normalized by the uncertainty. If there are multiple matches, the sum of the Mahalanobis distances between parameters (difference normalized by sum of variances) is used to select the best match. Updating segment parameters is performed by the update equations of the Kalman filter.

Performing stereo correspondence on the flow model provides cleaner data for stereo matching. In particular, this technique also permits the system to function in the presence of simple occlusions.

15.5 Correspondence Matching Using Dynamic Programming

This section describes the stereo correspondence matching process. The process "locks on" to correspondences by feeding the previously discovered disparity for each segment pair into the matching process.

15.5.1 Process Overview

The stereo matching process is illustrated in Figure 15.6. Line segments from the flow models from the left and right images are combined with predictions from the previous match to produce a new stereo correspondence list. The contents of the stereo correspondence list gives a list of vectors each containing 4 values: the left ID, right ID, disparity, and CF.

The left ID and right ID are the IDs of the matching segments. The disparity is the most recently observed horizontal disparity in pixels. The CF is the number of times that this correspondence has been found in the last five images. IF CF goes to zero, then the correspondence is removed from the list.

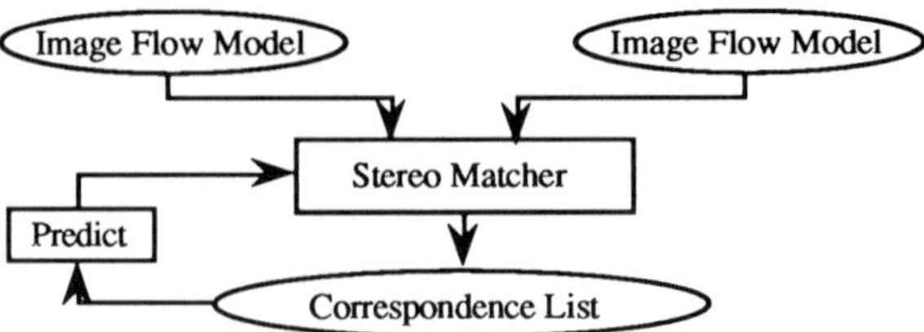

FIGURE 15.6. The stereo matching process maintains a correspondence list.

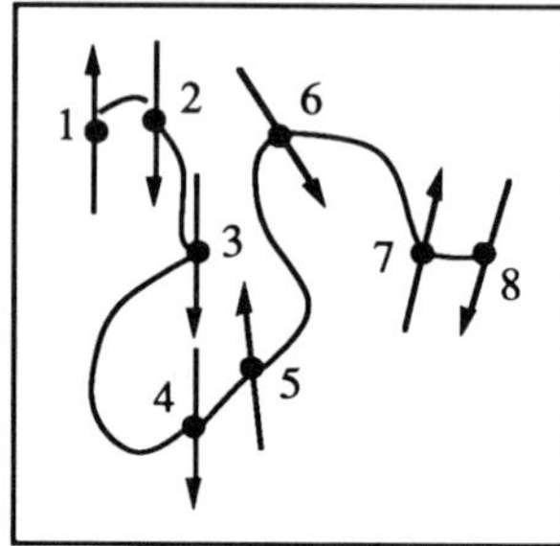

FIGURE 15.7. The segments in the flow models are sorted by row and column.

15.5.2 Matching by Dynamic Programming

Stereo matching is performed by a dynamic programming algorithm [9]. The algorithm operates on the lists of edge segments from the left and right flow models. The flow model segments are maintained in a list sorted by column of the center point and, for segments on the same column, by row. This is illustrated in Figure 15.7.

The dynamic programming algorithm calculates the best global match provided that the order of the segments is the same in the two images. The algorithm works by propagating matching costs in a grid. The columns of the grid correspond to the segments from the left image, and the rows correspond to the segments from the right image. Paths through the grid correspond to possible matches of segments.

A "lawn-mower" style algorithm is then used to propagate the cumulative cost for every possible path based on a cost function for each pair of matches. After propagating the cost from the bottom to the top of the grid, the least cost path is traced back to the far corner of the grid (Figure 15.8). The cells in this path provide the most likely globally consistent set of correspondences of segments from the left and right flow models.

15.5.3 Cost Functions

The individual cost of matching a segment from the left image to a segment in the right image is based on the Mahalanobis distance between the attributes

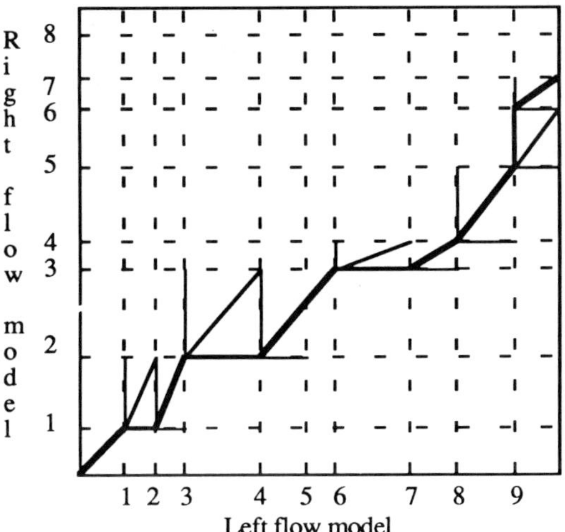

FIGURE 15.8. The cost of matching is propagated through a grid of possible matches. The least cost path provides a most likely globally consistent match.

of the segments. This cost function includes a term for the expected disparity. Let us refer to S_n as the nth segment from the left flow model and S_m as the mth segment from the right flow model. The individual cost for matching S_n to S_m is given by

$$C(n, m) = C_o(n, m) + C_\theta(n, m) + C_d(n, m).$$

The term $C_o(n, m)$ is based on the overlap of the segments, as computed from the parameters d and h. If the segments do not overlap, then the normalized difference is greater than one and the match is rejected by setting the cost to infinity.

Overlap

Cost $\qquad C_o(n, m) = \dfrac{\text{ABS}(d_n - d_m)}{(h_n + h_m)}.$

Rejection $\qquad \text{IF}[C_o(n, m) > 1] \quad \text{THEN} \quad C_o(n, m) = \infty.$

The term $C_\theta(n, m)$ is based on the similarity of orientation of the segments, as computed from the parameter θ. An experiment was performed in which it was observed that the largest difference in angles occurred when a 3-D line segment is tilted away from the cameras by $45°$. In this case, the observed difference in angle is $10°$. Thus, the cost for the difference in image plane orientation between the left and right images is given by the difference normalized by $10°$.

Orientation

cost $\qquad C_\theta(n, m) = \dfrac{(\theta_n - \theta_m)}{10°}.$

Rejection $\qquad \text{IF } C_\theta(n, m) > 2 \quad \text{THEN} \quad C_\theta(n, m) = \infty.$

The third component in the cost is the difference between an expected disparity and the observed disparity. A nominal disparity is initially determined from a fixation distance. This distance is a parameter that can be dynamically controlled during matching. Whenever a stereo match that exists from the previous frame has been determined to have a CF > 1, the nominal disparity is reset to this previous disparity. In this way, the process is biased to prefer existing matches. The cost is determined by dividing the difference from the fixation disparity by an uncertainty, σ_0, which is also a parameter that can be controlled in the process.

Disparity (with respect to a large uncertainty σ_0, around a fixation disparity, D_0)

$$\text{cost} \qquad C_d(n, m) = \frac{(c_n - c_m - D_0)}{\sigma_0}.$$

$$\text{Rejection} \qquad \text{IF} \quad [C_d(n, m) > 2] \quad \text{THEN} \quad C_d(n, m) = \infty.$$

The cost of skipping a match is equal to the cost from a 1 standard deviation difference on all 3 measures, that is $C_{\text{skip}} = 3$.

15.5.4 Cost Propagation

The cost propagation algorithm considers each pair of matches in a raster-scan manner. At each grid cell (n, m), the algorithm computes three cumulative costs:

C_{01}, the cost of a path that skips a correspondence for the left segment,
C_{10}, the cost of a path that skips a correspondence for the right segment, and
C_{11}, the cost of a path that includes the correspondence of the left segment to
    the right segment.

At each grid point (n, m), the cumulative cost C_c is given by the minimum of

$$C_{01}(n, m) = C_c(n - 1, m) + C_{\text{skip}},$$

$$C_{10}(n, m) = C_c(n, m - 1) + C_{\text{skip}},$$

$$C_{11}(n, m) = C_c(n - 1, m - 1) + C(n, m).$$

The cumulative cost is determined by

$$C_c(n, m) = \text{Min}\{C_{01}(n, m), C_{10}(n, m), C_{11}(n, m)\}.$$

15.5.5 Example of Matching Results

The system has been regularly operated using live images during debugging over the last few months. In a typical experiment, a sequence of 5 to 10 pairs of stereo images is a made as the mobile robot moves in a straight line. Matching statistics have been improved from early results of around 80% correct to nearly 100% correct by improving the stability of edge segments that are

FIGURE 15.9. The inference of 3-D form from the the correspondence of edge segments.

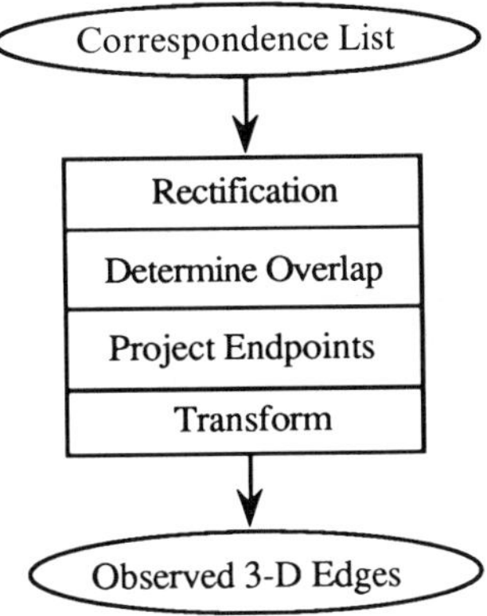

detected. This additional stability was achieved by increasing the degree of smoothing from $m = 2$ to $m = 4$, as well as by enlarging the tolerance for recursive line fitting to 2 pixels.

15.6 Recovery of 3-D Position from Stereo Information

For each stereo correspondence, a 3-D vertical edge segment is obtained. These 3-D segments are then integrated into a 3-D composite scene model as an independent observation. This section describes the process of 3-D inference.

The 3-D inference process is described in Figure 15.9. It begins by rectification of the position of segments that are found to correspond. Segments are then limited to their overlapping parts, followed by the calculation of the depth and uncertainty for the endpoints in a camera-centered coordinate system. Segment parameters are then computed from the endpoints, and the representation is transformed to scene-centered coordinates. To describe this process, we begin by defining our notation.

15.6.1 Coordinate Systems

To develop 3-D reconstruction, we will require the following notation. Let us refer to a point in the scene expressed in scene coordinates as $P_s = (x_s, y_s, z_s)$. The same point expressed in a camera-centered coordinate system is $P_c = (x_c, y_c, z_c)$. The projection of this point on the retina is written as $P_r = (x_r, y_r, 0)$. This point corresponds to a pixel in the image, $P_i = (i, j)$. The intrinsic parameters of a camera are the size of a pixel, (d_i, d_j) in millmeters/pixel, the optical center of the image, (c_i, c_j) in pixels, the the focal length, F, in millimeters. These are illustrated in Figure 15.10.

For calibration, we use a technique developed by Faugeras and Toscani

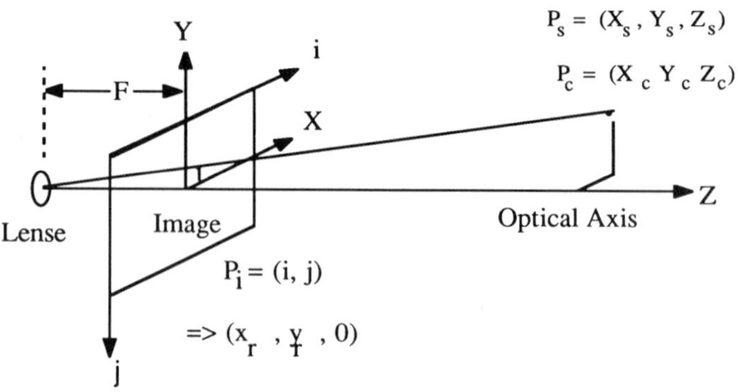

FIGURE 15.10. Image, camera, and scene coordinates.

[10]. This technique does not directly provide the pixel size or focal length, but simply the ratios of size to focal length along two orthogonal axes, $(d_i/F, d_j/F)$. Using these ratios, we project pixels to a vertual retina at focal length 1. It is this retina that is then rectified to produce a second virtual retina. The transformation of an image point (i, j) to a point on the virtual retina $(F = 1)$ is given by

$$x_r = \frac{(i - c_i)d_i}{F}, \qquad y_r = \frac{(c_j - j)d_j}{F}.$$

15.6.2 Rectification

When the retina's of stereo cameras are coplaner, the depth equations reduce to a trivial form. However, it is impossible to mechanically mount cameras such that their retinas are sufficiently close to parallel for this simplification to apply. This problem is avoided by calibrating a transformation that projects points in stereo images to a pair of virtual coplanar retinas. We call these the virtual rectified retinas. We use a technique inspired by Ayache [1] to perform this transformation.

Our stereo cameras are rigidly mounted on a rectangular plate. The mounting permits a rotation about a vergence angle, α, as illustrated in Figure 15.11. The point of rotation is somewhere near the optical axis. A point, p_r, on the unrectified virtual retina is projected to the virtual receified retina. This transformation uses the vergence angle α and the translation T, as well as the intrinsic camera parameters obtained by calibration.

15.6.3 Depth from Coplanar Stereo Cameras

The depth equation falls directly from a difference in similar triangles, as illustrated in Figure 15.12. Let us refer to a scene point, which is observed by

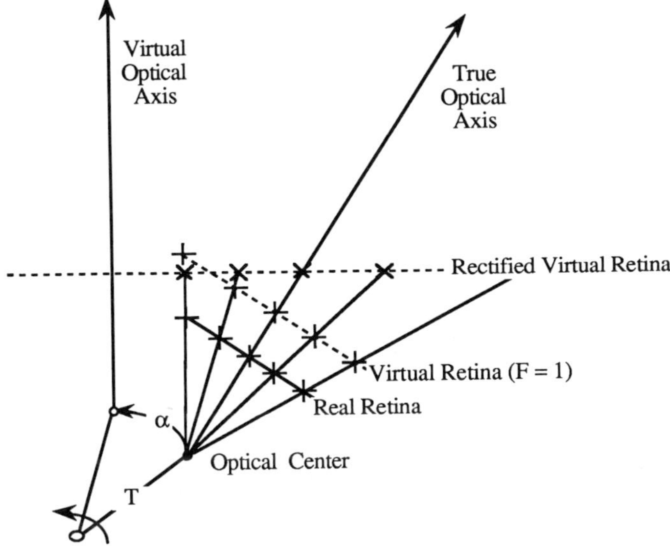

FIGURE 15.11. Cameras can be mounted with a vergence angle α. Rectification corrects for this angle by projecting back to virtual coplanar retinas.

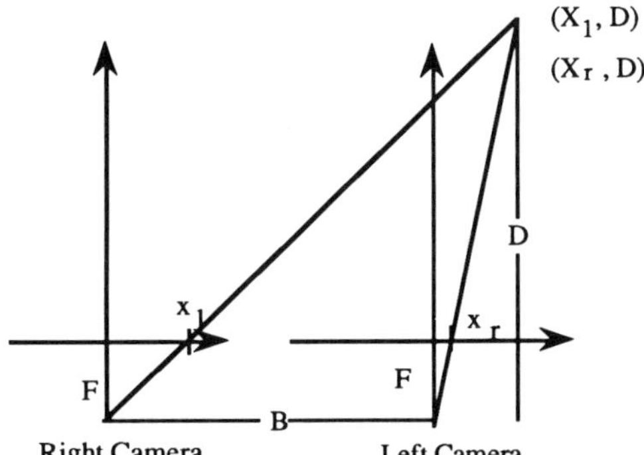

FIGURE 15.12. Calculation of depth from disparity along coplanar retinas.

two coplanar cameras. In the coordinates of the left camera, this point occurs at (x_{cl}, D), while in the coordinates of the right camera, its position is written (x_{cr}, D). If the focal lengths of the right and left cameras are both equal to F, then we can write two equivalence relations by similar triangles:

$$x_1/F = x_{cl}/D, \qquad x_r/F = x_{cr}/D.$$

By taking the difference of these two relations, we obtain

$$x_1/F - x_r/F = x_{cl}/D - x_{cr}/D.$$

If we also note that the translation from x_c to x_{cr} is given by

$$x_{cl} = x_{cr} + B,$$

then by a small amount of algebra we can arrive at the equation

$$D = BF/(x_r - x_l).$$

If we note disparity as $\Delta x = (x_l - x_r)$ and we observe that our edge lines are rectified to a virtual retina where $F = 1$, then this equation may be expressed as

$$D = B/\Delta x.$$

The linear term B is determined as a final step in calibration by calculating the distance between the optical centers of the left and right cameras as provided by calibration.

15.6.4 Projecting Stereo Matches to 3-D

Stereo projection is performed on the segment endpoints. Because segment length is not always reliable, we must first determine the overlapping part of the corresponding segments. Let us refer to the segment endpoints as P_1 and P_2, such that P_1 is the upper endpoint and P_2 is the lower endpoint. Thus, the left segment is given by $P_{1l} = (x_{1l}, y_{1l}, P_{2l} = (x_{2l}, y_{2l})$, and the right segment is given by $P_{1r} = (x_{1r}, y_{1r})$, $P_{2r} = (x_{2r}, y_{2r})$. Segments are expressed in the right-hand coordinate system of the virtual retina (up is positive y, right is positive x).

Since the segments are nearly vertical, we need only be concerned with the vertical position for the endpoints. For the upper endpoint of the left and right endpoints, we replace the y value with the maximum of the two values. For the lower endpoint, we replace the row number with the minimum of the y values, as illustrated in Figure 15.13.

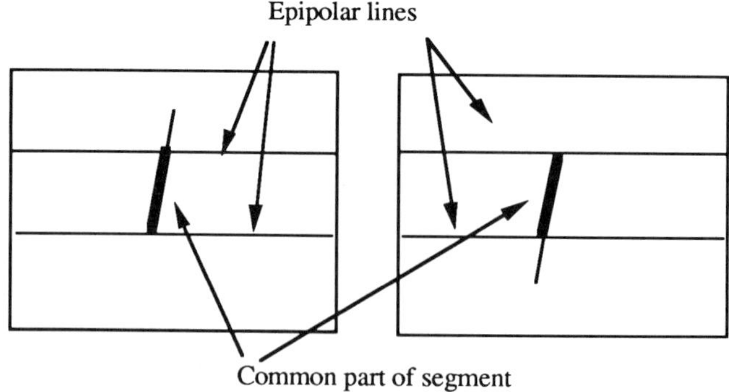

FIGURE 15.13. Determining the common part of a segment.

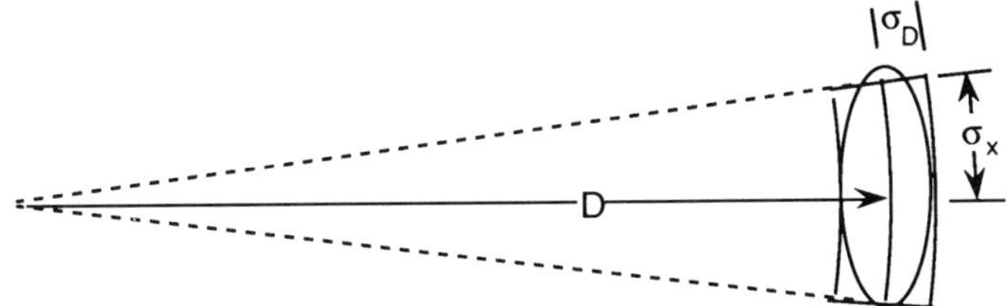

FIGURE 15.14. The covariance in recovered depth.

$$y_{1r} = y_{1l} = \text{MAX}(y_{1r}, y_{1l})$$

$$y_{2r} = y_{2l} = \text{MIN}(y_{2r}, y_{2l})$$

$$D = B/\Delta x.$$

We then compute the corresponding point in the scene in the coordinate system of the stereo pair of cameras as

$$x_c = D x_{rr}\left(\frac{d_i}{F}\right) + \frac{B}{2},$$

$$y_c = D y_{rr}\left(\frac{d_j}{F}\right),$$

$$z_c = D.$$

The uncertainty of each recovered point is modeled as having two independent components, an uncertainty in x, σ_x, and an uncertainty in D, σ_D, as shown in Figure 15.14.

The term along the x axis is dependent on the uncertainty in the angle to which the pixel from the left retina is projected. Errors in this angle may come from errors in calibration of the intrinsic camera parameters d_i/F, and c_i and errors in calibration of the rectification angles α. The term in the D direction is dependent on the precision of the size of a pixel. By performing multiple observations of the calibration pattern and then reconstructing the grid, we can obtain estimates of this error as a variance.

15.7 Generating 3-D Vertical Segments in Scene Coordinates

The 3-D position of segments is initially recovered in a coordinate system centered on a point halfway between the pair of stereo cameras. Projected line segments are expressed in a parametric representation that is the 3-D analog of the 2-D MDL representation used for edge segments. Segments in this representation are then transformed to scene coordinates.

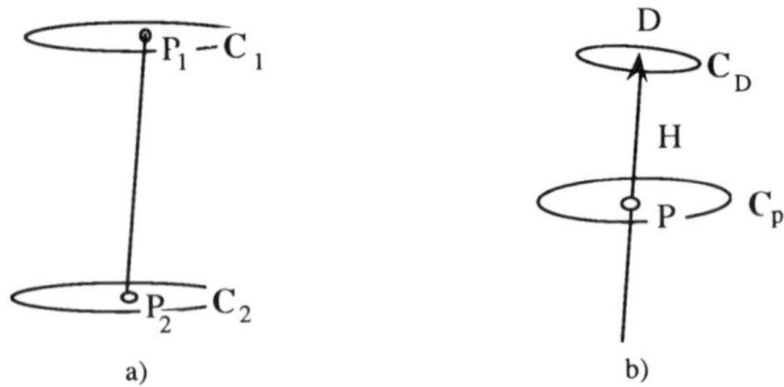

FIGURE 15.15. Representation for a 3-D segment and its uncertainties: (a) endpoint representation; (b) midpoint, direction, and length in 3-D.

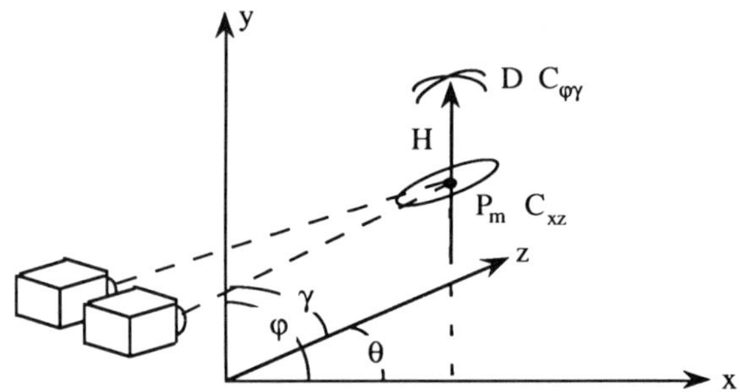

FIGURE 15.16. The stereo–camera-centered coordinate system.

15.7.1 Transformation to Midpoint, Direction, and Length

Segments are obtained in an endpoint representation, illustrated in of Figure 15.15(a). This representation is composed of two 3-D points and their covariances.

Top Endpoint $P_1, C_1,$
Bottom Endpoint $P_2, C_2.$

To facilitate matching and integration, segments are transformed to a representation in terms of midpoint, direction, and length, as shown in Figure 15.15(b).

The camera-centered coordinates, with the camera axis parallel to the z axis, are shown in Figure 15.16. Because we are restricted to vertical segments, segment representation is simplified. For segment position, the y component of the uncertainty may be considered to be modeled by the length of the segment. Thus, the uncertainty in position is restricted to the x and z axes in the camera-centered coordinate system. For the same reason, the set of orientations can be expressed as the angle y, the tilt angle in the y–z plane, and φ can be expressed as the roll angle in the x–y plane. The uncertainties in the angles can be expressed as covariances in these two angles.

Thus, segments are transformed to a representation composed of the following parameters:

Midpoint	$P_m = (x, y, z)$
Direction	$D = (\varphi, \gamma)$
Half-length	H
Covariance of midpoint	C_{xz}
Covariance of direction	$C_{\varphi\gamma}$

15.7.2 Transformation to Scene Coordinates

The difference between camera-centered coordinates and scene coordinates is limited to a rotation about the vertical axis, a permutation of axes, and a translation to the scene coordinates origin. Both the rotation and the translation with respect to the scene coordinates are provided by the robot vehicle controller.

To pass from the coordinate system of the camera pair to the coordinate system of the scene requires the position and orientation of the camera with respect to the robot and the position of the mobile robot with respect to the scene. The position of the cameras with respect to the robot is constant. The cameras are aligned looking forward from the robot at a height of 1.6 m. They are placed 10 cm in front of the axis of rotation of the robot vehicle.

The robot vehicle controller [5] returns the estimated position P_R and orientation θ of the vehicle accompanied by an error boundary in the form of a covariance in position, C_{PR}, and a variance in orientation, C_θ. These values are used to transform the position and covariances of the recovered 3-D point to a common scene-centered coordinate system. The scene-centered coordinate system is illustrated in Figure 15.17.

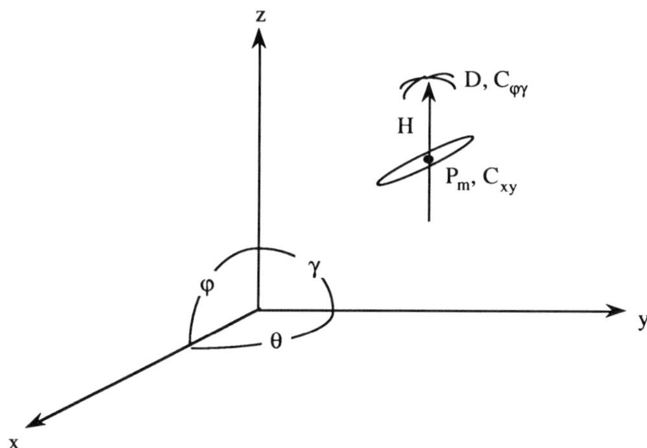

FIGURE 15.17. The scene centered coordinate system.

The transformation from segment position in camera coordinates, P_c, to the segment position to scene coordinates, P_s, begins by transposing the y and z axes by multiplication with a matrix $\mathbf{Q}$, followed by rotation in the x–y plane using the matrix $\mathbf{R}_3(\theta)$. The robot position, P_R, is then added:

$$P_s = \mathbf{R}_3(\theta)\mathbf{Q}P_c + P_R.$$

The uncertainty in position is represented only in the coordinates (x, y). The uncertainty is rotated in the x–y plane by the angle θ, expressed by the 2×2 rotation matrix $\mathbf{R}_2(\theta)$. To this, we add the uncertainty in the robot's position, C_{pR}:

$$C_{xy} = \mathbf{R}_2(\theta)C_{xz}\mathbf{R}_2(\theta) + C_{pR}.$$

The roll and tilt angles are also rotated to be expressed as angles in the x–z and y–z planes:

$$D_s = \mathbf{R}_2(\theta)D_c.$$

Finally, the uncertainty in the segment orientation is transformed to the scene-centered coordinate system:

$$C_{D_s} = \mathbf{R}_2(\theta)C_D\mathbf{R}_2(\theta).$$

15.7.3 An Example of 3-D Reconstruction

Figure 15.18. shows an example of stereo reconstruction in our laboratory. The upper part of the figure shows a stereo pair of images of a corner of the lab. The middle part shows the segments that were extracted. The bottom part shows the correspondences between segments that were obtained.

Figure 15.19. shows an overhead view of the 3-D position of the recovered vertical segments. Each segment is accompanied by its uncertainty in position, shown as an ellipse. Figure 15.20. shows a side view of the superposition of the 3-D segments reconstructed from the vertical line stereo system and the horizontal segments reconstructed from the ultrasonic range sensor.

15.8 Conclusions and Perspectives

This system is the subject of continuing experimentation and improvement. In particular, a significant effort has been expended on the calibration and rectification procedure, and at the time of this writing, we had recovered reasonable 3-D segment positions, with precisions on the order of a centimeter by a few millimeters at a distance of 3 m. The current rigid camera mount will soon be replaced by a controllable vergence mechanism. This will necessitate the use of algorithms for rectification and 3-D recovery that accept a vergence parameter. The system has proven to be both simple and reliable. The real-time implementation that is currently being developed is expected to become a standard component of our mobile robot navigation system.

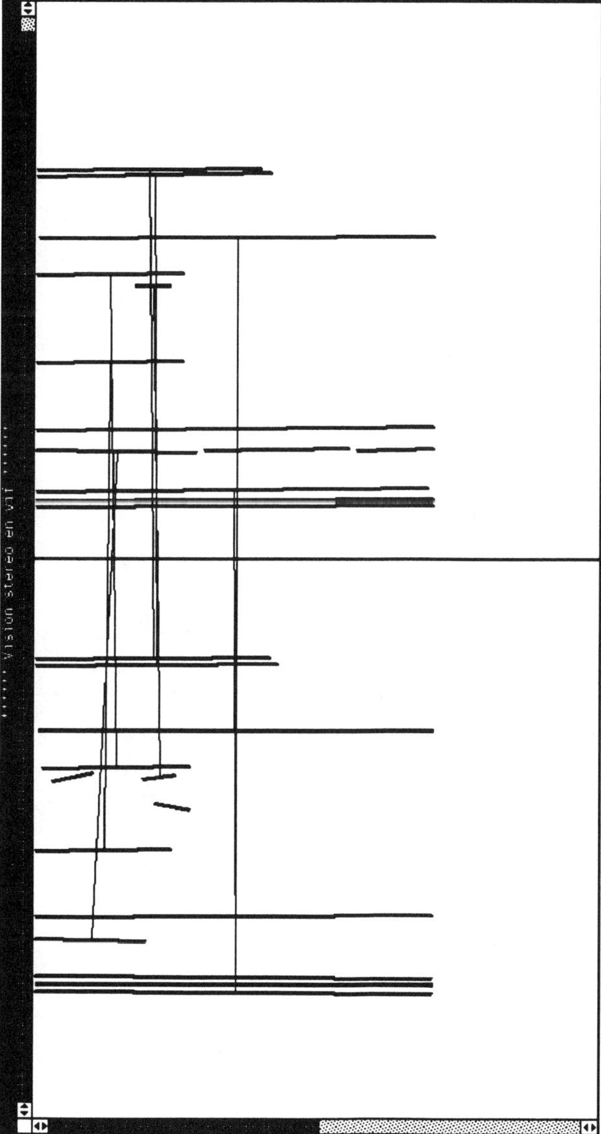

FIGURE 15.18. Stereo correspondence for segments extracted from the stereo images shown in Figure 15.4. Correspondences are illustrated by a horizontal line between segment midpoints. Because the correspondence lines were displayed with an exclusive OR, when the lines overlap, they cancel. This is the second pair from a set of 5 pairs taken at 10-cm displacements. Matches were 100% correct in 4 of the 5 pairs in this sequence.

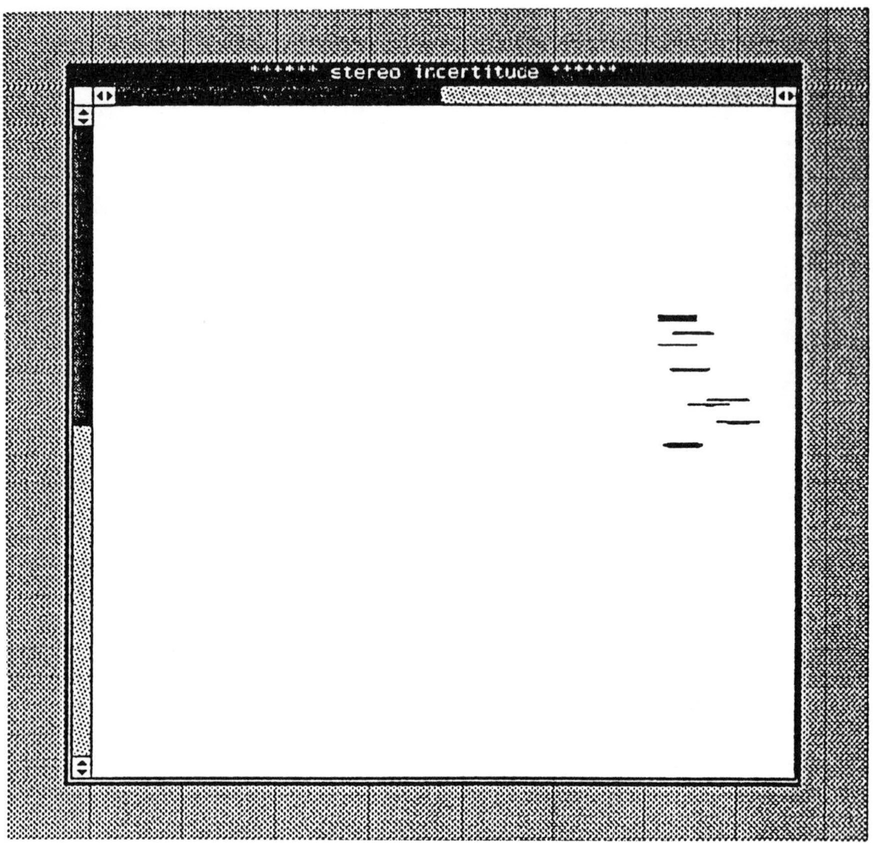

FIGURE 15.19. This figure shows the overhead view of the vertical segments reconstructed from the correspondences in Figure 15.18. The segments are perfectly vertical and thus appear as points. Each point is accompanied by the error ellipse about its midpoint. These error ellipses have major axes of a few centimeters and minor axes of a few meters.

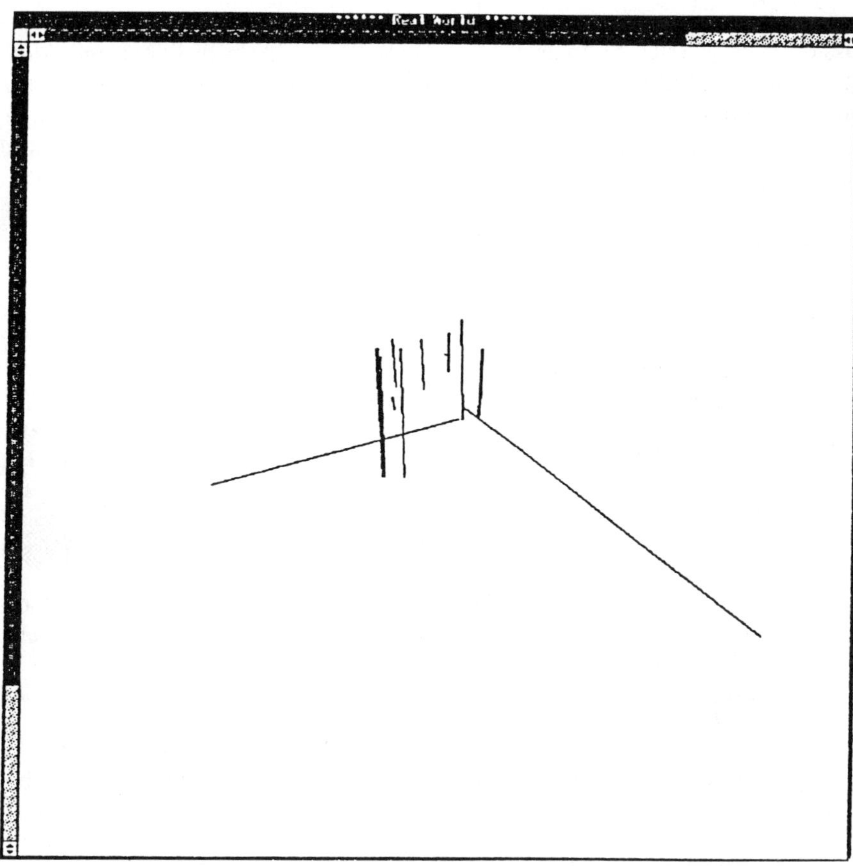

FIGURE 15.20. Side view of superposition of 3-D vertical lines constructed by stereo system and 3-D horizontal lines recovered from ultrasonic ranging.

Acknowledgments. The architecture for this system was developed during discussion with Patrick Stelmaszyk and Haon Hien Pham of ITMI and Alain Chehikian of LTRIF, INPG. The edge detection and chaining procedures are based on codes written by Per Kloor of the University of Linkoping during a post-doctoral visit at LIFIA. The early version of the segment extraction and tracking process was contructed by Stephan Mely and Michel Kurek. This system is the result of a continuous team effort, and the authors thank all involved for their mutual support and encouragement.

References

[1] Ayache, N. (1988). "Construction et Fusion de Représentations Visuelles 3D." Thèse de Doctorat d'Etat, Universtié Paris-Sud, centre d'Orsay.

[2] Stelmaszyk, P., Discours, C., and Chehikian, A. (1988). "A Fast and Reliable Token Tracker." In *IAPR Workshop on Computer Vision, Tokyo, Japan, October, 1988.*

[3] Crowley, J. L., and Ramparany, F. (1987). "Mathematical Tools for Manipulating Uncertainty in Perception." *AAAI Workshop on Spatial Reasoning and Multi-Sensor Fusion,* October, Kaufmann Press.

[4] Crowley, J. L., Stelmaszyk, P., and Discours, C. (1988). "Measuring Image Flow by Tracking Edge-Lines." *ICCV 88: 2nd International Conference on Computer Vision, Tarpon Springs, Florida, December 1988.*

[5] Crowley, J. L. (1989). "Control of Translation and Rotation in a Robot Vehicle." *1989 IEEE Int. Conf. on Robotics and Automation, Scottsdale, Arizona, May.*

[6] Discours, C. (1989). "Analyse du Mouvement par Mise en Correspondance d'Indices Visuels." Thèse de Doctorat de nouveau régime, INPG, Novembre 1989.

[7] Duda, R. O., and Hart, P. E. (1973). *Pattern Classification and Scene Analysis.* Wiley, New York.

[8] Kriegman, D. J., Triendl, E., and Binford, T. O. (1989). "Stereo Vision and Navigation in Buildings for Mobile Robots." *IEEE Transactions on Robotics and Automation* 5(6).

[9] Ohta, Y., and Kanade, T. (1985). "Stereo by Intra and Inter Scanline Search Using Dynamic Programming." *IEEE Trans. on PAMI,* 7, 139–154.

[10] Faugeras, O. D., and Toscani, G. (1986). "The Calibration Problem for Stereo," *IEEE Conf. on C.V.P.R.,* Miami Beach, Fl, June.

Index